The Enzymes of Glutamine Metabolism

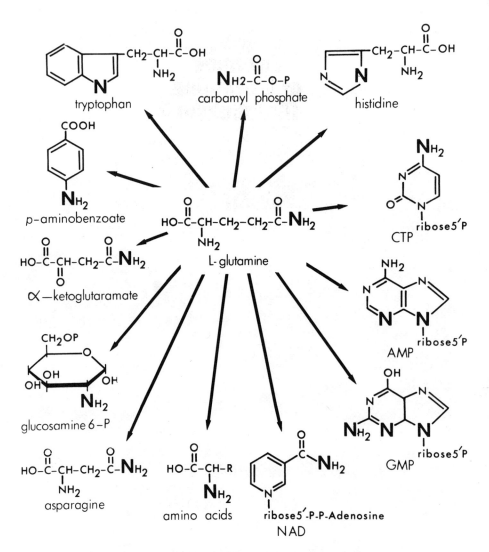

METABOLIC FATES OF THE AMIDE
NITROGEN OF L-GLUTAMINE

The Enzymes of Glutamine Metabolism

EDITED BY

Stanley Prusiner

Laboratory of Biochemistry
National Heart and Lung Institute
National Institutes of Health
Bethesda, Maryland
and
Department of Neurology
University of California
School of Medicine
San Francisco, California

Earl R. Stadtman

Laboratory of Biochemistry
National Heart and Lung Institute
National Institutes of Health
Bethesda, Maryland

Based on a Symposium Held at the 164th National Meeting of the American Chemical Society in New York City on August 30-31, 1972

ACADEMIC PRESS (AP) New York and London 1973

ACADEMIC PRESS, INC.
111 Fifth Avenue, New York, New York 10003

United Kingdom Edition published by
ACADEMIC PRESS, INC. (LONDON) LTD.
24/28 Oval Road, London NW1

LIBRARY OF CONGRESS CATALOG CARD NUMBER: 72-88378

PRINTED IN THE UNITED STATES OF AMERICA

The whole is equal to
the sum of its parts.

—EUCLID

From a glance at a drawing condensing
what is now known of cellular metabolism we can tell that even if at each step
each enzyme carried out its job perfectly,
the sum of their activities could only be
chaos were they not somehow inter-
locked so as to form a coherent system.
We do indeed have the most manifest
evidence of the extreme efficiency of the
chemical machinery of living beings, from
the simplest to the most complex.

— JACQUES MONOD,
Chance and Necessity

CONTENTS

CONTENTS

CONTENTS

PARTICIPANTS AND CONTRIBUTORS

Diane Albrycht, Department of Medicine, Section of Hematology, University of Chicago, Chicago, Illinois 60637

Bruce N. Ames, Department of Biochemistry, University of California, Berkeley, California 94720

M. A. Berberich, Laboratory of Biochemistry, National Heart and Lung Institute, National Institutes of Health, Bethesda, Maryland 20014

Jean E. Brenchley,* Department of Biology, Massachusetts Institute of Technology, Cambridge, Massachusetts 02139

C. M. Brown, Department of Microbiology, Medical School, University of Newcastle upon Tyne, England

John M. Buchanan, Department of Biology, Massachusetts Institute of Technology, Cambridge, Massachusetts 02139

A. J. L. Cooper, Department of Biochemistry, Cornell University Medical College, New York, New York 10021

Norman P. Curthoys,† Department of Pharmacology, Washington University School of Medicine, St. Louis, Missouri 63110

Thomas F. Deuel, Department of Medicine, Section of Hematology, University of Chicago, Chicago, Illinois 60637

William C. Dolowy, University of Washington School of Medicine, Seattle, Washington 98105

Leonard Estis, Department of Biochemistry, Cornell University Medical College, New York, New York 10021

---

*Present address: Department of Microbiology, Pennsylvania State University, University Park, Pennsylvania 16802

†Present address: Department of Biochemistry, University of Pittsburgh School of Medicine, Pittsburgh, Pennsylvania 15213

A. Ginsburg, Laboratory of Biochemistry, National Heart and Lung Institute, National Institutes of Health, Bethesda, Maryland 20014

N. Glansdorff, Laboratoire de Microbiologie, l'Université Libre de Bruxelles, Brussels, Belgium

M. Grenson, Laboratoire de Microbiologie, l'Université Libre de Bruxelles, Brussels, Belgium

Standish C. Hartman, Department of Chemistry, Boston University, Boston, Massachusetts 02215

Rudy H. Haschemeyer, Department of Biochemistry, Cornell University Medical College, New York, New York 10021

John S. Holcenberg, University of Washington School of Medicine, Seattle, Washington 98105

H. Holzer, Biochemisches Institut, Universität Freiburg, Freiburg, West Germany

B. L. Horecker, Roche Institute of Molecular Biology, Nutley, New Jersey 07110

Bernard Horowitz, Department of Biochemistry, Cornell University Medical College, New York, New York 10021

T. Katsunuma, Department of Enzyme Chemistry, School of Medicine, Tokushima University, Tokushima, Japan

N. Katunuma, Department of Enzyme Chemistry, School of Medicine, Tokushima University, Tokushima, Japan

Alfred Lerner, Department of Medicine, Section of Hematology, University of Chicago, Chicago, Illinois 60637

Alexander Levitzki, Department of Biophysics, The Weizmann Institute of Science, Rehovot, Israel, and Department of Biochemistry, University of California, Berkeley, California 94720

Oliver H. Lowry, Department of Pharmacology, Washington University School of Medicine, St. Louis, Missouri 63110

Boris Magasanik, Department of Biology, Massachusetts Institute of Technology, Cambridge, Massachusetts 02139

J. L. Meers, Agricultural Division, I.C.I. Ltd., Billingham, Teeside, England

Alton Meister, Department of Biochemistry, Cornell University Medical College, New York, New York 10021

Richard E. Miller,* Laboratory of Biochemistry, National Heart and Lung Institute, National Institutes of Health, Bethesda, Maryland 20014

Gregory Milman, Department of Biochemistry, University of California, Berkeley, California 94720

Alan Peterkofsky, Laboratory of Biochemical Genetics, National Heart and Lung Institute, National Institutes of Health, Bethesda, Maryland 20014

Charles F. Phelps, Department of Biochemistry, University of Bristol, Bristol, England

A. Piérard, Laboratoire de Microbiologie, l'Université Libre de Bruxelles, Brussels, Belgium

Lawrence M. Pinkus, Department of Biochemistry, Cornell University Medical College, New York, New York 10021

Michael J. Prival,† Department of Biology, Massachusetts Institute of Technology, Cambridge, Massachusetts 02139

Stanley Prusiner,‡ Laboratory of Biochemistry, National Heart and Lung Institute, National Institutes of Health, Bethesda, Maryland 20014

Joseph Roberts, University of Washington School of Medicine, Seattle, Washington 98105

H. Schutt, Biochemisches Institut, Universität Freiburg, Freiburg, West Germany

Robert W. Sindel, Department of Pharmacology, Washington University School of Medicine, St. Louis, Missouri 63110

*Present address: Division of Metabolic Disease, Department of Medicine, University of California at San Diego, La Jolla, California 92034

†Present address: Center for Science in the Public Interest, 1179 Church Street, N.W., Washington, D. C.

‡Present address: Department of Neurology, University of California School of Medicine, San Francisco, California 94122

David Smotkin, Department of Biochemistry, University of California, Berkeley, California 94720

P. R. Srinivasan, Department of Biochemistry, Columbia University College of Physicians and Surgeons, New York, New York 10032

Earl R. Stadtman, Laboratory of Biochemistry, National Heart and Lung Institute, National Institutes of Health, Bethesda, Maryland 20014

Suresh S. Tate, Department of Biochemistry, Cornell University Medical College, New York, New York 10021

D. W. Tempest, Microbiological Research Establishment, Porton, Salisbury, England

David C. Tiemeier, Department of Biochemistry, University of California, Berkeley, California 94720

I. Tomino,* Department of Enzyme Chemistry, School of Medicine, Tokushima University, Tokushima, Japan

T. Towatari, Department of Enzyme Chemistry, School of Medicine, Tokushima University, Tokushima, Japan

Paul P. Trotta, Department of Biochemistry, Cornell University Medical College, New York, New York 10021

Vaira P. Wellner, Department of Biochemistry, Cornell University Medical College, New York, New York 10021

J. M. Wiame, Laboratoire de Microbiologie, l'Université Libre de Bruxelles, Brussels, Belgium

Peter J. Winterburn, Department of Biochemistry, University College, Cardiff, England

Bernard Witholt,† Department of Chemistry, University of California at San Diego, La Jolla, California 92037

*Present address: Department of Nutrition, Jikei University of Medicine, Tokyo, Japan.

†Present address: Biochemistry Laboratory, University of Groningen, Groningen, The Netherlands

R. M. Wohlhueter, Biochemisches Institut, Universität Freiburg, Freiburg, West Germany

James B. Wyngaarden, Department of Medicine, Duke University Medical Center, Durham, North Carolina

Howard Zalkin, Department of Biochemistry, Purdue University, Lafayette, Indiana

PREFACE

The amino acid, glutamine, lies at the center of cellular nitrogen metabolism. The amide and amino nitrogens of glutamine may be utilized in the biosynthesis of amino acids while the amide nitrogen may also be used in the synthesis of nucleotides, amino sugars, and cofactors.

In recent years significant advances in the biology of glutamine metabolism have been made. The pathways of glutamine synthesis and degradation have been elucidated and their regulation studied in detail. The enzymes which catalyze the formation and breakdown of glutamine have been purified and characterized extensively. Much of the information available on these enzymes lies scattered, because glutamine participates in many important biological reactions and consequently has attracted numerous scientists from a variety of disciplines. The interconversion of glutamate and glutamine has been extensively studied in mammals and micro organisms as described in the first section of this volume. In the second section, the various biosynthetic reactions in which glutamine donates its amide nitrogen are examined.

These studies are presented in a single volume with the hope that such a collection may help to organize and integrate our knowledge of the cellular control of nitrogen assimilation and elimination, the regulation of multi-enzyme systems with common substrates, and the interaction between biosynthetic and catabolic processes in metabolism. In addition, this unique collection of papers may interest scientific investigators in such diverse fields as cellular regulation, nitrogen metabolism, cancer biology and the neurosciences, and may also be useful in advanced courses of study in cellular metabolism and biochemistry. Lastly, we would hope that these studies will stimulate the discovery of new enzymatic reactions which utilize the amide nitrogen of glutamine. The biosynthesis of oxytocin, vasopressin and vitamin B_{12} may possibly involve such reactions (see contribution by Tate and Meister).

Because the use of trivial names for enzymes is often confusing, an appendix is included which lists the systematic and trivial names and the enzymatic reactions of the enzymes of glutamine metabolism.

We are indebted to the American Chemical Society for allowing us to hold this symposium on the Enzymes of Glutamine Metabolism as part of the 164th annual meeting in New York City on August 30-31, 1972. Also, we are most appreciative of the travel funds for foreign scientists which were provided by a grant from the National Institute of General Medical Sciences of the National Institutes of Health. We thank Linda Alvord for help in organizing the symposium and Sandra Prusiner for editorial assistance.

Stanley Prusiner
Earl R. Stadtman

The Enzymes of Glutamine Metabolism

A NOTE ON THE SIGNIFICANCE OF GLUTAMINE IN INTERMEDIARY METABOLISM

E.R. Stadtman

National Institutes of Health
Bethesda, Md. 20014

Introduction

The large amount of time devoted to a discussion of "The Enzymes of Glutamine Metabolism" is not unrealistic, considering the great importance of glutamine in intermediary metabolism and the wealth of basic information that is available on the mechanisms of glutamine dependent reactions as well as on the enzymes that catalyze them, to say nothing of the vast knowledge that has accumulated concerning the cellular regulation of glutamine metabolism.

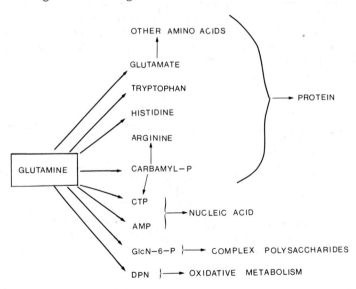

Figure 1. The Role of glutamine in intermediary metabolism.

1

The central role of glutamine in nitrogen metabolism is
illustrated in Fig. 1, which shows those pathways in which
the amide group of glutamine is utilized as a preferred
source of nitrogen for the biosynthesis of various amino
acids, purine and pyrimidine nucleotides, glucosamine-6-P,
and DPN. It has been known for sometime that the indole
nitrogen of tryptophan (Srinivasan, 1959), and an imidazole
nitrogen atom of histidine (Neidle, and Waelsch, 1959), are
derived from the amide group of glutamine; moreover, the
potential importance of glutamine in biosynthesis of other
amino acids was suggested by the discovery that the alpha
amino groups of glutamine can be transferred to various
α-ketoacids to produce the corresponding amino acids
(Meister, 1962). However, an even greater role of glutamine
in the biosynthesis of amino acids was disclosed by the discovery by Tempest et al. (1970) of an enzyme in Aerobacter
aerogenes that catalyzes a TPNH linked reductive amination
of α-ketoglutarate (reaction 1).

Glutamine + α-ketoglutarate + TPN → 2 glutamate + TPN   (1)

As is shown in Fig. 2, when coupled with glutamine synthetase and various glutamate-α-ketoacid transaminases, this
new enzyme provides a heretofore unrecognized pathway of
amino acid biosynthesis, wherein glutamine serves on the one
hand as the primary product of ammonia assimilation and on
the other hand as the principle source of the α-amino group
for synthesis of all amino acids.

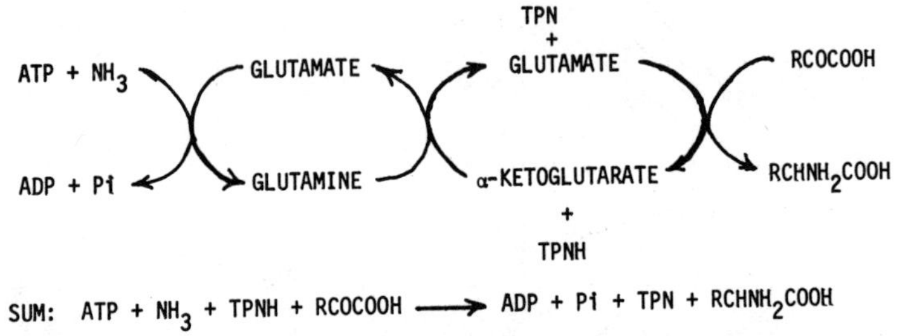

Figure 2. The glutamine dependent mechanism of amino acid
biosynthesis.

The glutamate synthase catalyzing reaction (1) is widely distributed among microbial species (Nagatani et al. 1971) and has been highly purified from extracts of Escherichia coli (Miller and Stadtman, 1972). Later discussions in this symposium by Miller and by Tempest are concerned with the properties and regulation of this important enzyme.

Prior to the discovery of reaction 1, it was generally assumed that the principle mechanism for amino acid synthesis in most organisms involves direct assimilation of ammonia via the action of glutamate dehydrogenase (reaction 2) coupled with transamination (reaction 3).

$$\text{TPNH} + \alpha\text{-ketoglutarate} + \text{NH}_4{}^+ \rightleftharpoons \text{glutamate} + \text{TPN}^+ \qquad (2)$$

$$\text{Glutamate} + \text{RCOCOOH} \rightleftharpoons \text{RCHNH}_2\text{COOH} + \alpha\text{-ketoglutarate} \qquad (3)$$

$$\text{TPNH} + \text{NH}_4{}^+ + \text{RCOCOOH} \rightleftharpoons \text{RHNH}_2\text{COOH} + \text{TPN}^+ \qquad (4)$$

This glutamate dehydrogenase dependent pathway is theoretically not as well suited for amino acid synthesis as is the glutamine dependent pathway depicted in Fig. 2, for two reasons: (1) The glutamine pathway is driven by ATP and is essentially irreversible under physiological conditions, thus favoring amino acid synthesis; whereas the glutamate dehydrogenase pathway is freely reversible and can lead to amino acid breakdown as well as to synthesis. (2) The affinity of glutamate dehydrogenase in most organisms is relatively low so that efficient synthesis of amino acids by reactions 2-4 will take place only in the presence of high intracellular ammonia concentrations. In contrast, glutamine synthetase has a high affinity for ammonia, and the glutamate synthase catalyzing reaction 1 has high affinities for both α-ketoglutarate and glutamine; thus, amino acid synthesis by the mechanism shown in Fig. 2 can occur readily even at low ammonia concentrations. For these reasons it is believed that the glutamine dependent mechanism is probably the most important mechanism for amino acid synthesis in microorganisms, except under conditions of very high ammonia availability when the glutamine synthetase is repressed (Woolfolk and Stadtman, 1964, 1966) and the glutamate dehydrogenase pathway can function in the biosynthetic direction.

In view of its role in the formation of amino acids, CTP, AMP, glucosamine-6-P and DPN, it is evident that glutamine is a key intermediate in the ultimate synthesis

3

of protein, nucleic acids and complex polysaccharides, and
is also of potential importance in the pyridine nucleotide
linked oxidation-reduction reactions. It is evident therefore that glutamine synthetase which catalyzes the synthesis of glutamine from ATP, glutamate and NH3 is an enzyme
of singular importance since it catalyzes the first step in
a highly branched metabolic pathway that leads ultimately
to the synthesis of nearly all of the important macromole-
cules of the cell.

Because of its central role in metabolism, it appeared
to us that glutamine synthetase should be a strategic target
for cellular control. Prompted by this consideration we
(Woolfolk and Stadtman, 1964, 1966) initiated a series of
studies in Bethesda to investigate the regulation of glutamine synthetase in E. coli. These early studies lead to the
discovery of a novel mechanism of cumulative feedback control (see Fig. 3) in which each metabolite derived from
glutamine could partially inhibit activity of glutamine
synthetase, and collectively they could inhibit its catalytic activity almost completely (Woolfolk and Stadtman,
1964).

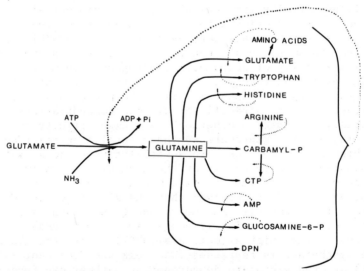

Figure 3. The regulation of glutamine metabolism. The
dotted lines indicate those steps or pathways that are
regulated by the indicated end product.

4

In the meantime continued investigations in Bethesda
(Stadtman, et al. 1968, 1970; Brown, et al. 1971) and in
Freiburg (Holzer, 1969) have led to the disclosure of
several other highly sophisticated mechanisms for the regulation of glutamine synthetase activity in E. coli.
Results of these studies will be summarized by Dr. A.
Ginsburg in her presentation later on in this symposium.
    Control of the first common enzymatic step in a highly
branched pathway must be reinforced by specific controls of
the first divergent steps that are uniquely concerned with
the synthesis of each particular end product. This is
necessary to prevent continued synthesis of individual end
products that are already in excess, and to insure continued
production of those that are not (Stadtman, 1966, 1970).
    As is illustrated in Fig. 3, in addition to its ability
to inhibit glutamine synthetase, each one of the various
end products of glutamine metabolism is able to regulate
the divergent pathway that is uniquely concerned with its
own synthesis. Although not specifically indicated in Fig.
3, this regulation involves inhibition, and/or repression
of the enzyme that catalyzes the first step in the
divergent pathway. In other words all of those reactions
in which glutamine is directly involved as a substrate are
inhibited by the ultimate end product of the pathway in
which they are involved.
    The enzymes catalyzing these glutamine dependent
reactions are of particular importance from the physiological point of view, and they offer the enzymologists and
protein chemists unique opportunities to investigate
ligand induced conformational changes of protein structure
and catalytic potential. In addition, as a class, these
enzymes catalyze rather complicated multi-substrate biochemical transformations and are therefore of particular
interest from the standpoint of reaction mechanism. It is
therefore not surprising that all of the glutamine dependent
reactions and the enzymes that catalyze them have been under
intensive investigation during the last decade. These
investigations have contributed richly to our understanding
of the mechanisms of enzyme catalysis and its cellular
regulation.
    In the organization of this symposium an effort was made
to include a presentation by at least one prominent
investigator in each area of glutamine biochemistry.
    These proceedings therefore constitute an authorative

5

comprehensive summary of all aspects of glutamine
biochemistry.

References

Brown, M.S., Segal, A. and Stadtman, E.R. (1971).
    Proc. Nat. Acad. Sci., U.S. 68, 2949.
Holzer, H. (1969) Advan. Enzymol. 32, 97.
Meister, A. (1972). In "The Enzymes" (P.D. Boyer, H. Lardy
    and K. Myrback, eds). Vol. VI. 2nd Edition, pp 196,
    Academic Press, New York.
Miller, R.M. and Stadtman, E.R. (1972). J. Biol. Chem.,
    in press.
Nagatani, H., Shimizu, M. and Valentine, R.C. (1971).
    Arch. Mikrobiol. 79, 1974.
Neidle, A. and Waelsch, H. (1959). J. Biol. Chem., 234, 586.
Srinivasan, P.R. (1959). J. Am. Chem. Soc. 81, 1772.
Stadtman, E.R. Shapiro, B.M., Kingdon, H.S., Woolfolk, C.A.
    and Hubbard, J.S. (1968) Advan. Enzym. Reg. 6, 257.
Stadtman, E.R., Ginsburg, A., Ciardi, J.E., Yeh, J.
    Advan. Enzyme. Reg. 8, 99.
Stadtman, E.R. (1966). Advan. Enzymol. 28, 41.
Stadtman, E.R. (1970). In "The Enzymes" (P.D. Boyer, Ed.)
    Vol. I, 3rd Edition, pp 297-459, Academic Press,
    New York.
Tempest, D.W., Meers, J.L. and Brown, E.M. (1970).
    Biochem. J. 117, 405.
Woolfolk, C.A. and Stadtman, E.R., (1966). Arch. Biochem.
    Biophy. 116, 177.
Woolfolk, C.A. and Stadtman, E.R. (1964). Biochem. Biophys.
    Res. Comm. 17, 313.

SECTION I

ENZYMATIC INTERCONVERSION OF GLUTAMATE AND GLUTAMINE

REGULATION OF GLUTAMINE SYNTHETASE IN <u>ESCHERICHIA</u> <u>COLI</u>

A. Ginsburg and E. R. Stadtman

Laboratory of Biochemistry, National Heart and
Lung Institute, NIH, Bethesda, Md. 20014

ABSTRACT

Glutamine synthetase has a central role in the
nitrogen metabolism of <u>E</u>. <u>coli</u>. A rigorous cellular control of glutamine synthetase activity in this microorganism
involves the following regulatory mechanisms: (1) repression; (2) feedback inhibition by end products of glutamine
metabolism; (3) environmental availability and type of
divalent cation; (4) a cascade system consisting of several
metabolite-regulated enzymes and a small regulatory protein, which together modulate the adenylylation and
deadenylylation of glutamine synthetase and thereby determine its catalytic potential, susceptibility to feedback
inhibition, and divalent cation specificity. The cascade
system includes: (a) uridylyltransferase catalyzing
uridylylation of the regulatory protein; (b) uridylyl
removing enzyme catalyzing a cleavage of 5'-UMP from the
regulatory protein; (c) adenylyltransferase catalyzing the
adenylylation of glutamine synthetase in the presence of
unmodified regulatory protein or deadenylylation of
adenylylated glutamine synthetase in the presence of
uridylylated regulatory protein. The site of adenylylation in glutamine synthetase is a specific tyrosyl residue
within each subunit polypeptide chain of the native
dodecamer of 600,000 mol. wt. Deadenylylation occurs by
a phosphorolytic cleavage of the stable 5'-adenylyl-O-
tyrosyl derivative. Heterologous interactions occur
between adenylylated and unadenylylated subunits in
hybrid glutamine synthetase molecules. Both heterologous
and homologous subunit interactions influence activity
expression, with each subunit of glutamine synthetase
being potentially active in catalysis.

SYMBOLS

$GS_{\bar{n}}$	= Glutamine synthetase with $\bar{n}$ average molar equiv ($\bar{n}$ = 0-12) covalently bound AMP groups
A	= Adenylylation
D	= Deadenylylation
ATase	= ATP:glutamine synthetase adenylyltransferase catalyzing A and D
P_{IIA}	= unmodified small protein component stimulating adenylylation (A)
UTase	= UTP:P_{IIA} uridylyltransferase
P_{IID}	= Uridylylated P_{IIA} stimulating deadenylylation (D)
UR enzyme	= Uridylyl removing enzyme catalyzing conversion of P_{IID} to P_{IIA}
PMPS	= p-chloromercuriphenylsulfonate
α-KG	= α-ketoglutarate (or 2-oxoglutarate)
GLN	= L-glutamine

INTRODUCTION

Glutamine synthetase catalyzes the synthesis of L-glutamine in the reaction:

$$L\text{-glutamate} + ATP + NH_4^+ \xrightarrow{Me^{2+}} L\text{-glutamine} + ADP + P_i \quad (1)$$

The metal ion (Me^{2+}) requirement depends on the form of the enzyme from E. coli (Wulff et. al., 1967; Kingdon et. al., 1967). Reaction (1) favors the biosynthesis of L-glutamine with a free energy change of about -5.2 kcal $mole^{-1}$ at pH 7 and 37° (Levintow and Meister, 1954; Alberty, 1968).

Because glutamine is an important intermediate in the assimilation of ammonia by E. coli, glutamine synthetase plays a central role in the nitrogen metabolism of this microorganism. The amide nitrogen of glutamine is utilized in the biosynthesis of AMP, CTP, tryptophan, histidine, glucosamine 6-phosphate, and carbamyl phosphate (Meister, 1962) and also of L-glutamate (Tempest et. al., 1970). In addition, glutamine synthetase may be coupled with glutamate synthase (Tempest et. al., 1970; R. E. Miller, this volume) and various transaminases to provide a pathway for ATP-dependent synthesis of most amino acids (Miller and Stadtman, 1971). A rigorous

cellular control of glutamine synthetase activity in
E. coli has been demonstrated; (see reviews of Holzer,
1969 and of Shapiro and Stadtman, 1970).

Some physical and chemical properties of E. coli
glutamine synthetase, together with references to
particular studies are summarized in Table I. These
studies have been reviewed elsewhere (Ginsburg, 1972).

Table I

E. Coli Glutamine Synthetase

		Ref.*
A. Native Protein at pH 7.		
Molecular weight	600,000	a
Apparent specific volume	~ 0.707ml/g	a
Sedimentation coefficient	20.3S	a
α-Helical structures	$\sim 36\%$	b
Reactive sulfhydryl groups	-	c
Stability in presence of Mn^{2+}	+	a,d,e
Removal of Mn^{2+} or Mg^{2+}	Inactivation	f
Appearance in electron-	Dodecamer:	
microscopy	Double hexagon	g
Hexagonal ring cross section	129Å	h
B. Subunits.		
Molecular weight	50,000	d,g
Molecular dimensions	45x45x53Å	g
Number of sulfhydryl groups	5 (no-S-S-)	c
N-Terminal amino acid	Serine	i
C-Terminal amino acid	Valine	i,j
Adenylylation site	R-Asp-Asn-Leu-Tyr-Asp-R'	
	AMP	k,l
(R = H_2NIle-His-Pro-Gly-Glu-Ala-Met-Lys-)		l
(R'= Leu-Pro-Pro-Glu-Gly-Glu-Ala-LysCOO⁻)		l

*Ref.: a/Shapiro and Ginsburg (1968). b/Hunt and Ginsburg
(1972). c/Shapiro and Stadtman (1967). d/Woolfolk et. al.
(1966). e/Woolfolk and Stadtman (1967b). f/Kingdon et. al.
(1968). g/Valentine et. al. (1968). h/Eisenberg et. al.
(1971). i/Lahiri et. al. (1972). j/Ginsburg (1972).
k/Shapiro and Stadtman (1968). l/Heinrikson and Kingdon (1971).

The native enzyme has twelve identical subunits which are molecularly arranged in two face to face hexagons. In hydrodynamic studies, the enzyme behaves as a spherical particle of 600,000 molecular weight. The dodecamer is very stable in the presence of the activating divalent cations, Mn^{2+} or Mg^{2+}; monovalent cations also specifically effect a stabilization of the quaternary structure. A specific tyrosyl residue of each of the twelve subunits can be adenylylated enzymatically to form a 5'-adenylyl-$\underline{O}$-tyrosyl derivative. Adenylylation and deadenylylation are catalyzed by a cascade-type of enzyme system that is itself regulated by intracellular metabolite concentrations (see below). As discussed below also, various catalytic properties of glutamine synthetase are altered drastically by adenylylation. Further, multiple molecular forms, differing in number and orientation of adenylylated subunits, may be produced by adenylylation. Subunit interactions in hybrid enzyme forms containing both adenylylated and unadenylylated subunits affect both stability and catalytic properties of glutamine synthetase (Stadtman et. al., 1970).

REGULATORY MECHANISMS

Four different mechanisms for the cellular control of glutamine synthetase activity in E. coli will be discussed. Emphasis will be on the novel adenylylation-deadenylylation mechanism of regulation since this dominates other control mechanisms.

1. ## Repression

The cellular synthesis of glutamine synthetase is repressed when E. coli cells are grown on media containing either readily available nitrogen or high concentrations of ammonium salts (Woolfolk et. al., 1966), in which cases ammonia can substitute for glutamine in biosynthetic pathways. Cell growth under conditions of limiting ammonia produces a twenty-fold derepression of glutamine synthetase, with the latter representing about 1% of the total soluble proteins of the cell. Glutamate synthase, which is another important enzyme in glutamine metabolism, is not appreciably repressed by ammonia levels that inhibit the synthesis of glutamine synthetase (Miller and Stadtman, 1971; 1972).

12

Prusiner et. al. (1972) have explored the effect of
cyclic AMP on the intracellular levels of five enzymes
concerned with the interconversion of glutamate and
glutamine in E. coli. Cyclic AMP was supplied to the
culture medium of a mutant of E. coli K-12 deficient in
adenyl cyclase. The exogeneously supplied cyclic AMP
caused a 2- to 3-fold increase in glutamine synthetase
and in glutamate dehydrogenase whereas intracellular
levels of glutamate synthase were decreased 2-fold. These
effects of cyclic AMP were not observed in mutants lacking
cyclic AMP receptor protein (Pastan and Perlman, 1970) or
when protein synthesis was inhibited in the mutant deficient in adenyl cyclase. Possibly, different molecular
mechanisms exist for the regulation of the synthesis of
these enzymes by cyclic AMP. The data of Berberich (1972)
show that there is no close genetic linkage between the
structural genes in E. coli K-12 coding for glutamate
synthase, glutamate dehydrogenase, and glutamine synthetase. However, indirect linkage in a broad scheme of
nitrogen metabolism was not excluded.

2. Divalent Cation Specificity

Certain divalent cations such as Mn^{2+} are involved in
the structural stabilization and activation of glutamine
synthetase. Divalent cations which induce specific
structural changes in glutamine synthetase are Mn^{2+}, Mg^{2+},
Ca^{2+}, Co^{2+}, and Zn^{2+}. The activation of glutamine synthetase in the biosynthetic reaction (1) by Mg^{2+}, Mn^{2+}, or
Co^{2+} is intimately linked to the adenylylation state of
the enzyme, and in this respect these metal ions also
could have a regulatory function.

One of the most dramatic changes induced in glutamine
synthetase by adenylylation, and one first noted by Wulff
et. al. (1967) and Kingdon et. al. (1967), is the inactivation of the enzyme in a Mg^{2+}-activated biosynthetic
assay. This is illustrated in Fig. 1, which also shows
that adenylylation produces a reciprocal, although less
pronounced, activation by Mn^{2+}. The changed metal ion
specificity in reaction (1) of glutamine synthetase upon
adenylylation was first observed by Kingdon et. al. (1967).
Moreover, the pH optimum with Mg^{2+} or Mn^{2+} is different.
Since the fully adenylylated form of glutamine synthetase

has no activity with Mg^{2+}, Mn^{2+} is concluded to be a specific activator of adenylylated subunits. Conversely, Mg^{2+} or Co^{2+} (Fig. 2) appears to specifically activate unadenylylated subunits of the dodecamer.

Unlike mixtures of unadenylylated and the fully adenylylated enzymes, different preparations of glutamine synthetase exhibit a nonlinear decrease in Mg^{2+}- or in Co^{2+}-, or a nonlinear increase in Mn^{2+}-, activated

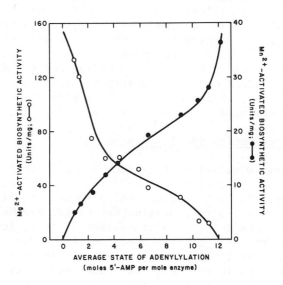

Fig. 1. The variation in Mg^{2+}- and Mn^{2+}-dependent biosynthetic activities (μmoles ADP formed per minute at 37° per milligram protein) as a function of the average extent of adenylylation of the glutamine synthetase preparation. Approximately optimal activities of different enzyme forms in Mg^{2+}-activated (O) or Mn^{2+}-activated (●) spectrophotometric assays at 37° are shown. The Mg^{2+}-activated assays at pH 7.6 contained 50 mM $MgCl_2$, 90 mM KCl, 50 mM NH_4Cl, 30 mM L-glutamate, and 5 mM ATP; Mn^{2+}-activated assays at pH 6.5 contained 6 mM $MnCl_2$, 90 mM KCl, 100 mM NH_4Cl, 100 mM L-glutamate, and 5 mM ATP. The data are from Ginsburg et. al. (1970).

14

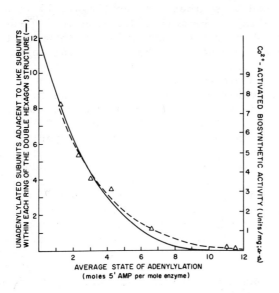

Fig. 2. Heterologous and homologous subunit interactions in hybrid glutamine synthetase molecules that affect Co^{2+}-activation in the biosynthetic reaction. The average number of unadenylylated subunits adjacent only to other unadenylylated subunits within each ring of the double hexagon structure was calculated at different average states of adenylylation (see text) and is plotted against the latter quantity in the continuous solid curve. The open triangles connected by the dashed line indicate the Co^{2+}-activated biosynthetic activities at different adenylylation states. The Co^{2+}-supported biosynthetic activity was determined by measuring inorganic phosphate release per minute at $37°$ in an assay mixture containing 25 mM imidazole-25 mM triethanolamine buffer at pH 7.0, 50 mM L-glutamate, 50 mM NH_4Cl, 7.5 mM ATP, and 7.5 mM $CoCl_2$. The data are from Segal and Stadtman, 1972a).

biosynthetic assays as a function of increasing adenylylation (Figs. 1 and 2). A more striking effect of subunit interactions affecting the catalytic potential of each

15

activated subunit of the enzyme has been obtained in the Co^{2+}-activated biosynthetic assay (Segal, 1971; Segal and Stadtman, 1972a). The marked decrease in Co^{2+}-dependent activity with increasing adenylylation shown in Fig. 2 can be described quite well by the theoretical solid curve of Fig. 2. This curve was constructed, assuming that Co^{2+}-activated biosynthetic activity is expressed only by the average number of unadenylylated subunits adjacent to other unadenylylated subunits within a ring of the double hexagon structure (Segal and Stadtman, 1972a). For these calculations, the number of unique, nonsuperimposable configurations of glutamine synthetase for each state of adenylylation was determined, assuming a randon distribution of adenylylated subunits within the enzyme molecule (Table I). The average number of unadenylylated subunits not in contact with adenylylated subunits was computed for all unique configurations at each state of adenylylation. A corollary to this analysis is that adjacent adenylylated subunits inhibit unadenylylated subunits from expressing Co^{2+}-dependent activity. There is little doubt that interactions between adenylylated and unadenylylated subunits in hybrid molecules containing both types of subunits modulate the response of glutamine synthetase to activation by Mg^{2+}, Mn^{2+}, or Co^{2+} in the biosynthetic reaction (1). Homologous interactions between adjacent like subunits are probably important also, as is illustrated by the activation by Co^{2+} which is augmented by homologous interactions between adjacent unadenylylated subunits.

Glutamine synthetase from E. coli also catalyzes a γ-glutamyl transfer reaction (Woolfolk et. al., 1966):

$$\text{L-glutamine} + NH_2OH \xrightarrow[\text{ADP, arsenate}]{Me^{2+}} \gamma\text{-glutamyl-}$$

$$\text{hydroxamate} + NH_4^+ \qquad (2)$$

The divalent cation specificity in reaction (2) is different from that in the biosynthetic reaction (1). Under certain assay conditions (with a mixed buffer of 2-methylimidazole,2,4-dimethylimidazole, and imidazole), there is an isoactivity point at pH 7.15, at which pH Mn^{2+} gives equivalent activation of glutamine synthetase forms that are adenylylated to different extents (Stadtman et. al., 1970). This is despite the fact that

the γ-glutamyl transferase activities of unadenylylated
and adenylylated glutamine synthetase have different pH
optima (Stadtman et. al., 1970), different sensitivities
to feedback inhibitors (Shapiro et. al., 1967) (see
below), and different responses to Mg^{2+} in the presence
of Mn^{2+} (Stadtman et. al., 1968). This latter property
has provided the following useful relationship for measuring the average state of adenylylation ($\bar{n}$) of glutamine
synthetase in either pure or impure enzyme preparations:

$$\bar{n} = 12 - 12(b)/(a) \qquad (3)$$

where $\bar{n}$ may vary between 0 and 12 moles 5'-adenylyl
groups per mole active enzyme; (a) is a measure of the
total transferase activity of both adenylylated and
unadenylylated subunits in the presence of 0.3 mM Mn^{2+} at
pH 7.15; (b) is the measure of the transferase activity
of only unadenylylated subunits in the presence of 0.3 mM
Mn^{2+} plus 60 mM Mg^{2+} at pH 7.15 (Stadtman et. al., 1970).
With 60 mM Mg^{2+} plus 0.3 mM Mn^{2+}, the activity of
unadenylylated glutamine synthetase is the same as it is
with Mn^{2+} alone, and the pH-activity profile is shifted so
that it peaks at the isoactivity pH 7.15. In contrast,
the presence of 60 mM Mg^{2+} in the transferase assay
completely inhibits adenylylated subunits from expressing
this activity. The linear relationship of Eq. (3) has
been shown by Stadtman et. al. (1970) to be valid for
most enzyme preparations tested, including many with $\bar{n}$
intermediate between $\bar{n} = 0.8$ and $\bar{n} = 12$. Heterologous
subunit interactions at intermediate stages of adenylylation that would influence (a) or (b) therefore are not
evident. Like Mg^{2+}, Zn^{2+} (Miller et. al., 1972b) also can
partially support the γ-glutamyl transferase activity of
unadenylylated, but not of adenylylated, glutamine
synthetase.

Glutamine synthetase is inactivated by a removal of
Mn^{2+} or Mg^{2+} (Table I). Consequently, fluctuations in the
environmental free concentration of specific divalent
cations in the cell could be important. The inactivation
of glutamine synthetase by the removal of Mn^{2+} causes a
conformational change (relaxation) in the protein structure
that leads to an exposure of sulfhydryl groups (Shapiro and
Stadtman, 1967), to an increased susceptibility to

17

dissociation (Woolfolk and Stadtman, 1967b), and (without a change in molecular weight) to an exposure of tryptophanyl and tyrosyl residues (Fig. 3) and a decrease in sedimentation rate of $\Delta s_{20,w}^{0} \simeq -0.6S$ (Shapiro and Ginsburg, 1968). Relaxation, however, does not appear to produce changes in the secondary structure of the enzyme, as judged by optical rotatory dispersion and circular dichroism measurements (Hunt and Ginsburg, 1972). A <u>tightening</u> of the metal ion-free enzyme by the addition of Mn^{2+}, Mg^{2+}, or Ca^{2+} reactivates glutamine synthetase (Kingdon <u>et</u>. <u>al</u>., 1968), and reverses the conformational changes produced by relaxation (Shapiro and Ginsburg, 1968). The unadenylylated enzyme has a significantly higher affinity than does

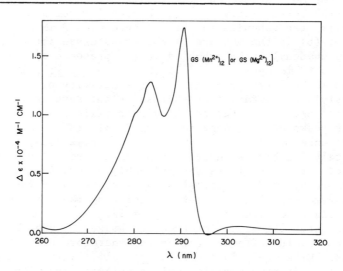

Fig. 3. Ultraviolet difference spectra at 25° for unadenylylated glutamine synthetase with 12 molar equiv Mn^{2+} (or Mg^{2+}) bound <u>vs</u> the metal ion-free enzyme at the same concentration in the reference compartment. The enzyme (2.8×10^{-6} M) was in 0.05 M Tris-0.1 M KCl buffer at pH 7.2. A 0-0.1 slidewire of a Cary Model 15 spectrophotometer was used for difference spectral measurements, which are shown after completion of time-dependent spectral changes produced by the addition of metal ions. The data are from Hunt and Ginsburg (1972), although the spectral perturbations of glutamine synthetase by Mn^{2+}, Mg^{2+}, and Ca^{2+} were first shown by Shapiro and Ginsburg (1968).

18

the adenylylated form for Mn^{2+} in either equilibrium binding (Denton and Ginsburg, 1969) or kinetic (Stadtman and Smyrniotis, 1972) studies. Associated with the tightening or activation process is the binding of one metal ion per enzyme subunit. For example, the activation of the unadenylylated enzyme at pH 7.2 (37^{0}) requires the binding of 12 molar equiv Mn^{2+} with $K' \simeq 2 \times 10^{6}$ M^{-1} (Denton and Ginsburg, 1969), Mg^{2+} with $K' \simeq 2 \times 10^{4}$ M^{-1} (Hunt and Ginsburg, 1972), or Ca^{2+} with $K' \simeq 5 \times 10^{4}$ M^{-1} (Hunt, 1972). The tightening process has an activation energy of 21 kcal $(mole\ subunit-Mn)^{-1}$ (Hunt and Ginsburg, 1972). The binding of either Mn^{2+}, Mg^{2+}, or Ca^{2+} to an enzyme subunit causes the release of two equiv protons; one proton is displaced from the enzyme instantaneously and the second proton is released in a slow first order process. The half-time of

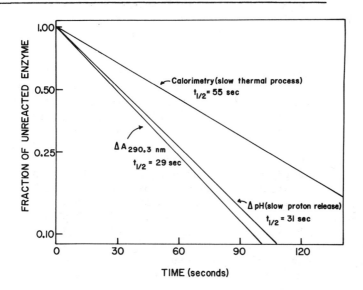

TIME (seconds)

Fig. 4. Idealized first order rate plots for the tightening process. These rate plots were constructed from an average of half-time values observed at 37^{0} and pH 7.2 in measurements of the slow thermal process (Hunt et. al., 1972), of the slow proton release by pH changes in dilute buffers, or of the ultraviolet absorbancy change at 290.3 nm (Hunt and Ginsburg, 1972) for the binding of 12 molar equiv Mn^{2+} (or Mg^{2+}) to glutamine synthetase.

the slow proton release was similar to that measured for
activation (Kingdon et. al., 1968) or for the reburial of
tryptophanyl-tyrosyl residues in ultraviolet difference
spectral measurements (Figs. 3 and 4) at 25°. Thus, the
interaction of Mn^{2+}, Mg^{2+}, or Ca^{2+} with glutamine synthetase occurs as a bimolecular reaction which is followed by
a slow first order reaction, involving an induced conformational change in the enzyme structure. Fig. 4 shows that
longer half-times were observed at 37° in calorimetric
measurements, in which a small endothermic contribution was
not resolved from the slow proton release in the apparently
first order slow thermal process (Hunt et. al., 1972). The
calorimetric experiments indicated that little net heat is
associated with the interaction of Mn^{2+} (or Mg^{2+}) with
glutamine synthetase. For the binding of each Mn^{2+} to
unadenylylated glutamine synthetase ($K_{eq} = K'/(H^+)^2$) at 37°
and pH 7.2 (standard state for hydrogen ions at activity
of $10^{-7 \cdot 2}$ M), $\Delta H \simeq + 3$ kcal mole^{-1} and $\Delta S \simeq + 38$ cal deg^{-1}
(mole subunit-Mn)$^{-1}$ (Hunt et. al., 1972).

The activity studies of Segal and Stadtman (1972b)
with pairs of divalent cations have demonstrated that two
different divalent cations can occupy sites that modulate
the expression of biosynthetic activity at the subunit
level and also affect subunit interactions in the dodecamer. For example, Cd^{2+} (which by itself cannot support
glutamine synthetase activity), when present with Mg^{2+} or
Co^{2+}, shifts the pH optimum and changes the saturation
functions for substrates from those observed with either
Mg^{2+} or Co^{2+} alone. The Mg^{2+}-dependent activity of
unadenylylated glutamine synthetase (with 5 mM $MgCl_2$-7.5 mM
ATP present in biosynthetic assays at pH 8) is inhibited
50% by 8×10^{-5} M Ca^{2+}, 7×10^{-6} M Cd^{2+}, or 2×10^{-5} M
Mn^{2+}, whereas 1 mM Co^{2+} stimulates the Mg^{2+}-dependent
activity about 2.5-fold under the same conditions.

The conformation at the active site of the glutamine
synthetase subunit therefore can be affected by at least
two metal ion binding sites per subunit. When only a
single species of metal ion is present, however, it is
uncertain whether one or two metal ion binding sites need
to be saturated for the expression of activity. The
expression of biosynthetic activity does not necessarily
have the same requirements as the induced ultraviolet

20

spectral perturbations with Mn^{2+}, Mg^{2+}, or Ca^{2+} (Fig. 3),
and as the relatively small spectral perturbations with
Co^{2+} (Segal and Stadtman, 1972b) or Zn^{2+} (Miller et. al.,
1972a), which require the saturation of each subunit with
only one metal ion.

3. Feedback Inhibition

Glutamine synthetase activity may be regulated
through feedback inhibition by the multiple end products
of glutamine metabolism. Woolfolk and Stadtman (1964,
1967a) have described the cumulative nature of the feedback
inhibition patterns observed with mixtures of alanine,
glycine, histidine, tryptophan, CTP, AMP, carbamyl phosphate, and glucosamine 6-phosphate. The effects of multiple inhibitors are cumulative, provided that each inhibitor
is present at a physiological concentration that produces
only partial inhibition by itself (Shapiro and Stadtman,
1970).

The sensitivity of glutamine synthetase toward feedback inhibition was found to be altered by adenylylation
(Kingdon and Stadtman, 1967; Kingdon et. al., 1967;
Shapiro et. al., 1967). Generally, Mn^{2+}-activated
adenylylated enzyme forms are more inhibited than is the
unadenylylated enzyme by AMP, L-tryptophan, L-histidine,
or CTP. Studies with mixtures of inhibitors are enormously complicated by the fact that the divalent cation
and substrate concentrations determine the response of
different enzyme forms to individual inhibitors (Ginsburg,
1969; Stadtman and Smyrniotis, 1972).

Kinetic evidence (Woolfolk and Stadtman, 1967a;
Stadtman and Smyrniotis, 1972), the effects of inhibitors
on the inactivation of glutamine synthetase by mercurials
(Shapiro and Stadtman, 1967) and binding studies (Ginsburg,
1969) suggest that the enzyme has separate sites for most,
if not all, of the feedback inhibitors. Separate sites for
AMP and L-tryptophan were demonstrated additionally by the
calorimetric results presented in Table II (Ross and
Ginsburg, 1969). In these studies, the sum of the heats (Q)
measured for the individual effectors was equal within
experimental error to that measured for a saturating
mixture of these inhibitors.

Table II

Calorimetric Studies[a]

On The Interaction Of Glutamine Synthetase With Inhibitors

Expt (Inhibitors present at saturating concentrations)	Q[b] (mcal/mg enzyme)
I. GS + L-tryptophan	- 0.147
II. GS + AMP	- 0.040
III. GS + L-tryptophan + AMP	- 0.192

a/Data of Ross and Ginsburg (1969). The calorimetric experiments were performed at 25°, using a buffer of 0.02 M imidazole-chloride-0.1 M KCl-0.001 M $MnCl_2$ at pH 7.07.

b/Q (the measured heat of binding) is corrected for a small endothermic heat of dilution of the enzyme (GS), which resulted from the effector solution being added to the enzyme in a batch-type calorimeter.

The stoichiometry of the binding of two of the feedback inhibitors (AMP and L-tryptophan) and of the substrate (ATP-Mn) to glutamine synthetase is shown in Fig. 5. Since 12 molar equiv of each ATP, AMP, and L-tryptophan are bound, each subunit of the enzyme appears to possess both inhibitor and substrate sites. The evidence presented above in Section 2 suggested that each subunit has a potential catalytic site, the divalent cation specificity of which is dictated by the absence or presence of a covalently bound 5'-adenylyl group to the subunit. Although not readily apparent from the plot in Fig. 5, a small negative-type interaction between glutamine synthetase and ATP-Mn was observed at low concentrations of this substrate (Denton and Ginsburg, 1970). The data of Fig. 5 indicate that the 12 AMP binding sites are independent and equivalent with $K' = 8 \times 10^3 M^{-1}$. However, L-tryptophan or L-glutamate appeared to lower the affinity of the enzyme for AMP (Ginsburg, 1969). Although the binding of AMP was slightly decreased by the presence of L-tryptophan, L-serine, L-histidine, L-alanine, and glycine were without effect.

22

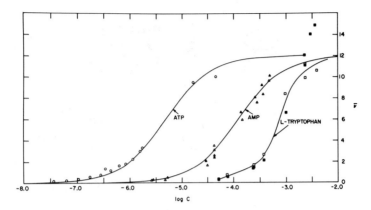

Fig. 5. Equilibrium binding of substrate (ATP-Mn) or inhibitors (AMP and L-tryptophan) to glutamine synthetase (GS$\overline{2.3}$) at 4°. $\bar{\nu}$ is the average number of moles of ^{14}-ATP(O), ^{32}P-AMP (△, ▲), or L-(^{14}C-methylene) tryptophan (□, ●) bound per mole of glutamine synthetase at the free concentration, C, of the compound present. The buffer in each case was 20 mM imidazole-chloride, 100 mM KCl (pH 7.4-7.5) with 5 mM MnCl$_2$ (O, ▲) or 1 mM MnCl$_2$ (■) or with 50 mM MgCl$_2$(△) or 1 mM MgCl$_2$ (□). The ATP-Mn binding data are from Denton and Ginsburg (1970), with the curve theoretical for 12 sites with K' = 2 x 10^5 M^{-1}. In the inhibitor-binding studies of Ginsburg (1969), the AMP curve was constructed for 12 independent sites with K' = 8 x 10^3 M^{-1}; the binding curve for L-tryptophan was drawn arbitrarily to fit the data.

The binding of L-tryptophan to the enzyme (Fig. 5) is cooperative, suggesting that homologous interactions occur between L-tryptophan binding sites of different subunits.

Fig. 6 schematically shows one subunit of the native dodecamer. Two binding sites for the activating and stabilizing divalent cation and an additional possibly specific site for a monovalent cation are shown. In addition, substrate and inhibitor sites are shown. Besides these sites, the subunit must have intersubunit contact with adjacent subunits in the same and in adjoining hexagonal rings. An inferred site of interaction for the adenylylating-deadenylylating enzyme system (see below) of E. coli with each subunit is indicated by the open circle of Fig. 6.

23

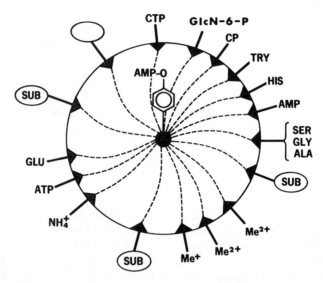

Fig. 6.   Diagrammatic representation of a single subunit of E. coli glutamine synthetase showing multiple binding sites. The picture is from Stadtman et. al. (1972). Each closed triangle on the circumference represents distinct binding sites for the inhibitors (CTP, glucosamine 6-P, carbamyl phosphate, L-tryptophan, L-histidine, and AMP, with a possible single site for L-serine, glycine, and L-alanine), for activating and stabilizing divalent and monovalent cations (Me^{2+}, Me^{+}), and for the substrates (L-glutamate, ATP, and NH$_4^+$). The encircled "SUB" indicates interaction sites between adjacent subunits. The open circle shows a binding site for the enzyme system that catalyzes adenylylation and deadenylylation. The solid center circle represents the site composed of the tyrosyl residue that undergoes adenylylation and deadenylylation. Interactions between ligand binding sites and the adenylylation site are indicated by the dashed lines.

At the center is shown the specific tyrosyl residue which undergoes covalent modification, with 5'-adenylic acid attached in phosphodiester linkage to the phenolic hydroxyl group (Shapiro and Stadtman, 1968). Since adenylylation

24

can affect the interactions of all ligands with the enzyme,
the interaction of the subunit with the adenylylating-
deadenylylating proteins, and subunit interactions within
the dodecamer, dashed lines are drawn between these sites
(peripheral closed triangles of Fig. 6) and the site of
adenylylation (center closed circle in Fig. 6).

Without regulatory subunits, the regulation of glutamine synthetase activity in E. coli occurs at the subunit
level. The direct and indirect structural alterations
induced by the novel, regulatory mechanism of adenylylation
provide an example of single-site modification affecting
both intra- and intersubunit interactions of an oligomeric
protein.

4. The Adenylylation-Deadenylylation Enzyme System:
   A Metabolite-Regulated Cascade Enzyme System

Both adenylylation (Eq. 4) and deadenylylation (Eq. 5)
reactions are catalyzed by adenylyltransferase in the
presence of Mg^{2+} or Mn^{2+} (Anderson et. al., 1970; Anderson
and Stadtman, 1971; Brown et. al., 1971).

$$12\ ATP + GS \xrightarrow[\text{10 mM } Mg^{2+};\ pH\ 7]{\text{ATase}} GS \cdot (AMP)_{12} + 12\ PP_i \qquad (4)$$

$$(\Delta G^o_{obs} = -1.0\ kcal/mole)$$

$$GS \cdot (AMP)_{12} + 12\ P_i \xrightarrow[\text{10 mM } Mg^{2+};\ pH\ 7]{\text{ATase}} GS + 12\ ADP \qquad (5)$$

$$(\Delta G^o_{obs} = -1.0\ kcal/mole)$$

In reaction (4), unadenylylated glutamine synthetase
can be adenylylated to the extent of 12 molar equiv AMP
groups attached (Kingdon et. al., 1967; Wulff et. al., 1967),
with a corresponding release of pyrophosphate. Mantel and
Holzer (1970) have shown that reaction (4) is reversible
and that the adenylyl-O-tyrosyl bond in glutamine synthetase is apparently energy rich, with $\Delta G^o_{obs} = -9.0$ kcal per
mole. (The free energy changes indicated were calculated
from the data of Mantel and Holzer (1970) and of Alberty
(1969) for ATP hydrolysis at pH 7 and $pMg^{2+} = 2$).

In reaction (5), deadenylylation of adenylylated
glutamine synthetase was shown by Anderson and Stadtman
(1970) to occur by a phosphorolytic cleavage of the

25

adenylyl-$\underline{O}$-tyrosyl bond. This reaction is exergonic by about 1 kcal/mole also, but is apparently irreversible.

Since both adenylylation and deadenylylation reactions are catalyzed by the same enzyme, coupling between them must be prevented by an appropriate control of each function. Otherwise, a futile cycle will exist in which glutamine synthetase fluctuates between adenylylated and unadenylylated forms and ATP will undergo phosphorolysis to ADP and pyrophosphate, as shown by the sum of reactions (4 and 5) in Eq. (6).

$$12 \text{ ATP} + 12 \text{ P}_i \longrightarrow 12 \text{ ADP} + 12 \text{ PP}_i \qquad (6)$$

$$(\Delta G^o_{obs} = -2.0 \text{ kcal/mole})$$

An aimless coupling of reactions (4 and 5) is prevented by an elaborate regulatory system involving metabolite control of the adenylyltransferase (Table III) and an interaction of adenylyltransferase with a small regulatory protein (P_{II}), which exists in two interconvertible forms (Fig. 7).

The scheme shown in Fig. 7 summarizes our present knowledge of the cellular control of the adenylylation and deadenylylation reactions. At least four discrete proteins and a cascade-type of enzyme regulation are involved. The interconversion of a small regulatory protein ($P_{IIA} \rightleftharpoons P_{IID}$) is catalyzed by enzymes that have recently been separated from ATase and P_{II} by column chromatography in unpublished experiments of Mangum and Magni (1972). A uridylyltransferase (which is activated by Mg^{2+} or Mn^{2+}, α-KG and ATP) covalently attaches UMP to P_{IIA} to form P_{IID} (Brown et. al., 1971; Mangum, 1972). The uridylylated regulatory protein (P_{IID}) is the form of P_{II} that stimulates ATase-catalyzed deadenylylation. The unmodified form of the regulatory protein (P_{IIA}) can be regenerated from P_{IID} by the action of a Mn^{2+}-specific uridylyl removing enzyme which catalyzes the hydrolytic cleavage of UMP from P_{IID} (Mangum and Magni, 1972). The unmodified regulatory protein (P_{IIA}) interacts with ATase to stimulate the adenylylation reaction. Glutamine and inorganic phosphate are potent inhibitors of the uridylylation reaction (Brown et. al., 1971). Preliminary studies of Mangum and Magni (1972) indicate that Mg^{2+}

26

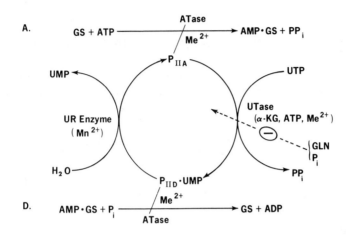

Fig. 7. Scheme of $P_{IIA} \rightleftharpoons P_{IID}$ Interconversion from the studies of Brown et. al. (1971) and Mangum and Magni (1972). The unmodified form of P_{II} (P_{IIA}) reacts with ATase to stimulate adenylylation (A). The uridylylated form of P_{II} (P_{IID}) interacts with ATase to activate deadenylylation (D). The uridylyltransferase UTase is activated by α-KG, ATP, and Mg^{2+} or Mn^{2+}, and is markedly inhibited by GLN or P_i. A uridylyl removing (UR) enzyme is specifically activated by Mn^{2+} in hydrolyzing UMP from P_{IID}.

cannot replace Mn^{2+} in the P_{IID} to P_{IIA} conversion. Specific metabolite controls (more of which may be discovered later) of the P_{IIA} - P_{IID} interconversion (Fig. 7) prevent a futile cycle of UTP hydrolysis to UMP and PP_i, which is the net reaction involved.

The adenylyltransferase and the regulatory protein (P_{II}), as well as the modifying enzymes of P_{II}, could exist as a complex in the cell. In this case, a cascade-type of regulation actually arises from the separate inputs by metabolites. Alternatively, the regulatory protein (P_{II}) could serve in other as yet unknown regulatory functions of the cell.

As indicated in Table III, the adenylyltransferase has an absolute requirement for either Mg^{2+} or Mn^{2+} in

Table III

Some Catalytic Properties Of Adenylyltransferase

Activity	Requirements	Positive Effectors	Negative Effectors	Ref.[*]
Adenylylation (Optimum pH = 8)	Mg^{2+} or Mn^{2+} ATP, GS	GLN, P_{IIA}	α-KG, UTP	a-g
Deadenylylation (Optimum pH = 7.2)	Mg^{2+} or Mn^{2+} P_i, GS·(AMP)	P_{IID} with α-KG, ATP	-	f-i
PP_i-ATP Exchange (1% of A-activity)	PP_i, ATP Mg^{2+}, GS, GLN	-	-	e,j

[*]Ref.: a/Wulff et. al. (1967). b/Kingdon et. al. (1967). c/Stadtman et. al. (1968). d/Ebner et. al. (1970). e/Oliver (1972). f/Brown et. al. (1971). g/Mangum (1972). h/Shapiro (1969). i/Anderson and Stadtman (1971). j/Wohlhueter et. al. (1972).

catalyzing the deadenylylation or adenylylation reactions. Deadenylylation is activated by P_{IID} (Fig. 10 below), ATP, and α-ketoglutarate. Adenylylation is stimulated by P_{IIA} and/or glutamine, with divalent cations also having a specific role (Fig. 8). In the experiments of Fig. 8, the ratio of activities: (ATase + P_{IIA})/ATase for P_{IIA} stimulation, is about 5 in a Mn^{2+}- supported assay without glutamine and about 3 in a Mg^{2+}-supported assay with or without glutamine present. In contrast to the stimulatory action of P_{IIA}, P_{IID} inhibits the Mg^{2+}-supported, ATase-catalyzed adenylylation reaction (Oliver, 1972). The data of Fig. 8 show that the Mg^{2+}-supported (P_{IIA}-stimulated) activity is extremely responsive towards activation by glutamine, whereas the Mn^{2+}-supported (P_{IIA}-stimulated) adenylylation rate is almost unchanged by varying glutamine concentrations. Glutamine has mixed activating and inhibitory effects in the Mn^{2+}-supported adenylylation assay. Some effects of modifying the sulfhydryl groups of the adenylyltransferase are discussed below.

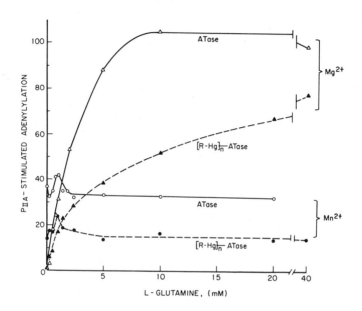

Fig. 8. The effects of glutamine on ATase-catalyzed (P_{IIA}-stimulated) adenylylation in the presence of Mg^{2+} or Mn^{2+}. An adenylyltransferase fraction which had been pretreated with PMPS is indicated by $[R-Hg]_n$—ATase. Assays for ^{14}C-AMP incorporation into GS contained in 50 μl: 50 mM each imidazole, 2-methylimidazole, and 2,4 dimethyl-imidazole buffer (pH 8.0), 1 mM ^{14}C-ATP (10^6 cpm/μmole), 300 μg $GS_{\overline{1}}$, either 20 mM $MgCl_2$ with 0.3 μg ATase or 5.4 mM $MnCl_2$ with 0.6 μg ATase, a saturating amount of P_{IIA} (8 μg), and the indicated concentration of L-glutamine. P_{IIA}-Stimulated adenylylation is expressed as nmoles ^{14}C-AMP incorporated into GS per minute at 37° per mg ATase. Without GLN, (ATase + P_{IIA}) is 8-fold more active in the Mn^{2+}- than in the Mg^{2+}-supported adenylylation assay. These data are from studies of Oliver (1972).

The supporting evidence for the scheme of Fig. 7 will be presented in 2 parts: (a) properties of the adenylyl-transferase; (b) the $P_{IIA} \rightleftharpoons P_{IID}$ interconversion. The reciprocal relationships between metabolites, which are indicated in Fig. 7 and Table III, will be considered last in part (c).

(a) Properties of the adenylyltransferase
     In the original studies of the deadenylylating enzyme system by Shapiro (1969), two apparently obligatory protein components (labeled P_I and P_{II}) were found to be easily separable by gel filtration (Fig. 9). In later studies of Anderson et. al. (1970), it was found that the larger component (P_I) of the deadenylylating enzyme system had adenylylating activity (Fig. 9). The size of this protein, with a molecular weight of about 130,000 $\pm$ 10,000 (Hennig et. al., 1970), was found to be about the same as that reported for the adenylyltransferase isolated by Ebner et. al. (1970). Deadenylylating and adenylylating activities of the P_I component (ATase) of Fig. 9 were shown by Anderson

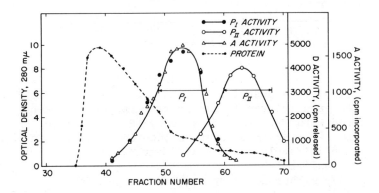

Fig. 9. Agarose gel filtration of a partially purified deadenylylating enzyme preparation from the paper of Anderson et. al. (1970). The enzyme fraction (32 ml) was applied to a column (4 x 80 cm) of Bio-Gel A-0.5m agarose equilibrated with 50 mM phosphate-10 mM 2-mercaptoethanol-0.25 mM $K_2MgEDTA$ (pH 7.2); fractions of 10 ml were collected. Activities are reported as either cpm released from ^{14}C-AMP-adenylylated GS (D ACTIVITY) or as cpm incorporated from ^{14}C-ATP into unadenylylated GS (A ACTIVITY). To determine D-activity of P_I or P_{II}, respectively, fractions from the P_I peak were mixed with P_{II} or fractions from the P_{II} peak were mixed with P_I.

et. al. (1970) to copurify, to comigrate in polyacrylamide
gel electrophoresis, and to be coinactivated by heat treatment (38-48°). The studies of Anderson and Stadtman (1971)
and of Brown et. al. (1971) also demonstrated that the
adenylyltransferase (P_I) by itself could slowly catalyze a
Mn^{2+}- or a Mg^{2+}-supported deadenylylation reaction. It is
concluded therefore that the adenylyltransferase possesses
both adenylylating and deadenylylating activities, as
presented above in reactions (4 and 5).

The adenylyltransferase is an acidic protein with an
isoionic point (pI) of pH 4.3 (Wolf et. al., 1972), as is
its substrate, glutamine synthetase, with pI = pH 4.9
(Ciardi et. al., 1972). Adenylyltransferase has an apparent
association constant for unadenylylated glutamine synthetase
of ~ 5 x 10^5 M^{-1} and an absolute specificity for the native,
dodecameric glutamine synthetase structure (Hennig and
Ginsburg, 1971; Wolf et. al., 1972).

Fluorescence measurements with a hydrophobic probe
(Holzer, 1969), as well as kinetic studies (Hennig and
Ginsburg, 1971), indicate that adenylyltransferase has one
or more allosteric sites for L-glutamine. The results of
Wolf and Ebner (1972) also show that the adenylyltransferase contains 5 sulfhydryl groups, 2 of which rapidly
react with sulfhydryl reagents without producing inactivation. The data of Fig. 8 show that a partial reaction of
adenylyltransferase with PMPS causes some desensitization
toward the activation by glutamine in a Mg^{2+}-supported
(P_{IIA}-stimulated) adenylylating reaction, although the
maximal activity in this case was almost unchanged. In
comparable studies, the Mg^{2+}-supported adenylylating activity
of ATase alone (which is about 3-fold lower than it is with
saturating P_{IIA}) shows this same effect of mercurial modification on the glutamine activation curve (Oliver, 1972).
Thus, the activating sites of adenylyltransferase for the
regulatory protein (P_{IIA}) and for glutamine are apparently
independent. In the case of Mg^{2+}-supported activity (with
or without P_{IIA}), which is extremely responsive toward
glutamine activation, high concentrations of glutamine
appear to reverse structural alterations in adenylyltransferase produced by mercurial treatment. In contrast, the
Mn^{2+}-supported (P_{IIA}-stimulated) activity of the mercurial
derivative is not restored by high concentrations of
glutamine.

Impure fractions of adenylyltransferase have significant glutamate- or glutamine-stimulated PP_i-ATP exchange

31

activity (Denton and Ginsburg, 1968; Anderson et. al., 1970).
However, the most purified preparations of adenylyltransferase exhibit a negligible PP_i - ATP exchange activity
unless glutamine and glutamine synthetase are also present,
in which case this activity still is only 1% of the adenylylating activity (Table III). Thus, a stable AMP-adenylyl-
transferase intermediate does not appear to be involved in
the adenylylation mechanism (Wohlhueter et. al., 1972).
Rather, a nonrandom or concerted mechanism involving a
ternary complex between ATase, ATP-Me^{2+}, and GS during
adenylylation appears to occur.

The subunit structure of adenylyltransferase is as yet
unknown. A smaller unit of about 70,000 molecular weight
possessing adenylylation, but not deadenylylation, activity
was isolated and characterized by Hennig and Ginsburg (1971).
Preliminary evidence suggested that the adenylyltransferase
with both adenylylating and deadenylylating activities is
composed of subunits of different size (Hennig et. al.,
1970). These conclusions are regarded as tentative, however,
until more thorough dissociation studies have been performed
on a completely homogeneous preparation of adenylyltransferase. Somewhat confusing are the results from sedimentation equilibrium studies in which apparently homogeneous
preparations of adenylyltransferase had molecular weights of
140,000 (assuming $\bar{V}$ = 0.73 ml/g) in the experiments of
Caban (1972) and of 115,000 in those of Wolf et. al. (1972).
In crude extracts, the apparent size of the adenylyltransferase is much larger (Wolf et. al., 1972; Adler, 1972),
which could mean that ATase is normally complexed to other
regulatory proteins in the scheme of Fig. 7. Thus, a
purification of the adenylyltransferase could result in a
varying loss of other regulatory proteins, one or more of
which may be necessary to maintain a conformation of this
enzyme capable of expressing both adenylylating and
deadenylylating activities.

(b) $P_{IIA} \rightleftharpoons P_{IID}$ Interconversion

The studies of Brown et. al. (1971) illustrated in
Fig. 10 first suggested the occurrence of a time dependent
conversion of the regulatory protein (P_{II}) to a form that
stimulated deadenylylation of adenylylated glutamine synthetase. With this time dependent activation of
deadenylylating activity, there was a reciprocal decrease
in the ability of P_{II} to stimulate the adenylylation of

32

glutamine synthetase. This transformation of the regulatory protein ($P_{IIA} \longrightarrow P_{IID}$) was shown in these studies to depend upon the presence of UTP, divalent cation, ATP, and α-ketoglutarate during the preincubation of P_{IIA} with the crude ATase fraction, which is known now to have contained also the uridylyltransferase. Glutamine and inorganic phosphate were found to be potent inhibitors of the P_{II} transformation illustrated in Fig. 10. The dashed lines of Fig. 10 are reference activities for the ATase-catalyzed

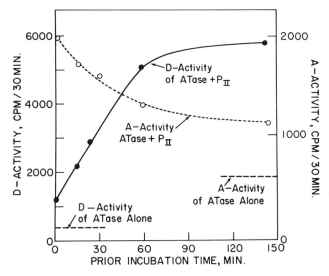

Fig. 10. Results of Brown et. al. (1971) showing a reciprocal effect of prior incubation of P_{II} with (ATase + UTase) on adenylylating (A) and deadenylylating (D) activities. A crude ATase fraction containing also UTase activity (0.21 mg) and P_{II} (0.10 mg) were incubated at 37° in 0.2 ml containing 0.1 mM ATP, 1.0 mM UTP, 1.0 mM α-KG, 1.0 mM $MnCl_2$, 1.0 mM dithiothreitol, and 50 mM 2-methyl-imidazole (pH 7.2). At the times indicated, 0.02 ml aliquots were assayed for Mg^{2+}-supported A- and D-activities at pH 7.2. A-Activity was measured by cpm incorporation into GS from ^{14}C-ATP with both negative and positive effectors present; D-activity was measured by cpm release from ^{14}C-AMP-GS_{12} in the presence of 20 mM P_i and effectors (Table III). The dashed lines show A- and D-activities of the ATase, preincubated without P_{II}.

reactions (4 and 5) in the absence of the regulatory protein (P_{II}); these activities were unaffected by the preincubation.

Brown et. al. (1971) also observed that if [14C]-UTP was used during the preincubation in an experiment similar to the one shown in Fig. 10, a partial radioactive labeling of the P_{II} component resulted. Later studies of Mangum (1972) have shown that the stoichiometry of [14C]-UMP or $[\alpha\text{-}^{32}P]$-UMP incorporation into P_{IID} is slightly more than one equiv per 50,000 g of P_{II}, during the P_{IIA} to P_{IID} conversion (Table IV). The experiments of Table IV show that

Table IV

Covalent Binding Of UMP During Conversion Of P_{IIA} To P_{IID}[*]

Radioactively labeled UTP added	Equivalents Radioactive Group Bound Per 50,000 g of P_{II}		
	Expt. 1	Expt. 2	Expt. 3
[14C] Uridine-P-P-P	0.97	-	1.17
Uridine-^{32}P-P-P	-	1.36	0.86
Uridine-P-^{32}P-^{32}P	0.02	-	-

[*]Data of Mangum (1972). In each experiment, the incubation mixture contained 50 mM 2-methylimidazole (pH 7.2), 1 mM dithiothreitol, 1 mM α-KG, 0.1 mM ATP, 10 mM $MgCl_2$, and in Expt. 1, 92 μg P_{IIA}, 400 μg (ATase + UTase) with 280 nmoles UTP containing [14C]-UTP (2×10^5 cpm/nmole) and $[\beta,\gamma\text{-}^{32}P]$-UTP ($3.4 \times 10^5$ cpm/nmole) in 1.5 ml; in Expt. 2, 124 μg P_{IIA}, 600 μg (ATase + UTase) with 300 nmoles UTP containing $[\alpha\text{-}^{32}P]$-UTP (4×10^5 cpm/nmole) in 2.0 ml; in Expt. 3, 62 μg P_{IIA}, 400 μg (ATase + UTase) with 280 nmoles UTP containing [14C]-UTP (2.4×10^5 cpm/nmole) and $[\alpha\text{-}^{32}P]$-UTP (2.4×10^5 cpm/nmole) in 1.0 ml. Each mixture was incubated for 10-12 hrs at 37° and then dialyzed for 72 hrs against several 500 ml volumes of 20 mM Tris-0.25 mM $Na_2MgEDTA$-0.5 mM dithiothreitol buffer at pH 7.2 prior to the determination of protein-bound radioactivity in a liquid scintillation counter.

radioactivity is incorporated into P_{IID} only when ^{14}C-UTP and/or $[\alpha\text{-}^{32}P]$-UTP were included in the preincubation mixture. The nonincorporation of radioactivity from $[\beta,\gamma\text{-}^{32}P]$-UTP indicates that pyrophosphate is released during the conversion of P_{IIA} to P_{IID}. For the calculations of Table IV, it was assumed that the P_{IIA} preparation was pure and 100% active, either of which may not be entirely correct. Recent studies of Mangum and Magni (1972) indicate that the regulatory protein (P_{II}) may exist as a 25,000 molecular weight species. If so, the stoichiometry of UMP attachment to P_{II} may well be two equiv per 50,000 g of P_{II}.

The results of Mangum (1972) illustrated in Fig. 11 further show that the adenylyltransferase included in the preincubation mixture is not labeled by $[\alpha\text{-}^{32}P]$-UTP during the P_{IIA} to P_{IID} conversion. All of the radioactivity from $[\alpha\text{-}^{32}P]$-UTP was associated with the P_{IID} peak of the elution profile shown in Fig. 11, with P_{IID} being identified by its

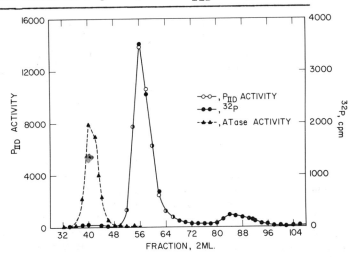

Fig. 11. Gel-filtration of $\alpha\text{-}^{32}P$-UMP-P_{IID} from Expt. 2 of Table IV (Mangum, 1972). After dialysis, the mixture of Expt. 2 of Table IV was loaded onto a Sephadex G-100 column (1.5 x 90 cm) and eluted (20 ml/hr) with 25 mM potassium phosphate-0.5 mM dithiothreitol-0.25 mM NaMgEDTA buffer at pH 7.2. Aliquots (0.05 ml) from the fractions collected were tested for ATase D-activity by supplementing each with 4 μg P_{IID}, and for P_{IID} activity by measuring the stimulation of ATase (18 μg) catalyzed D-activity.

ability to stimulate ATase-catalyzed deadenylylation. The results given in Table IV and Fig. 11 strongly suggest that UMP is covalently attached to the regulatory protein (P_{IID}) with a corresponding liberation of PP_i.

Recent experiments of Mangum and Magni (1972) have shown also that $\underline{E}$. $\underline{coli}$ has a Mn^{2+}-specific enzyme (UR enzyme) which is capable of reversing the P_{IIA} to P_{IID} conversion. The results of Fig. 12 show that the loss in the ability of P_{IID} to stimulate the ATase-catalyzed

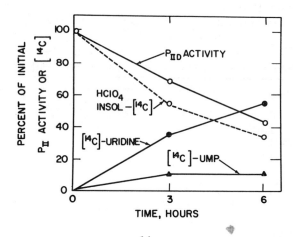

Fig. 12. A conversion of ^{14}C-UMP-P_{IID} to P_{IIA}, showing the time dependent loss in P_{IID} activity and in protein-bound radioactivity together with the release of ^{14}C-UMP and ^{14}C-uridine. The radioactively labeled ^{14}C-UMP-P_{IID} (15 μg) was incubated at 37° with a crude ATase fraction containing the UR enzyme, 1 mM $MnCl_2$, 50 mM 2-methylimidazole (pH 7.2), and 1 mM dithiothreitol in a volume of 0.3 ml. At the times indicated, 0.02 ml aliquots were analyzed for free ^{14}C-UMP and ^{14}C-uridine by thin layer chromatography; 0.005 ml aliquots were each supplemented with 18 μg ATase and assayed for D-activity; 0.025 ml aliquots (with 310 μg bovine serum albumin added to each) was precipitated by adding 0.075 ml of 6% $HClO_4$ and the radioactivity in each supernatant was subtracted from the total radioactivity in the aliquot. The data are from studies of Mangum and Magni (1972).

deadenylylation reaction parallels the loss in protein-bound radioactivity and the release of ^{14}C-UMP + ^{14}C-uridine. The predominance of ^{14}C-uridine in this latter measurement seemed to arise from a hydrolysis of the enzymatically released ^{14}C-UMP by contaminating phosphatases present in the crude enzyme fraction used for the experiment of Fig. 12. Although not illustrated in Fig. 12, the loss in the capacity of P_{IID} to stimulate the deadenylylation reaction corresponded to an increased ability of the treated protein to activate the adenylylation reaction (Mangum and Magni, 1972). Thus, the transformation shown in Fig. 12 represents a P_{IID} to P_{IIA} conversion. The UR enzyme in the crude ATase fraction appeared to specifically require Mn^{2+}, since a substitution of 10 mM Mg^{2+} for Mn^{2+} in an incubation as in Fig. 12 resulted in no change in the protein-bound radio-activity or in the catalytic properties of P_{IID}. Also, UMP was found to strongly inhibit the UR-enzyme-catalyzed hydrolysis of the uridylylated regulatory protein (Mangum and Magni, 1972).

(c) Some reciprocal effects of metabolites
     As indicated in Table III and in the scheme of Fig. 7, there is a reciprocal relationship between metabolites. When E. coli is grown on an environment of excess ammonia, the intracellular level of glutamine increases (see R. M. Wohlhueter et. al., this volume) and activates the adenylylation of glutamine synthetase, while indirectly inhibiting deadenylylation by blocking the modification of the regulatory protein (Fig. 7). When ammonia is limiting in the growth medium, an increase in the intracellular level of α-ketoglutarate will activate deadenylylation and inhibit the adenylylation reaction (4). UTP is required for the formation of P_{IID} from P_{IIA} (Fig. 7), and UTP also has the capacity to inhibit adenylylation.
     Using conditions that allow both adenylylation and deadenylylation to occur simultaneously, Segal et. al. (1972) have studied the effects of various metabolites on the state of adenylylation ($\bar{n}$) of glutamine synthetase in vitro. When glutamine synthetase preparations of different states of adenylylation are incubated under these conditions, glutamine synthetase forms are rapidly converted to the same state of adenylylation (Figs. 13 and 14). The steady state level of adenylylation was found in these studies to be specified by the relative concentrations of

ATP, α-ketoglutarate, glutamine, Mg^{2+}, Mn^{2+}, and by the form of the regulatory protein (P_{IIA} or P_{IID}) that are present. Increasing concentrations of α-ketoglutarate, with glutamine kept constant at 0.3 mM, in Fig. 13a led to a lower final state of adenylylation of glutamine synthetase. In a reciprocal manner (as shown in Fig. 13b), increasing concentrations of glutamine with α-ketoglutarate kept

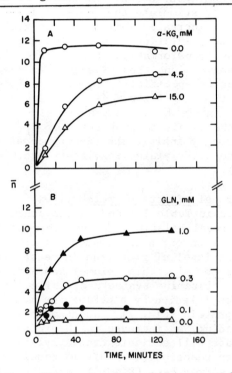

Fig. 13. The effect of α-ketoglutarate and glutamine on the state of adenylylation ($\bar{n}$) of glutamine synthetase. Incubations at 37° contained in 0.1 ml: crude fraction of ATase (32 μg) containing UTase, $P_{IIA} + P_{IID}$ (5 μg), GS$\overline{0.8}$ (95 μg), 50 mM 2-methylimidazole (pH 7.2), 1.0 mM dithiothreitol, 20 mM $MgCl_2$, 20 mM K-PO$_4$, 1.0 mM ATP, 1.0 mM UTP, and, unless varied, 15 mM α-ketoglutarate in (B) and 0.3 mM glutamine in (A). At the times indicated, aliquots were removed and the state of adenylylation of glutamine synthetase was determined by the assay method given in expression (3). These data are from studies of Segal et. al. (1972).

38

constant at 15 mM led to a higher final state of adenylylation of glutamine synthetase. Increasing concentrations of
ATP, which is required as an allosteric effector for
deadenylylation and as a substrate for adenylylation, also
resulted in a higher final state of adenylylation of
glutamine synthetase. The constant state of adenylylation
achieved was independent of the concentration of glutamine
synthetase over a 20-fold range (0.2 - 4.0 mg/ml); the
concentration of adenylyltransferase only affected the rate
at which the final state of adenylylation was attained.

The interconversion of the regulatory protein also is
a part of the adenylylating-deadenylylating enzyme system
that'regulates the state of adenylylation of glutamine
synthetase. The experiments of Fig. 13 were performed with
a mixture of P_{IIA} and P_{IID}; the presence of phosphate during
the incubations of Fig. 13 blocked the conversion of P_{IIA}
to P_{IID} (Fig. 7). Fig. 14 shows results obtained by

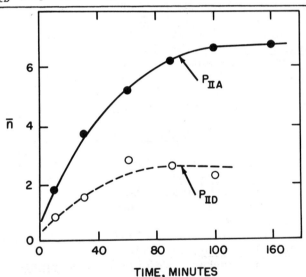

Fig. 14. Effect of P_{IIA} or P_{IID} on the state of
adenylylation ($\bar{n}$) of glutamine synthetase from the data
of Segal et. al. (1972). Either P_{IIA} (2.5 μg) or P_{IID}
(2.5 μg) were included in an incubation mixture as given in
the legend to Fig. 13 with 2 mM α-ketoglutarate and 0.3 mM
glutamine. The assay method of expression (3) was used to
determine the state of adenylylation of glutamine synthetase
at the different times.

Segal et. al. (1972) using a constant ratio of
α-ketoglutarate (2 mM) to glutamine (0.3 mM) and either
the unmodified (P_{IIA}) or the uridylylated (P_{IID}) regulatory
protein. It is evident that the presence of P_{IIA} favors
the high, whereas P_{IID} favors the low, final state of
adenylylation of glutamine synthetase. A rigorous cellular
control of the $P_{IIA} \rightleftharpoons P_{IID}$ interconversion therefore is
important. Metabolites function in generating either the
unmodified or the uridylylated regulatory protein while
simultaneously acting as effectors or substrates of
adenylylation and deadenylylation reactions. Thus, metabolites have a complex role in determining the state of
adenylylation of glutamine synthetase.

REFERENCES

Adler, S. P. (1972). Unpublished data.
Alberty, R. A. (1968). J. Biol. Chem. 243, 1337.
Alberty, R. A. (1969). J. Biol. Chem. 244, 3290.
Anderson, W. B., and Stadtman, E. R. (1970). Biochem.
    Biophys. Res. Commun. 41, 704.
Anderson, W. B., and Stadtman, E. R. (1971). Arch. Biochem.
    Biophys. 143, 428.
Anderson, W. B., Hennig, S. B., Ginsburg, A., and
    Stadtman, E. R. (1970). Proc. Nat. Acad. Sci. U. S.
    67, 1417.
Berberich, M. A. (1972). Biochem. Biophys. Res. Commun.
    47, 1498.
Brown, M. S., Segal, A., and Stadtman, E. R. (1971). Proc.
    Nat. Acad. Sci. U. S. 68, 2949.
Caban, C. E. (1972). Unpublished data.
Ciardi, J. E., Cimino, F., and Stadtman, E. R. (1972).
    Unpublished data.
Denton, M. D., and Ginsburg, A. (1968). Fed. Proc. Fed.
    Amer. Soc. Exp. Biol. 27, 783. Abstr.
Denton, M. D., and Ginsburg, A. (1969). Biochemistry 8,
    1714.
Denton, M. D., and Ginsburg, A. (1970). Biochemistry 9,
    617.
Ebner, E., Wolf, D., Gancedo, C., Elsässer, S., and
    Holzer, H. (1970). Eur. J. Biochem. 14, 535.
Eisenberg, D., Heidner, E. C., Goodkin, P., Dastoor, M. N.,
    Weber, B. H., Wedler, F., and Bell, J. D. (1971).
    Cold Spring Harbor Symp. 36, 291.

Ginsburg, A. (1969). Biochemistry 8, 1726.

Ginsburg, A. (1972). Advan. Protein Chem. 26, 1.

Ginsburg, A., Yeh, J., Hennig, S. B., and Denton, M. D. (1970). Biochemistry 9, 633.

Heinrikson, R. L., and Kingdon, H. S. (1971). J. Biol. Chem. 246, 1099.

Hennig, S. B., and Ginsburg, A. (1971). Arch. Biochem. Biophys. 144, 611.

Hennig, S. B., Anderson, W. B., and Ginsburg, A. (1970). Proc. Nat. Acad. Sci. U. S. 67, 1761.

Holzer, H. (1969). Advan. Enzymol. 32, 297.

Hunt, J. B. (1972). Unpublished data.

Hunt, J. B., and Ginsburg, A. (1972). Biochemistry 11, 3723.

Hunt, J. B., Ross, P. D., and Ginsburg, A. (1972). Biochemistry 11, 3716.

Kingdon, H. S., and Stadtman, E. R. (1967). J. Bacteriol. 94, 949.

Kingdon, H. S., Shapiro, B. M., and Stadtman, E. R. (1967). Proc. Nat. Acad. Sci. U. S. 58, 1703.

Kingdon, H. S., Hubbard, J. S., and Stadtman, E. R. (1968). Biochemistry 7, 2136.

Lahiri, A. K., Noyes, C., Kingdon, H. S., and Heinrikson, R. L. (1972). Fed. Proc., Fed. Amer. Soc. Exp. Biol. 31, 473. Abstr.

Levintow, L., and Meister, A. (1954). J. Biol. Chem. 209, 265.

Mangum, J. H. (1972). Unpublished data.

Mangum, J. H., and Magni, G. (1972). Unpublished data.

Mantel, M., and Holzer, H. (1970). Proc. Nat. Acad. Sci. U. S. 65, 660.

Meister, A. (1962). In "The Enzymes" (P. D. Boyer, H. Lardy, and K. Myrbäck, eds.), Vol. 6, p. 193. Academic Press, New York.

Miller, R. E., and Stadtman, E. R. (1971). Fed. Proc., Fed. Amer. Soc. Exp. Biol. 30, 1067. Abstr.

Miller, R. E., and Stadtman, E. R. (1972). J. Biol. Chem. (in press)

Miller, R. E., Shelton, E., and Stadtman, E. R. (1972a). Unpublished data.

Miller, R. E., Smyrniotis, P. Z., and Stadtman, E. R. (1972b). Unpublished data.

Oliver, E. J. (1972). Unpublished data.

Pastan, I., and Perlman, R. (1970). Science 169, 339.

Prusiner, S., Miller, R. E., and Valentine, R. C. (1972).
Proc. Nat. Acad. Sci. U. S. (in press).
Ross, P. D., and Ginsburg, A. (1969). Biochemistry 8, 4690.
Segal, A. (1971). Fed. Proc., Fed. Amer. Soc. Exp. Biol.
30, 1175. Abstr.
Segal, A., and Stadtman, E. R. (1972a). Arch. Biochem.
Biophys. 152, 356.
Segal, A., and Stadtman, E. R. (1972b). Arch. Biochem.
Biophys. 152, 367.
Segal, A., Brown, M. S., and Stadtman, E. R. (1972).
Unpublished data.
Shapiro, B. M. (1969). Biochemistry 8, 659.
Shapiro, B. M., and Ginsburg, A. (1968). Biochemistry 7,
2153.
Shapiro, B. M., and Stadtman, E. R. (1967). J. Biol. Chem.
242, 5069.
Shapiro, B. M., and Stadtman, E. R. (1968). J. Biol. Chem.
243, 3769.
Shapiro, B. M., and Stadtman, E. R. (1970). Annu. Review
Microbiol. 24, 501.
Shapiro, B. M., Kingdon, H. S., and Stadtman, E. R. (1967).
Proc. Nat. Acad. Sci. U. S. 58, 642.
Stadtman, E. R., and Smyrniotis, P. Z. (1972). Unpublished
data.
Stadtman, E. R., Shapiro, B. M., Ginsburg, A., Kingdon,
H. S., and Denton, M. D. (1968). Brookhaven Symp.
Biol. 21, 378.
Stadtman, E. R., Ginsburg, A., Ciardi, J. E., Yeh, J.,
Hennig, S. B., and Shapiro, B. M. (1970). Advan.
Enzyme Regulation 8, 99.
Stadtman, E. R., Ginsburg, A., Anderson, W. B., Segal, A.,
Brown, M. S., and Ciardi, J. E. (1972). PAABS Symp.
"Molecular Basis of Biological Activity" (K. Gaede,
B. L. Horecker, and W. J. Whelan, eds.), Vol. 1,
p. 127. Academic Press, New York.
Tempest, D. W., Meers, J. L., and Brown, C. M. (1970).
Biochem. J. 117, 405.
Valentine, R. C., Shapiro, B. M., and Stadtman, E. R.
(1968). Biochemistry 7, 2143.
Wohlhueter, R. M., Ebner, E., and Wolf, D. H. (1972).
J. Biol. Chem. 247, 4213.
Wolf, D. H., and Ebner, E. (1972). J. Biol. Chem. 247, 4208.
Wolf, D. H., Ebner, E., and Hinze, H. (1972). Eur. J.
Biochem. 25, 239.

Woolfolk, C. A., and Stadtman, E. R. (1964). Biochem.
    Biophys. Res. Commun. 17, 313.
Woolfolk, C. A., and Stadtman, E. R. (1967a). Arch. Biochem.
    Biophys. 118, 736.
Woolfolk, C. A., and Stadtman, E. R. (1967b). Arch. Biochem.
    Biophys. 122, 174.
Woolfolk, C. A., Shapiro, B. M., and Stadtman, E. R. (1966).
    Arch. Biochem. Biophys. 116, 177.
Wulff, K., Mecke, D., and Holzer, H. (1967). Biochem.
    Biophys. Res. Commun. 28, 740.

REGULATION OF GLUTAMINE SYNTHESIS
IN VIVO IN E. COLI[+]

R. M. Wohlhueter, H. Schutt, and H. Holzer

Biochemisches Institut der Universität Freiburg
and Institut für Biochemie der Gesellschaft für
Strahlen- und Umweltforschung, München

Abstract

The regulation of glutamine synthetase by
adenylylation is a phenomenon common to several
bacteria. The mechanism allows for a rapid readjustment of nitrogen metabolism in response to
changes in the nature of the exogenous source of
nitrogen. In E. coli adenylylation/deadenylylation
is catalyzed by a system composed of two or more
proteins, which, judging from sucrose-density-
gradient centrifugation of sonicated cells,
exists in the cell as a structural unit.

The rapid readjustment to new steady-state
concentrations of glutamine, glutamate, and ATP
following addition of 10 mM NH_4^+ to cells growing
in a proline-glycerol medium is interpreted as
follows. In proline-cultured cells ammonium is
the limiting factor for the synthesis of glutamine; the added ammonium causes a burst of glutamine synthesis imposing a drastic drain on ATP.
The accumulated glutamine, however, activates the
adenylyltransferase, which, in turn effects the
inactivation of glutamine synthetase. Thus the
consumption of ATP is throttled, and repletion of
the ATP pool allowed. Glutamine functions, there-

[+] Unless otherwise cited, data in this paper are
taken from the doctoral dissertation of H.Schutt,
University of Freiburg, 1972.

fore, as a potent feedback inhibitor of its own
synthesis, not via a conventional allosteric feedback loop, but amplified by way of its action on
adenylyltransferase. The relationship between
initial and final steady-states indicates that, in
the presence of NH_4^+, the cell is able to meet
its growth requirements with a reduced demand for
glutamine.

Introduction

The regulation of glutamine synthetase in
Escherichia coli is an elaborate process. Expression of the enzyme is controlled at the genetic
level (Mecke and Holzer, 1966; Woolfolk,Shapiro and
Stadtman, 1966; Meers and Tempest, 1970), its
activity is inhibited by no less than eight end-
products (reviewed by Stadtman et al., 1968), and
it exists in chemically interconvertible, catalytically distinct forms, arising as permutations
among 12 subunits, each of which may be adenylylated or not.

This interconversion has been the subject of
extensive, enzymological studies which allow a
fairly detailed description of it (for recent reviews see Shapiro and Stadtman, 1970; Holzer and
Duntze, 1971). Each of the 12 identical subunits
of glutamine synthetase may bear an AMP group in
phosphodiester linkage to the hydroxyl of a specific tyrosine residue (Shapiro and Stadtman,
1968; Heinrikson and Kingdon, 1970). Adenylylation of the enzyme effects a variety of changes
in its catalytic properties, most notably a
decrease of its physiologically important activity (the Mg^{2+}-dependent synthesis of glutamine)
that is roughly proportional to the extent of
adenylylation. The adenylylation reaction is catalyzed by a specific enzyme (Ebner et al., 1970;
Hennig and Ginsburg, 1971; Wolf et al., 1972;
Wohlhueter et al., 1972), whose activity is stimulated by glutamine. This enzyme, furthermore,
appears to be under the influence of a regulator
protein, which, in the presence of α-ketogluta-

rate and UTP, enables it to catalyze also the phosphorylytic deadenylylation of glutamine synthetase (Anderson and Stadtman, 1970; Brown et al., 1971). The entire system thus completes a novel sort of control loop in which the concentrations of glutamine, product of the glutamine synthetase reaction, and of α-ketoglutarate, a precursor, modulate glutamine synthetase activity. This control is superimposed upon feedback control exerted directly on glutamine synthetase by end products of pathways utilizing glutamine amide nitrogen.

We address ourselves here to the task of relating this body of enzymatic knowledge to the cell, to the questions of how this system functions in vivo and of the metabolic consequences of adenylylation of glutamine synthetase.

Biological Occurence of Adenylylation

A first step in our inquiry is to establish in which organisms the adenylylation/deadenylylation system exists. Perhaps the simplest screening technique for microorganisms is to test for the rapid inactivation of glutamine synthetase after adding NH_4^+ to cells growing in glutamate medium. By this criterion, all of four Enterobacteriaceae were positive for adenylylation, including species of Escherichia, Salmonella, Shigella, and Klebsiella (Gancedo and Holzer, 1968). Evidence for adenylylation on the Gram-negative Pseudomonas putida has been cited Shapiro and Stadtman (1970), namely that putida glutamine synthetase is adenylylated in vitro by coli adenylyltransferase.

Outside the Gram-negative bacteria there have been no demonstrations of regulation of glutamine synthetase by adenylylation. Among Gram-positive bacteria Sarcina, Bacillus, and Lactobacillus species have been tested and found negative (Gancedo and Holzer, 1968; Deuel and Stadtman, 1970). Ferguson and Sims (1971) report an inactivation of glutamine synthetase in the

47

yeasts Candida utilis, Saccharomyces cerevisiae
and Torulopsis candida following addition of ammonia to glutamate-grown cells. This inactivation,
though less rapid than in bacteria, occurs too
rapidly to be explained by repression, and is reversible. The evidence for adenylylation is in
these cases negative, but points rather to a
reversible, effector-controlled disaggregation of
the decamer to pentamers (Sims, personal communication).

There have been no reports of adenylylation
of glutamine synthetase in mammals (for example,
see Tate and Meister, 1971).

Adenylylation Functions In Vivo

The use of ammonium-induced inactivation of
glutamine synthetase as a criterion for adenylylation wants some elaboration. It was in fact this
phenomenon which led to the discovery of the interconversion of glutamine synthetase (Mecke and
Holzer, 1966). When intact E. coli cells grown on
glutamate were sudenly confronted with ammonium,
glutamine synthetase activity fell to about 25 %
of initial activity within two minutes (37°).
Crucial to the interpretation of this inactivation as a chemical interconversion à la glycogen
phosphorylase were the observations that the
γ-glutamyl transferase activity of glutamine synthetase was relatively unaltered by this treatment, and that the synthetase activity was regenerated upon removal of NH_4^+, even in the presence of chloramphenicol (Mecke and Holzer,
1966). In vitro studies soon established that
inactivation of synthetic activity was correlated
with the incorporation into glutamine synthetase
of the AMP-moiety of ATP (Shapiro et al., 1967;
Wulff et al., 1967). This reaction, however, was
stimulated by glutamine and not at all by NH_4^+.

That the ammonium-induced inactivation observed in vivo is actually due to adenylylation
has been demonstrated by isolating $[^{14}C]$AMP-glut-
amine synthetase from a purine-requiring mutant
fed $[^{14}C]$adenine (Heinrich and Holzer, 1970).
That ammonium serves as trigger for adenylylation

48

in intact cells probably means, simply, that
addition of NH_4^+ to the medium is about the most
rapid way to elevate the intercellular glutamine
pool (see figure 4 below). Glutamine can also
trigger the inactivation, but first at a concentration about 100-times higher than that necessary for NH_4^+ (Holzer et al., 1967). Exogenous
glutamine appears rapidly in the cell's glutamate
pool (Hartman, 1968), quite possibly the prey of
a periplasmic glutaminase activity (Wade et al.,
1971). 6-Diazo-5-oxo-L-norleucine (DON) is nearly
as effective in vitro in stimulating the adenylyltransferase reaction. We have found the maximal
reaction velocity with DON 61 % of that obtained
with L-glutamine, with half-maximal velocity at
1.1 mM compared to 1.5 mM for glutamine. DON,
supplied exogenously to cells growing in glutamate medium, unleashes the inactivation of glutamine synthetase, even at concentrations as low
as 10^{-6} M.

A natural example of adenylylation operating
in vivo is given by a batch culture in transition
to the stationary phase (fig. 1). Under the assay
conditions employed here the synthetase/transfe-
rase ratio is not strictly proportional to state
of adenylylation, but does serve as a rough
gauge. At the end of the logarithmic growth phase
in glutamate-glycerol medium, ammonium begins to
accumulate in the medium. Concurrently the synthetase activity of glutamine synthetase rapidly
decreases; the transferase activity decreases
less rapidly, that is glutamine synthetase is
converted to the adenylylated form.

The Functions of Glutamine in E. coli

To proceed with our task of putting the
chemical interconversion of glutamine synthetase
into its cellular context, we are obliged to
state what it is glutamine synthetase does for
the cell. While in brain and uricotelic livers
glutamine synthetase may excercise specialized
functions in γ-aminobutyrate metabolism (Tate and

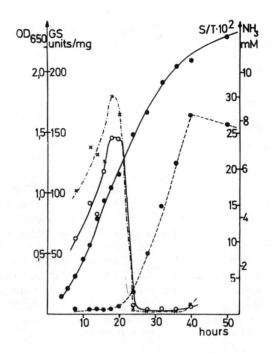

Fig. 1. <u>Changes in State of Adenylylation of
Glutamine Synthetase During Growth in Batch Cul-
ture</u> (taken from Holzer et al., 1968). <u>E. coli</u> B
was grown aerobically in 18 l of a salts medium
with glycerol (21 mM) and glutamate (19 mM).
Symbols: ●——● cell growth as optical density at
650 nm; ●----● ammonium in medium; o——o synthase
activity; x-•-x ratio of synthetase to transferase activity.

Meister, 1971) or ammonium detoxification (Vorhaben and Campbell, 1972), in <u>E. coli</u> its sole
purpose seems to be to supply glutamine. The
γ-glutamyl transferase activity of glutamine synthetase, which responds differently to adenylyl-

ation than does the synthetic activity, has no
known physiological function.

And to what end glutamine? The classical
answer would itemize the roles of glutamine as
amide-nitrogen donor in the synthesis of histidine, tryptophan, purines, pyrimidines and hexosamines. Since 1970, with the discovery of glutamate synthetase by Tempest and his colleagues (Tempest et al., 1970; Meers et al., 1970), one has
also to weigh glutamine's potential role as a
nitrogen donor in the synthesis of glutamate from
α-ketoglutarate. Glutamate synthetase exists in
E. coli (Miller and Stadtman, 1971), and the
route of nitrogen assimilation via glutamine to
glutamate is of demonstrable importance in B.me-
gaterium (Elmerich, 1972). While we are not in a
position to assess quantitatively the importance
of this route in E. coli, we want to make some

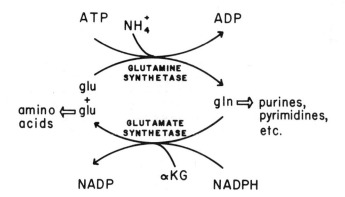

Fig. 2. Cyclic Pathway for the Synthesis of
Glutamate from Glutamine and α-Ketoglutarate.

points about it pertinent to an understanding of
the consequences in vivo of glutamine synthetase
inactivation. First is that the precursor-product
relationships between glutamate and glutamine become complex, presumably dependent on cellular
ammonium concentration, and may, in the extremes,
be reversed.

The second has to do with the relative demands on glutamine by the cell. On the basis of
the bulk composition of E. coli cells[1], we calculate that 1 kg dry weight contains ca. 1.3 g-
atoms of nitrogen originating as glutamine amido-
nitrogen and incorporated into amino acids, amino
sugars, purines and pyrimidines via those routes
usually utilizing glutamine (excepting the Tempest pathway). Total nitrogen content is on the
order of 11 g-atoms/kg dry weight. Thus nearly a
10-fold variation in the flux through glutamine
synthetase might be expected, depending on the
amount of glutamine diverted into the Tempest
pathway. We note that factor 10 is roughly the
variation on glutamine synthetase activity seen
in Gram-negative bacteria equipped for adenylylation, and that the specific activity of glutamine synthetase in these organisms after inactivation is similar to that in other bacteria and
yeast not possessing the inactivation machinery
(see table 1). The synthetic capacity is commen-
surate with these estimated requirements. The
activity content of 1 kg dry weight of derepressed cells would be sufficient to produce on the
order of 10 moles of glutamine per generation
time (ca. 60 min). Thus the potentially high activity and its ability to rapidly and efficiently
modulate this activity over a wide range by means
of adenylylation of the enzyme, would enable this
bacterium to cope with large fluctuations in its
nitrogenous environment.

Organization of the Adenylylating/Deadenyl-
ylating Enzyme System In Vivo

Although examination of typical purification

Table 1. Changes in Activity of Glutamine Synthetase in Several Microorganisms Following Addition
of Ammonium to Cells Growing in Glutamate. (Taken
from Gancedo and Holzer, 1968).

Microorganism	Specific glutamine synthetase activity	
	Glutamate culture	After NH_4^+ treatment
	(units/mg protein)	
Escherichia coli K$_{12}$	122	13
Salmonella typhimurium	132	12
Shigella flexneri	160	12
Klebsiella sp.	264	25
Sarcina citrea	32	32
Bacillus subtilis	16	16
Saccharomyces cerevisiae	30	27
Candida utilis	29	27
Lactobacillus plantarum	5	6

tables for glutamine synthetase and adenylyltransferase indicates that the two enzymes may
be present in nearly equimolar quantities, it
must be recognized that the ratio between them is
contingent upon growth conditions. Glutamine synthetase is considered a repressible enzyme
(Mecke and Holzer, 1966; Woolfolk et al., 1966;
Meers and Tempest, 1970; reviewed by Shapiro and
Stadtman, 1970), whereas adenylyltransferase
appears to be constituitive, as seen in table 2.
There is, of course, a certain affinity between
the two proteins, as follows from kinetics of the
adenylylation reaction (Hennig and Ginsburg,
1971; Wohlhueter et al., 1972), and as has been
measured by affinity chromatography (Hennig and
Ginsburg, 1971). The variable stoichiometry between the two proteins, however, and, as we shall
see, the relative ease of separating them, argue
against considering glutamine synthetase and
adenylyltransferase to constitute a multi-enzyme
complex. Also the two proteins are genetically

Table 2. Specific Activities of Adenylyltransferase and Glutamine Synthetase in E. coli under Various Growth Conditions.

Nitrogen source	Generation time	Adenylyl-transferase	Glutamine synthetase
	(min)	(cpm/min·mg protein)	(mU/mg protein)
Ammonium sulfate	72	594	19
Glutamate	66	642	55
Aspartate	60	632	77
Alanine	66	638	64
Glycine	114	438	140
Proline	90	900	144
Nutrient broth (0.4%)	90	520	-

Cells were grown in a salts medium with 15 mM glycerol and the indicated nitrogen source at 15 mM. Cells were harvested in logarithmic phase by centrifugation, sonicated and dialyzed against 0.1 M imidazole/HCl, pH 7.6, over night. Adenylyltransferase assay contained (0.1 ml, 37°): 194 µg glutamine synthetase, 1 mM $[8-^{14}C]$ATP, 0.5 µCi, 20 mM L-glutamine, 25 mM $MgCl_2$, and 100 mM imidazole/HCl, pH 7.6. Samples were removed at 5, 10 and 15 min for assay of protein-bound radioactivity according to Mans and Novelli (1961). Glutamine synthetase was assayed according to Ebner et al. (1970).

independent in the sense that a glutamine synthetase minus mutant of E. coli has adenylyltransferate activity comparable to that of the wild type (Varricchio and Holzer, 1969).

Adenylyltransferase, however, quite probably exists as part of a complex. As we isolate it at Freiburg (Ebner et al., 1970), adenylyltransferase has a molecular weight of 115,000 daltons (equilibrium sedimentation, Wolf et al., 1972). It possesses per se no phosphorolytic deadenylyl-

ating activity, but expresses such an activity
when combined with another protein fraction which,
alone, has neither adenylylating nor deadenylylating activity (unpublished observations). These
facts all conform to the model propounded by
Stadtman's group that adenylyltransferase ("P_I"
in his nomenclature) has a functional associate,
"P_{II}", which determines expression of the former's adenylylating or phosphorolytically deadenylylating activities, as well as its sensitivities to effectors glutamine and α-ketoglutarate
(Shapiro, 1968; Anderson et al., 1970; Brown et
al., 1971).

In vitro, deadenylylation may be accomplished
by reversing the adenylyltransferase reaction,
that is, by pyrophosphorolytic deadenylylation
(Mantel and Holzer, 1970; Wohlhueter, 1971). This
reaction we regard as unphysiological, however, on
the grounds that the requisite pyrophosphate is
not available in vivo, and that it makes no regulatory sense.

Alone the behavior of adenylyltransferase
and P_{II} upon purification suggested that their
functional association was also a structural one
(Anderson and Stadtman, 1971). Centrifugation of
briefly sonicated E. coli cells in sucrose gradients confirmed this supposition. Figure 3 shows
the results of such an experiment; the adenylyltransferase activity from sonicated cells sediments with an apparent molecular weight of ca.
220,000 daltons. Pure glutamine synthetase and
adenylyltransferase, with the effectors of the
adenylylation reaction, Mg^{2+} and glutamine, present throughout the gradient, sediment apparently
without interaction, corresponding in molecular
weights to those determined by other methods.
Existence of the "heavy" adenylyltransferase is
dependent, apparently, on the presence of small-
molecular ligands; thus if one omits Mg^{2+} from
the gradient, or dialyzes the sonicate prior to
centrifugation, adenylyltransferase turns up at
its "light" position.

We have some reservation as to whether the
"heavy" peak contains also P_{II}. The assay for

55

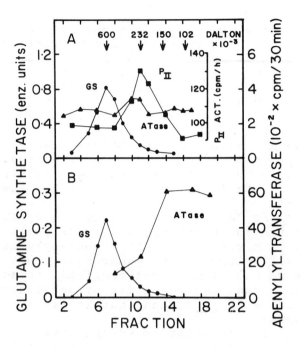

Fig. 3. **Sucrose Density Gradient Centrifugation of Glutamine Synthetase and Adenylyltransferase.** Centrifugation was carried out in a Beckman SW-56 rotor for 3.5 h at 56,000 rpm and 1° with a linear sucrose gradient from 5 to 20 % (w/v), containing 10 mM $MgCl_2$, 20 mM glutamine and 0.1 M imidazole/HCl, pH 7.6, throughout. Adenylyltransferase (ATase) was tested as in table 2, but at 30° and 30 min. with 480 µg glutamine synthetase in

assay mixture. Deadenylylating activity (P_{II}) was
assayed at 30° for 60 min. according to Shapiro
(1968), with 100 µg glutamine synthetase (12 moles AMP/mole, 23,800 cpm). Glutamine synthetase
(GS) was measured with the transferase assay of
Mecke and Holzer (1966). A. Cells from 1 l sus-
pension were collected by centrifugation, sus-
pended in 2 ml 0.1 M imidazole/HCl, pH 7.6, and
sonicated (Bronson, ca. 9 amp) for 3 min in 1/2
min sessions. After centrifuging the sonicate
(49,000 g for 20 min), supernate containing 2.3
mg protein was placed on the gradient. P_{II} activities are taken from a parallel run. B. 0.4 mg
pure glutamine synthetase (6 moles AMP/mole enzyme) and 0.07 mg pure adenylyltransferase were
placed on the gradient. Gradients were calibrated
in a parallel run with glutamine synthetase, bovine liver catalase, yeast hexokinase, and yeast
alcohol dehydrogenase.

---

deadenylylating activity in these gradients is of
questionable reliability, but does indicate an
association of P_{II} with the heavy adenylyltransferase peak. The 220,000 daltons would be somewhat too heavy for a 1:1 complex of adenylyltransferase with P_{II}, taking the molecular weight
of the latter as 50,000 (Brown et al., 1971), and
may indicate a more complex structure. That the
"heavy" peak represents dimerization of adenylyltransferase dependent on ligands originating in
the cell sap is made unlikely by the observation
that ultrafiltered cell sonicate does not affect
sedimentation of pure adenylyltransferase.

Consequences of Adenylylation of
Glutamine Synthetase In Vivo

The P_I-P_{II} complex doubtless bears closer
resemblance to the physiological regulatory entity than does naked adenylyltransferase. Therefore
the properties of this complex must be considered
the more relevant for the cell. The catalytic

behavior of this complex as studied in vitro is
an extremely complex function of the relative
concentrations of α-ketoglutarate, glutamine and
ATP (Brown et al., 1971). Our knowledge of the
cellular milieu is not sufficient to allow us to
describe the dependence of state of adenylylation
of glutamine synthetase in vivo on the relationships among the concentrations of these key effectors. We have recourse, however, to another
tactic, namely to perturb the glutamine metabolism of the cell, and then to observe the pattern of readjustment of these metabolites to
their new steady states.

The results of such a study are presented in
figure 4. In this case cells growing logarithmi-
cally in proline-glycerol medium were confronted
with NH_4^+ at a concentration of 10 mM. Glutamine
synthetase activity dropped to about 15 % within
30 sec, and then further decreased to a level
which represents essentially full adenylylation.

The burst of glutamine formation is at least
partially responsible for the adenylylation of
glutamine synthetase. The change from 0.4 mM to
8 mM encompasses the $S_{0.5}$ for glutamine stimulation of adenylyltransferase activity (1.5 mM,
Schutt and Holzer, 1972). It is evident from the
work of Brown et al. (1971) that the physiological adenylylation complex ("P_I-P_{IIAT}") is also
very sensitive to changes in glutamine concentrations in this range, although the extent is dependent upon α-ketoglutarate concentration. It
has been technically impossible to measure changes in α-ketoglutarate on such a short time
scale. Measuring α-ketoglutarate fluorometrically
(Nairns and Passoneau, 1970) in cells concentrated by ultrafiltration, we have found that, under
the conditions of the present experiment, the concentration of α-ketoglutarate decreases from
about 0.1 to 0.03 mM 5 min after addition of NH_4^+
(compare Lowry et al., 1971).

The rapid expansion of the glutamine pool
following addition of NH_4^+ to intact cells we
take as evidence, first that NH_4^+ penetrates the
cells rapidly, and second, that the concentration

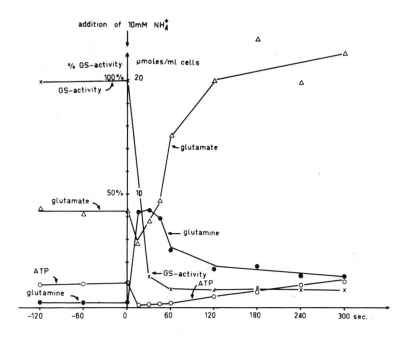

Fig. 4. Changes in Pools of Glutamine, Glutamate, and ATP following Inactivation of Glutamine Synthetase (taken from Schutt and Holzer, 1972). E. coli cells were grown in a proline medium to which was added 10 mM NH_4^+ at zero time. Metabolites were rapidly sampled directly into perchloric acid. For details, see the above reference.

of NH_4^+ is rate-limiting for glutamine synthesis. The latter conclusion raises a serious problem in interpretation of such in vivo experiments. The total concentrations of the substrates of glutamine synthetase, as measured here before the addition of NH_4^+, are: NH_4^+ (measured in parallel experiments) = 3.8 mM, glutamate = 8.4 mM, ATP = 2.2 mM. Comparing these to the respective K_M values of glutamine synthetase, K_{M,NH_4^+} = 1.8 mM, $K_{M,glu}$ = 2.4 mM, $K_{M,ATP}$ = 0.7 mM (Woolfolk et al.,

1966), one might conclude that the maximal possible increase in rate of glutamine synthesis over that in the pre-ammonium steady state is about 1/3. Such an increase could hardly account for the burst in glutamine pool size. Solution of the paradox, we believe, lies in the difference between total- und free-metabolite concentrations. We tally at least 40 enzymes which bind NH_4^+, 20 which bind glutamate, and 100 which bind ATP, all, of course, having different affinities and present at various concentrations. The total cellular concentrations of these metabolites are certaintly poor overestimates of their chemical activities in the cytoplasm.

The final steady-state concentration of glutamine, 3 mM, is some six times higher that the initial steady-state concentration: either its rate of synthesis is higher, or its rate of consumption lower (or both). While it is conceivable that influx of glutamine through the "inactivated" glutamine synthetase is higher than that through "activated", thanks to more saturating substrate concentrations, it seems more probable that the final steady state reflects decreased efflux from the pool. Growth in ammonium medium is somewhat more rapid than in proline medium. A reduced biosynthetic demand on glutamine, however, may arise 1) from the substitution of ammonium for glutamine as nitrogen donor in several biosynthetic enzymes, and 2) from a markedly decreased influx of glutamine into the Tempest pathway.

A necessary correlate to a burst of glutamine synthesis is a drain on ATP. In fact, the amount of glutamine accumulated here in 30 sec corresponds to 3 times the total initial ATP pool. The exhaustion of the ATP pool is thus understandable. Remarkable is the slow recovery. Turnover time of ATP in growing bacteria has been estimated at 0.13 to 1.2 sec (Holms et al., 1972; Harrison and Maitra, 1969), yet readjustment here to a new steady state is not complete by 300 sec, and this inspite of the fact that glutamine synthetase has been throttled. One can imagine that without benefit of the adenylylation system, reestablishment of viable ATP levels would be impossible. High activity of glutamine synthetase is a potential hazard. The nine microorganisms surveyed by Gancedo and Holzer (1968) fall into two classes of high and low specific glutamine synthetase activity, separated by nearly an order of magnitude. In every case, those organisms in the "high" category are equipped with the machinery of inactivation.

The movements of the glutamate pool following ammonium addition reciprocate those of glutamine. A transient depletion of glutamate coincides with the burst of glutamine, corroborating the concept that the enlarged glutamine pool is due to synthesis. The final steady state is approached quickly, at more than twice the concentration of the initial. By our previous logic, this increase in pool size probably does not reflect diminished biosynthetic demands. We suppose that increased synthesis via glutamate dehydrogenase is one component of it.

Anderson, W. B., and Stadtman, E. R. (1970).
    Biochem. Biophys. Res. Commun. 41, 704.
Anderson, W. B., and Stadtman, E. R. (1971).
    Arch. Biochem. Biophys. 143, 428.
Anderson, W. B., Hennig, S. B., Ginsburg, A., and
    Stadtman, E. R. (1970). Proc. Nat. Acad. Sci.
    US 67, 1417.
Brown, M. S., Segal, A., and Stadtman, E. R.
    (1971). Proc. Nat. Acad. Sci. US 68, 2949.
Deuel, F. T., and Stadtman, E. R. (1970). J. Biol.
    Chem. 245, 5206.
Ebner, E., Wolf, D., Gancedo, C., Elsässer, S.,
    and Holzer, H. (1970). European J. Biochem.
    14, 535.
Elmerich, C. (1972). European J. Biochem. 27, 216.
Ferguson, A. R., and Sims, A. P. (1971). J. Gen.
    Microbiol. 69, 423.
Gancedo, C., and Holzer, H. (1968). European J.
    Biochem. 4, 190.
Harrison, D. E. F., and Maitra, P. K. (1969).
    Biochem. J. 112, 647.
Hartman, S. C. (1968). J. Biol. Chem. 243, 853.
Heinrich, C. P., and Holzer, H. (1970). Arch.
    Mikrobiol. 73, 97.
Heinrickson, R. L., and Kingdon, H. S. (1970).
    J. Biol. Chem. 245, 138.
Hennig, S. B., and Ginsburg, A. (1971). Arch.
    Biochem. Biophys. 144, 611.
Holms, W. H., Hamilton, J. D., and Robertson, A.
    G. (1972). Arch. Mikrobiol. 83, 95.
Holzer, H., and Duntze, W. (1971). Ann. Rev.
    Biochem. 40, 345.
Holzer, H., Mecke, D., Wulff, K., Liess, K., and
    Heilmeyer, L.Jr. (1967). Advan. Enz. Regula-
    tion 5, 211.
Holzer, H., Schutt, H., Mašek, Z., and Mecke, D.
    (1968). Proc. Nat. Acad. Sci. US 60, 721.
Lowry, O. H., Carter, J., Ward, J. B., and
    Glaser, L. (1971). J. Biol. Chem. 246, 6511.
Luria, S. E. (1960). In "The Bacteria" (I.C.Gun-
    salus and R.Y.Stanier, ed.), Vol.I, pp 1-34.
    Academic Press, New York.
Mans, R. J., and Novelli, G. D. (1961). Arch.
    Biochem. Biophys. 94, 48.

Mantel, M., and Holzer, H. (1970). Proc. Nat. Acad. Sci. US 65, 660.

Mecke, D., and Holzer, H. (1966). Biochim. Biophys. Acta 122, 341.

Meers, J. L., and Tempest, D. W. (1970). Biochem. J. 119, 603.

Meers, J. L., Tempest, D. W., and Brown, C. M. (1970). J. Gen. Microbiol. 64, 187.

Miller, R. E., and Stadtman, E. R. (1971). Fed. Proc. 30, 1067 Abs.

Nairns, R. G., and Passonneau, J. V. (1970). In Methoden der Enzymatischen Analyse" (H.U. Bergmeyer, ed.) Vol.II, pp 1540-1543. Verlag Chemie, Weinheim/Bergstrasse.

Salton, M. R. J. (1960). In "The Bacteria" (I.C. Gunsalus and R.Y.Stanier, eds.) Vol.I, pp 97-151. Academic Press, New York.

Schutt, H., and Holzer, H. (1972). European J. Biochem. 26, 68.

Shapiro, B. M. (1968). Biochemistry 8, 659.

Shapiro, B. M., and Stadtman, E. R. (1968). J. Biol. Chem. 243, 3769.

Shapiro, B. M., and Stadtman, E. R. (1970). Ann. Rev. Microbiol. 24, 501.

Shapiro, B. M., Kingdon, H. S., and Stadtman, E. R. (1967). Proc. Nat. Acad. Sci. US 58, 642.

Stadtman, E. R., Shapiro, B. M., Kingdon, H. S., Woolfolk, C. A., and Hubbard, G. S. (1968). Advan. Enz. Regulation 6, 257.

Sueoka, N. (1960). In "The Bacteria" (I.C.Gunsalus and R.Y.Stanier, eds.) Vol.V, pp 419-443. Academic Press, New York.

Tate, S. S., and Meister, A. (1971). Proc. Nat. Acad. Sci. US 68, 781.

Tempest, D. W., Meers, J. L., and Brown, C. M. (1970). Biochem. J. 117, 405.

Varricchio, F., and Holzer, H. (1969). FEBS-Letters 3, 263.

Vorhaben, J. G., and Campbell, J. W. (1972). J. Biol. Chem. 247, 2763.

Wade, H. E., Robinson, H. K., and Phillips, B. W. (1971). J. Gen. Microbiol. 69, 299.

Wohlhueter, R. M. (1971). European J. Biochem. 21, 575.

Wohlhueter, R. M., Ebner, E., and Wolf, D. H. (1972). J. Biol. Chem. 247, 4213.

Wolf, D., Ebner, E., and Hinze, H. (1972). European J. Biochem. 25, 239.

Woolfolk, C. A., Shapiro, B. M., and Stadtman, E. R. (1966). Arch. Biochem. Biophys. 116, 177.

Wulff, K., Mecke, D., and Holzer, H. (1967). Biochem. Biophys. Res. Commun. 28, 740.

Footnote:

[1] Our calculations are based on figures given by Luria (1960) for water, protein, nucleic acid, and total nitrogen content, as well as for protein composition, by Salton (1960) for cell wall composition, and by Sueoka (1960) for nucleic acid base composition. They further assume an average amino acid residue weight of 120 daltons, and that 50 % of the chromatographically measured protein-glutamate was actually glutamine.

GLUTAMINE SYNTHETASE, REGULATOR OF THE SYNTHESIS OF GLUTAMATE-FORMING ENZYMES

Boris Magasanik, Michael J. Prival, and Jean E. Brenchley

Department of Biology
Massachusetts Institute of Technology
Cambridge, Massachusetts 02139

The intracellular level of glutamine synthetase is higher in a bacterium growing on a growth-rate limiting nitrogen source than in a bacterium growing on an excess of ammonia (Holzer *et. al.*, 1968). Recently, Tempest *et. al.* (1970) described an enzyme, glutamine (amide): α-ketoglutarate (NADP) amidotransferase oxido reductase (glutamate synthetase), which converts glutamine and α-ketoglutarate in the presence of NADPH to two molecules of glutamate. It is clear that this enzyme acting in concert with glutamine synthetase, produces a molecule of glutamate from one molecule of α-ketoglutarate and one of ammonia at the cost of the conversion of one molecule of ATP to ADP, and of one molecule of NADPH to $NADP^+$. It is obvious that the overall reaction has the same result as that catalyzed by glutamate dehydrogenase, except for the involvement of ATP. The other difference is that the affinity of glutamine synthetase for NH_3 is many times greater than that of glutamate dehydrogenase. Thus, the cell has two mechanisms for the essential assimilation of ammonia: the ATP dependent glutamine synthetase - glutamate synthetase system, which can function when the exogenous concentration of NH_3 is below 1mM, and the ATP independent glutamate dehydrogenase which is only effective at higher concentrations of ammonia. It is therefore not surprising that the cell responds to ammonia starvation by increasing its store of glutamine synthetase; it is also not surprising that under the same condition the cell dispenses with the synthesis of the now useless glutamate dehydrogenase. The level of this

65

enzyme is more than ten times higher in cells growing with an excess of ammonia than in those growing on a growth-rate limiting nitrogen source (Meers *et. al.*, 1970).

Many organisms, for example *Klebsiella aerogenes*, can obtain glutamate not only by assimilation of NH_3, but also by the degradation of an exogenously supplied amino acid, such as histidine (Magasanik and Bowser, 1955). The formation of the enzymes of histidine degradation in a medium containing glucose as the major source of carbon re-quires the presence of histidine and the absence of ammonia (Neidhardt and Magasanik, 1957). It is apparent that the cell regulates the formation of these enzymes according to its needs: without histidine the enzymes have no function to perform and in a medium containing glucose and ammonia their function is not required. The controls by induction through histidine and by repression through glucose and ammonia function independently: mutants constitutive for the *hut* (histidine utilization) system are still repressed by glucose-ammonia (Prival and Magasanik, 1971).

We can see that nitrogen limitation has opposite effects on the *hut* system and on glutamate dehydrogenase: derepression of the former and repression of the latter. We now wish to propose that the cellular element respon-sible for both regulatory effects is glutamine synthetase.

We have isolated glutamine-requiring mutants of a *hut* constitutive strain of *K. aerogenes*. These mutants appear to be located at two different sites on the chromosome, *glnA* and *glnB*, as shown by the fact that phage PW 52 lysates of the wild type organism and of the *glnB3* mutant strain MK-93 produce the same number of transductants when allowed to act on the *glnA6* mutant MK-104; phage lysates of other glutamine-requiring mutants such as MK-103 pro-duce no transductants when allowed to act on MK-104. Apparently, these strains carry mutations in the *glnA* site. Extracts of *glnA* or *glnB* mutants have very low or no measurable glutamine synthetase activity.

It is possible to distinguish *glnA* and *glnB* mutants by their revertants. Strain MK-103 (*glnA5*) yields revertants able to grow, like the wild strain, without glutamine at temperatures up to 42°. In addition, it yields temperature-sensitive revertants, able to grow with-out glutamine at 32°, but requiring glutamine at 42°. The glutamine synthetase activity measured after partial purification of the cell extract of one of the temperature-

sensitive revertants appears to be more labile at higher
temperatures than the glutamine synthetase of the wild
strain. On the basis of this evidence we would tentatively
identify *glnA* as the structural gene for glutamine synthetase (A. DeLeo and B. Magasanik, unpublished observation).
   No temperature-sensitive revertants of the *glnB3*
mutant MK-93 could be obtained, despite repeated attempts.
However, temperature insensitive revertants were readily
obtained; these differ from the wild type in their regulation of the formation of glutamine synthetase. This
enzyme is constitutive in *glnB* revertants: its level is
high whether the cells have been grown on excess ammonia
or on a growth-rate limiting nitrogen source. It is of
particular interest that the mutation which has restored
the ability of the *glnB* mutant to make glutamine synthetase,
but to produce it constitutively (*glnC*) is linked to the
glnA site, and not the *glnB* site. This is shown by the
fact that glutamine-independent transductants obtained by
having phage lysates of the revertants act on the *glnA*
mutant MK-104 produce glutamine synthetase constitutively.
   The glutamine-requiring mutants, both *glnA* and *glnB*,
differ from the parent strain not only by their inability
to produce glutamine synthetase, but also by alterations
in the regulation of the formation of glutamate dehydrogenase and histidase. These organisms contain a high level
of glutamate dehydrogenase but only a low level of histidase when grown in a medium containing glucose as major
carbon source and glutamine as the sole growth-rate
limiting nitrogen source. Under these conditions their
gln[+] *hut*-constitutive parent contains a very low level of
glutamate dehydrogenase and a high level of histidase.
   The opposite effects on glutamate dehydrogenase and
histidase are observed in *glnC* mutants: cells of these
strains produce essentially no glutamate dehydrogenase,
but contain histidase at a high level even when grown in a
medium containing an excess of NH_3.
   As shown in Table I, a high level of glutamine synthetase is always correlated with a high level of histidase
and a low level of glutamate dehydrogenase; conversely a
low level of glutamine synthetase is always correlated
with a low level of histidase and a high level of glutamate
dehydrogenase.

TABLE I

Response of glutamine synthetase, histidase,
and glutamate dehydrogenase to limitation
of the nitrogen source.

Strain	Pertinent Genotype	Nitrogen Source	Enzyme level in units per mg protein		
			Gln-S	Glut-D	Histidase
MK-53	gln^+	excess	270	244	81
		limiting	950	49	986
MK-104	$glnA6$	limiting	0	258	103
MK-93	$glnB3$	limiting	0	518	83
MK-94	$glnC4$	excess	1,400	0.1	580
		limiting	840	1.2	971

Glutamine synthetase (Gln-S) was determined in sonically disrupted cells by the transferase assay of Stadtman *et. al.* (1970); this assay measures the sum of adenylated and deadenylated enzyme. The *glnC* mutation did not affect adenylation. Glutamate dehydrogenase was determined in sonically disrupted cells as described by Meers *et. al.* (1970). Histidase was determined in detergent-treated whole cells as described by Smith *et. al.* (1971). Enzyme units are nmoles of product formed per minute. Glutamate synthetase activity was also measured, but found not to vary significantly.

This coordination is only found in media containing glucose, the source of catabolites capable of causing the repression of histidase. For example, in a medium containing citrate as sole source of carbon and an excess of ammonia, the cellular level of glutamine synthetase is low and that of histidase is high.

Mutants can be isolated which are specifically deficient in glutamate dehydrogenase. In these mutants glutamine synthetase is made under normal control.

We would like to propose the following hypothesis to account for our results. The expression of the structural

gene for glutamine synthetase, which we consider to be
glnA, is regulated by the product of the *glnB* gene.   In
the absence of this product the *glnA* gene cannot function,
and thus the *glnB* mutant fails to produce glutamine synthetase.   The *glnB* product can be inactivated by reacting
with glutamine, accounting for the repression of glutamine
synthetase.   A mutation in the *glnC* site, closely linked
to *glnA* renders the expression of *glnA* independent of the
glnB product, accounting for the ability of the *glnC*
mutants to produce glutamine synthetase constitutively,
irrespective of a mutation in the *glnB* site.

    We further postulate that glutamine synthetase
represses the formation of glutamate dehydrogenase and
overcomes the repression by glucose of enzyme systems that
produce glutamate from other amino acids, such as histidine.   We ascribe this effect to glutamine synthetase
itself, rather than to the product of its action, glutamine,
since addition or deprival of glutamine in the various
mutants of glutamine synthetase did not alter the expression of histidase or glutamate dehydrogenase.   Moreover, we have isolated a mutant in the *glnA* site, incapable of producing glutamine synthetase, but producing
histidase at a high rate in the presence of glucose, excess
ammonia, and glutamine.   This mutant may produce an
altered, enzymatically inactive glutamine synthetase,
still capable of overcoming the catabolite repression of
histidase.

    Our findings suggest the *K. aerogenes* has evolved an
elegant control system for the synthesis of enzymes
capable of supplying the cell with glutamate.   The key
element is glutamine synthetase.   Its synthesis is
controlled by repression: a decrease in the level of
glutamine or ammonia leads to rapid synthesis of this
enzyme.   The presence of a large amount of this enzyme in
the cell arrests the formation of glutamate dehydrogenase,
an enzyme incapable of contributing to the formation of
glutamate when the exogenous concentration of ammonia is
low.   At the same time, the large amount of glutamine
synthetase in the cell relieves enzymes capable of forming
glutamate from exogenous amino acids, such as histidine or
proline, from catabolite repression.   These enzymes are
inducible and can now be formed when the appropriate
amino acid is present in the medium.

    This work was supported by research grants #AM13894
and #GM07446 from the National Institutes of Health and
grant #GB32509 from the National Science Foundation. Preliminary reports have appeared (Prival and Magasanik, 1972;
Brenchley and Magasanik, 1972). A complete account is in
preparation. The present address of M.J. Prival is Center
for Science in the Public Interest, 1179 Church Street,
N.W., Washington, D.C.; and of J.E. Brenchley, Department
of Microbiology, Pennsylvania State University, University
Park, Pennsylvania.

References

Brenchley, J.E. and Magasanik, B. (1972). Amer. Soc.
    Microbiol. Abst. of Papers, p. 170.

Holzer, H., Schutt, H., Masek, Z., and Mecke, D. (1968).
    Proc. Nat. Acad. Sci. U.S. 60, 721.

Magasanik, B. and Bowser, H.R. (1955). *J. Biol. Chem. 213*,
    571.

Meers, J.L., Tempest, D.W., and Brown, C.M. (1970).
    J. Gen. Microbiol. 64, 187.

Neidhardt, F.C. and Magasanik, B. (1957). *J. Bacteriol. 73*,
    253.

Prival, M.J. and Magasanik, B. (1971). *J. Biol. Chem. 246*,
    6288.

Prival, M.J. and Magasanik, B. (1972). *Fed. Proc. 31*, 498.

Smith, G.R., Halpern, Y.S., and Magasanik, B. (1971).
    J. Biol. Chem. 246, 3320.

Tempest, D.W., Meers, J.L., and Brown, C.M. (1970).
    Biochem. J. 117, 405.

GENETIC STUDIES ON GLUTAMATE DEHYDROGENASE, GLUTAMATE SYNTHASE AND GLUTAMINE SYNTHETASE IN E̲. COLI̲ K_{12}

M. A. Berberich

Laboratory of Biochemistry, National Heart and
Lung Institute, NIH, Bethesda, Md. 20014

ABSTRACT

The genetics of the enzymes which catalyze glutamate and glutamine synthesis have been examined in E̲. coli̲ K_{12}. The possibility that an operon-type relationship exists for this system of enzymes can be eliminated.

INTRODUCTION

In E̲. coli̲ growing on minimal media, with NH_4^+ serving as the source of nitrogen, nitrogen is incorporated as shown in equation 1.

$$2 \text{ oxo-glutarate} \xrightarrow{NH_4^+} \text{glutamate} \xrightarrow{NH_4^+} \text{glutamine} \quad (1)$$

Glutamate may be synthesized from 2 oxo-glutarate by 3 different reactions: a) transamination as in equation 2; b) a NADPH-dependent reductive amination catalyzed by glutamate dehydrogenase, as in equation 3; and c) a NADPH-dependent $\in$ amino transfer reaction catalyzed by glutamate synthase, as in equation 4.

$$2 \text{ oxo-glutarate} + RCHNH_2COOH \rightleftharpoons \text{glutamate} + RCOCOOH \quad (2)$$

$$2 \text{ oxo-glutarate} + NH_4^+ + NADPH \rightleftharpoons \text{glutamate} + NADP \quad (3)$$

$$2 \text{ oxo-glutarate} + \text{glutamine} + NADPH \rightleftharpoons 2 \text{ glutamate} + NADP \quad (4)$$

The conversion of glutamate to glutamine is catalyzed by glutamine synthetase, as in equation 5.

$$\text{glutamate} + NH_4^+ + ATP \xrightarrow{M^{2+}} \text{glutamine} + ADP + P_i \quad (5)$$

71

The physical-chemical properties of each of these enzymes
as well as the physiological interrelationships which
control the level of their activities are discussed elsewhere in this volume. This paper will summarize some
studies, still in progress, which attempt to clarify
whether additional regulatory interrelationships for
glutamate dehydrogenase, glutamate synthase and glutamine
synthetase exist on the genetic level.

ENZYME ASSAYS

Cells were grown in a minimal salts-glucose medium
(Prusiner and Stadtman, 1971), containing 30 mM NH_4Cl.
Cultures were harvested in mid-log phase, the cells washed
once with distilled H_2O and resuspended in pH 7.5 buffer
containing 20 mM imidazole-Cl and 10 mM $MgCl_2$. Extracts
were prepared by sonication, clarified by centrifugation
and assayed. The biuret method was used for protein determinations (Layne, 1957). A mix was prepared containing
final concentrations of 50 mM HEPES, 1 mM EDTA, 5 mM
2-oxoglutarate, and adjusted to pH 7.5 with KOH. For
glutamate synthase activity 5 mM glutamine was added to
the above mix. For glutamate dehydrogenase activity 100 mM
NH_4Cl was added to the above mix. Assays were started by
the addition of 0.16 mM $NADPH_2$ and extract. Enzyme
activity in each case was followed by the reduction of
$NADPH_2$ at 340 mμ in a Coleman recording spectrophotometer.
Rates were corrected by substracting activity in mix minus
glutamine and NH_4Cl. Glutamine synthetase was measured by
a γ-glutamyl transferase assay which is independent of the
state of adenylylation (Stadtman et. al., 1970).
Glutaminase B was measured by the production of radioactive
glutamate from labelled glutamine (Prusiner and Stadtman,
1971).

RESULTS AND DISCUSSION

Glutamine synthetase.
The initial approach was to collect appropriate mutants,
defective in one or more of these enzyme activities, and to
localize the genes controlling the synthesis of the respective proteins on the chromosomal map of E. coli. A
prototrophic strain of E. coli K_{12} (strain 28 of Meselson)
was selected for the production of the glutamine synthetase
mutants. The method involved mutagenesis with DES followed

by cyclical penicillin selections for inability to grow in
the absence of exogenous glutamine in the presence of high
concentrations of NH_4^+ salts. A group of glutamine-requiring
mutants was obtained (Berberich, in preparation) whose
extracts are devoid of glutamine synthetase activity. The
activities of the other enzymes involved in NH_3 assimilation are identical to wild type. When these extracts were
examined in an Ouchterlony diffusion system, all the orignal
isolates ($\sim$ 6) showed cross-reactive material vs. rabbit
anti-purified glutamine synthetase (n_0 - n_{12}). Recombina-
tion via P_1 transduction (Rosner, 1972) localized the
structural gene for glutamine synthetase at $\sim$ 74 - 74.5 min
corresponding to a locus probably between ilv and met E.
Linkage to ilv is 65%; to met E, 83%; to met B (76.4'),
11.4%. Transductants selected simultaneously for the ilv$^-$
character and restoration of growth in the absence of
exogenous glutamine appear to distribute into 2 groups. One,
comprising approx. 10% of the total recombinants exhibits
marked inhibition of growth by glutamate. This observation
may be significant with respect to the hybrid (recombinant)
structure of glutamine synthetase or perhaps, more interestingly, with respect to the adenylylation system. It is
currently being pursued.

Glutamate dehydrogenase and glutamate synthase.
        Because of some irregularities observed in preliminary
crosses with a glutamate-requiring strain obtained from
Sanwal, several of the donor strains which were auxotrophic
for various amino acids were chosen at random and examined
for four of the enzymes involved in the assimilation of
NH_3 (see Table I). Strains AB 2550 and AB 1450 were found
to be defective for glutamate dehydrogenase and glutamate
synthase, respectively, although neither strain requires
glutamate for growth. The strain requiring glutamate for
growth, CB100, is defective for both the synthase and the
dehydrogenase. All glutamine synthetase and glutaminase B
levels correspond to those of wild type. Conversely, the
glutamine synthetase-minus mutant, gln 67, has normal levels
of dehydrogenase and synthetase.

        Subsequent genetic studies determined that the loci
for glutamate synthase and glutamate dehydrogenase are
unlinked. The double mutant, CB100, was used as the
recipient in transductions with P_1 phage grown on wild-type
cells. Colonies representing a random 10% sample of

Table I[*]

Comparison of Growth Requirements and Enzymic Activities[a]
of Bacterial Strains[b] Used in This Study

gln/glut[*] reqt.	Activities (units/mg protein)			
	GAT	GDH	GS	GB
none[1]	0.17	0.08	0.18	0.12
glut[2]	0.01	0.01	0.19	0.17
gln[3]	0.17	0.08	0.01	0.11
none[4]	0.13	0.01	0.16	0.16
none[5]	0.01	0.05	0.21	0.15
none[6]	0.08	0	0.18	0.09

[*]These data are from Berberich, 1972; gln = glutamine; glut = glutamate.

[a]GAT: glutamate synthase; units equal μm NADP formed/min. GDH: glutamate dehydrogenase; units equal μm NADP formed/min. GS: glutamine synthetase; units equal μm γ glutamyl hydroxamate formed/min. GB: glutaminase B; units equal μm glutamate formed/min.

[b]Sources and genotypes of strains.

[1] W3102 from Rosner; Sm^r, F^- (wild type); equals strain 28 of M. Meselson described in Gottesman and Yarmolinsky (1968).

[2] CB100 from Sanwal Sm^r, F^-, thr, leu, B_1, his, glut; selected on the basis of a glutamate growth requirement.

[3] gln 67 derived from W3102 by Berberich (1972) on the basis of a glutamine growth requirement.

[4] AB2550 from the collection of Adelberg; Sm^r, F^-, ilv, met E, thi, pro, trp, his.

[5] AB1450 from the collection of Adelberg; Sm^r, F^-, thi, ilv, met B, his.

[6] W1317G from Rickenberg; see Vender and Rickenberg (1964).

transductants growing in the absence of glutamate were cloned and re-tested for growth on appropriate media. Extracts were prepared as described in ENZYME ASSAYS. When the extracts were examined, 40% of the transductants were glutamate synthase[+], and 60% were glutamate dehydrogenase[+]. None of these transductants have both enzyme activities (Berberich, 1972). Both types of transductants have similar colony size and morphology, and in the absence of glutamate, both types grow with a doubling time of 120 min and show similar responses to changes in NH_4Cl concentration in the growth media, at least in batch cultures (Berberich, unpublished results).

Neither glutamate dehydrogenase nor glutamate synthase is linked to glutamine synthetase. Preliminary mapping studies (Berberich, unpublished results) using previously characterized Hfr donors, show that the time of entry of the gene for glutamate synthase activity corresponds to a location between 55' and 66' whereas entry of the gene for glutamate dehydrogenase activity corresponds to a location between 4 and 25 min on the E. coli chromosomal map.

Concluding remarks.
These initial studies indicate that the enzymes involved in NH_3 assimilation do not constitute a contiguous genetic unit such as an operon. However, a more thorough assessment of the possibilities for interrelationships at the genetic level, perhaps via a common regulatory molecule, awaits the isolation of regulatory mutants for one or more of the enzymes. In this regard, it is interesting that, upon addition of exogenous cyclic AMP to the growth media, the synthesis of glutamate dehydrogenase increases while the synthesis of glutamate synthase decreases (Prusiner et. al., 1972). It could also be significant that the dehydrogenase[-] strain AB 2550 and the synthase[-] strain AB 1450 were originally isolated on the basis of amino acid auxotrophy other than glutamate. This observation invites speculation that at least the two glutamate enzymes may be indirectly linked to each other in a greater organization concerned with the co-ordination of nitrogen metabolism.

REFERENCES

Berberich, M. A. (1972) Biochem. Biophys. Res. Commun. 47, 1498.

Gottesman, M. E. and Yarmolinsky, M. B. (1968) J. Mol. Biol. 41, 487.

Layne, E. (1957) in S. P. Colowick and N. O. Kaplan (ed.) Methods in Enzymology, Vol. 3, pp. 447-454. Academic Press, New York.

Prusiner, S. and Stadtman, E. R. (1971) Biochem. Biophys. Res. Commun. 45, 1474.

Prusiner, S., Miller, R. E., and Valentine, R. C. (1972) Proc. Nat. Acad. Sci. U. S. (in press).

Stadtman, E. R., Ginsburg, A., Ciardi, J. E., Yeh, J., Hennig, S. B., and Shapiro, B. M. (1970) Advances in Enzyme Regulation, Vol. 8, pp. 99-118. Pergamon Press, Oxford and New York.

Vender, J. and Rickenberg, H. V. (1964) Biochem. Biophys. Acta 90, 218.

GLUTAMINE SYNTHETASES OF MAMMALIAN LIVER AND BRAIN

Suresh S. Tate and Alton Meister

Department of Biochemistry
Cornell University Medical College
New York, N. Y. 10021

INTRODUCTION

Glutamine is of major metabolic importance not only because it is a required building block of most proteins, but also because it serves as a source of nitrogen for a variety of significant biosynthetic pathways (Figure 1).

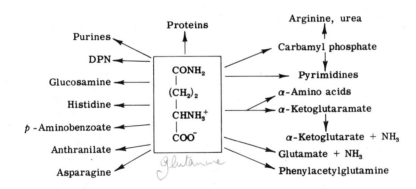

Fig. 1 Summary of Glutamine Metabolism

Thus, the amide nitrogen atom of glutamine is used for the synthesis of the amide nitrogen atoms of DPN and asparagine, nitrogen atoms 3 and 9 of the purine ring, the amino groups of glucosamine, guanine, cytidine, p-aminobenzoic acid, the nitrogen atom of carbamyl phosphate (which in turn used for the synthesis of arginine, urea, and nitrogen atom 1 of the pyrimidine ring), nitrogen atom 1 of the imidazole ring of histidine, and the pyrrole nitrogen atom of tryptophan[1]/. In serving these functions, glutamine seems to play a role that could conceivably be

77

fulfilled by ammonia itself, but for evolutionary and
metabolic reasons not yet fully understood, this is not
the case. However, it is notable that virtually all of
the glutamine amidotransferases can utilize ammonia in
place of glutamine at least to some extent, and it is significant that most of these enzymes exhibit at least some
glutaminase activity. Glutaminase, which may also be considered as a glutamine amidotransferase, is of metabolic
significance in certain cells in relation to the production of ammonia. It is important to note that the glutamine molecule is equipped with two nitrogen atoms[2/] and
that the α-amino group of glutamine is extensively utilized for the synthesis of amino acids by transamination.
In this pathway the amide nitrogen atom of glutamine is
released as ammonia in an indirect manner. Thus, transamination of glutamine with α-keto acids leads to the formation of α-ketoglutaramic acid, which is hydrolyzed to α-
ketoglutarate and ammonia by ω-amidase, an enzyme that
does not exhibit hydrolytic activity toward glutamine itself.

In view of the metabolic significance of glutamine,
it is not surprising to find that glutamine is the amino
acid present in highest concentration in mammalian blood.
It is also present in relatively high concentrations in
many mammalian tissues such as the heart and brain.
While glutamine is needed for the growth of several microorganisms and by certain mammalian cells when grown in
tissue culture, glutamine is not required in the diet of
animals; many animal tissues, plants, and microorganisms
can synthesize glutamine from glutamate and ammonia. The
reaction catalyzed by glutamine synthetase is a major
pathway for the utilization of free ammonia.

In this lecture, we will review published data on the
structure, mechanism of action, and regulation of mammalian glutamine synthetases, especially those of brain and
liver. Earlier reviews (Meister, 1962b, 1968a, b and c)
and the literature cited therein should be consulted for
historical information[3/] and for other details. The wide
variety of the metabolic functions of glutamine suggests
that glutamine synthetase must be subject to regulatory
mechanisms, which can control the supply of glutamine
available for different purposes. It is evident also that

glutamine serves different functions in different cells.
Thus, E. coli utilizes glutamine for the synthesis of
histidine, tryptophan, and p-aminobenzoic acid, while animals do not. There is substantial utilization of glutamine
in man (and in a few close relatives) for the synthesis of
phenylacetyglutamine; this reaction does not occur in other
animal species. While we know much about the metabolism
and enzymology of glutamine, less is known about the relationship between enzymatic events and physiological processes; yet it is evident that these must be closely connected. Thus, there is considerable reason to believe
that glutamine serves a special function in the central
nervous system related undoubtedly to the metabolism of
glutamate and γ-aminobutyrate, which are, respectively,
excitatory and inhibitory neurotransmitters. The high
concentration of glutamine in heart muscle still requires
study at both physiological and biochemical levels. These
considerations suggest that one must expect to find significant differences between the glutamine synthetases of
different cells with respect to structure, regulatory
mechanisms, and other properties. On the other hand, since
all of the glutamine synthetases catalyze the same chemical
reaction, it is to be expected that some properties of the
glutamine synthetases isolated from various sources will be
virtually the same, especially if one considers the probability that the catalytic sites of the enzyme are very
similar. The mammalian glutamine synthetases thus far
studied all have 8 subunits, exhibit similar substrate
specificity, and are irreversibly inhibited by methionine
sulfoximine. While the bacterial enzymes now known differ
from the mammalian enzymes in molecular weight and number
of subunits (and in some cases in being adenylylated), it
is notable that the subunit molecular weights of all the
glutamine synthetases thus far isolated range between
44,000 and 50,000. There are major differences between
mammalian and the bacterial glutamine synthetases with respect to the mechanisms by which they are controlled.
Nevertheless, examination of the available data indicates
some interesting parallels. For example, while α-ketoglutarate directly activates liver glutamine synthetase, the
effect of α-ketoglutarate in increasing glutamine formation in E. coli is mediated by stimulating deadenylylation
leading to the production of a more active enzyme, and also
by inhibiting adenylylation (Stadtman et al., 1968a, 1968b,

1970; Shapiro and Stadtman, 1968; Shapiro, 1969). Another example, discussed in detail below, concerns the inhibition of glutamine synthetases from a variety of sources
by carbamyl phosphate; this compound, which cannot be considered an allosteric inhibitor, apparently binds directly
to the acyl phosphate site at the catalytic center of the
enzymes (Tate et al., 1972).

Mechanism of the Reaction

Studies carried out over a period of several years
have shown that glutamine synthetase can catalyze a number
of reactions in addition to the physiologically significant
synthetase reaction:

(1) Glutamate + NH_3 + ATP $\overset{*}{\rightleftharpoons}$ Glutamine + ADP + Pi

(Functions with L- and D-glutamate; when NH_3 is replaced by NH_2OH, γ-glutamylhydroxamate is formed. Other
NH_3 analogs, including CH_3NHNH_2, NH_2NH_2, CH_3NH_2, glycine
ethyl ester, are also active. $^*Mg^{++}$, Mn^{++} or CO^{++} are required).

(2) L-Glutamine + NH_2OH $\overset{*}{\longrightarrow}$ L-γ-glutamylhydroxamate +
$$NH_3$$

(*ATP or ADP, Pi or Asi, and Mg^{++} or Mn^{++} are
required).

(3) L-Glutamine + H_2O $\overset{*}{\longrightarrow}$ L-glutamate + NH_3

(*ADP, Mg^{++} or Mn^{++}, and Asi are required).

(4) Glutamate + ATP $\overset{*}{\longrightarrow}$ pyrrolidone carboxylate + ADP +
$$Pi$$

(Functions with L- and D-glutamate; *Mg^{++} required; NH_3
must be excluded).

(5) L-Methionine-S-sulfoximine + ATP $\overset{*}{\longrightarrow}$ L-methionine-S-
$$\text{sulfoximine phosphate + ADP}$$

(*Mg^{++} or Mn^{++} required; products are enzyme-bound;

see the text).

(6) β-Glutamyl phosphate + ADP $\longrightarrow$ ATP + β-glutamate

(7) Carbamyl Phosphate + ADP $\longrightarrow$ ATP + CO_2 + NH_3

    (Functions also with the acetyl phosphate).

(8) Cycloglutamate* + ATP $\xrightarrow{\text{Mg}^{++}}$ cycloglutamylphosphate +

$$ADP + Pi$$

    (*<u>cis</u>-1-amino-1,3-dicarboxycyclohexane).

These reactions have been studied in detail with ovine brain glutamine synthetase; the available data indicate that they are also catalyzed by glutamine synthetases from other sources.

    Early studies showed that the glutamine synthetase reaction involves a step in which glutamate is activated. This was initially suggested by data obtained in studies on the synthesis of the L- and D-isomers of glutamine and γ-glutamylhydroxamate (Levintow and Meister, 1953). Thus, the synthesis of L-glutamine occurs about three times more rapidly than does that of D-glutamine, but when hydroxylamine is substituted for ammonia, the corresponding γ-glutamylhydroxamate isomers are formed at about the same rate. These findings suggested that the first step in glutamine synthesis is a relatively optically non-specific activation reaction, and that the second step is a more specific reaction of the activated intermediate with ammonia. Since γ-glutamyl compounds generally exhibit a tendency to cyclize to yield pyrrolidone carboxylic acid, it was reasoned that an activated γ-glutamyl derivative, such as that formed by glutamine synthetase, might cyclize rapidly. Such expectation was realized in the finding that glutamine synthetase catalyzes the formation of pyrrolidone carboxylic acid in the absence of ammonia (reaction [4]). It is notable that the rate of pyrrolidone carboxylic acid formation from L-glutamate is about the same as that from D-glutamate; cyclization appears to reflect formation of

81

an activated glutamate intermediate (Krishnaswamy et al., 1962).

Ultracentrifugation and ultrafiltration studies were carried out with labeled glutamate and labeled ATP. These experiments showed that ATP and magnesium ions are needed for the binding of glutamate to the enzyme, that the binding of glutamate to the enzyme is associated with its activation, that activation of glutamate is coupled with cleavage of ATP, and that the ADP and phosphate formed in the course of binding remain attached to the enzyme. In a series of pulse-labeling experiments, the enzyme was incubated with ATP, magnesium ions, and a small amount of labeled glutamate. After brief incubation, a large excess of unlabeled glutamate and hydroxylamine were added together. The mixture was then deproteinized and the formation of labeled γ-glutamylhydroxamate was determined. These studies showed that there is preferential conversion of labeled glutamate to glutamylhydroxamate indicating that equilibration between labeled and unlabeled glutamate does not occur under the experimental conditions employed (Krishnaswamy et al., 1962). These experiments, which provide indirect information about the activation of glutamate, when considered in relation to earlier studies which showed that the synthesis of glutamine was accompanied by a transfer of oxygen from glutamate to inorganic phosphate (Kowalsky et al., 1956; Boyer et al., 1956), are in accord with the hypothesis of an intermediate γ-glutamyl phosphate. Such an acyl phosphate intermediate had been demonstrated directly in the case of glutathione synthetase, which catalyzes a similar reaction. Thus, in studies on this enzyme it was possible to isolate the dipeptidyl phosphate intermediate after incubation of the enzyme with dipeptide, ATP, and magnesium ions (Nishimura et al., 1963, 1964). However, such an isolation effort in the case of γ-glutamyl phosphate was hindered by the marked tendency of this compound to cyclize.

An interesting approach to this problem became evident when the curious optical specificity of the enzyme was again considered. Since both D-glutamate and L-glutamate are substrates, it is apparent that the amino group of the substrate does not need to be in a specific position. Therefore, β-glutamic acid, in which the amino group is attached to the third possible position of the glutaric acid

82

carbon chain, was synthesized and tested. It proved to be
a good substrate. The β-aminoglutaryl phosphate intermediate that might be postulated in the enzymatic synthesis of
β-glutamine would be expected to be relatively stable as
compared to γ-glutamyl phosphate. Therefore, β-aminoglutaryl phosphate was synthesized, and it was found that the
enzyme can utilize this compound in the presence of ADP for
the synthesis of ATP (Khedouri et al., 1964). The ability
of glutamine synthetase to react with certain acyl phosphate compounds is further illustrated by its interaction
with carbamyl phosphate and acetyl phosphate as discussed
below. Additional evidence in support of the acyl phosphate hypothesis was obtained in studies with another analog of glutamate, cis-1-amino-1,3-dicarboxycyclohexane.
When this compound (which cannot cyclize to form an analog
of pyrrolidone carboxylic acid) was incubated with the enzyme in the presence of ATP and magnesium ions, but in the
absence of ammonia, evidence was obtained for the formation
of the corresponding acyl phosphate derivative (Tsuda et
al., 1971).

Further support for the intermediate participation of
γ-glutamyl phosphate in the synthesis of glutamine arose
from studies designed to elucidate the mechanism of inhibition of glutamine synthetase by methionine sulfoximine.
The enzyme was found to be irreversibly inhibited after incubation with methionine sulfoximine ATP, and metal ions.
Such inhibition is associated with the tight binding of
methionine sulfoximine to the enzyme and it was subsequently found that the inhibitor is bound to the enzyme in
the form of methionine sulfoximine phosphate (Ronzio and
Meister, 1968; Ronzio et al., 1969a; Rowe et al., 1969)
(Figure 2). Thus far, attempts to reactivate methionine
sulfoximine-inactivated enzyme have been unsuccessful, but
when the inactivated enzyme is denatured by heating at 100°
or by treatment with perchloric acid, methionine sulfoximine phosphate is released from the enzyme. In these
studies it was also found that the ADP formed remains attached to the enzyme. Methionine sulfoximine phosphate
was prepared by chemical synthesis and this compound, when
added to the enzyme, produces marked inhibition especially
in the presence of ADP and metal ion. The phosphorylation
of methionine sulfoximine by glutamine synthetase appears
to be analogous to the formation of γ-glutamyl phosphate

and the findings with methionine sulfoximine are therefore in accord with the acyl phosphate hypothesis.

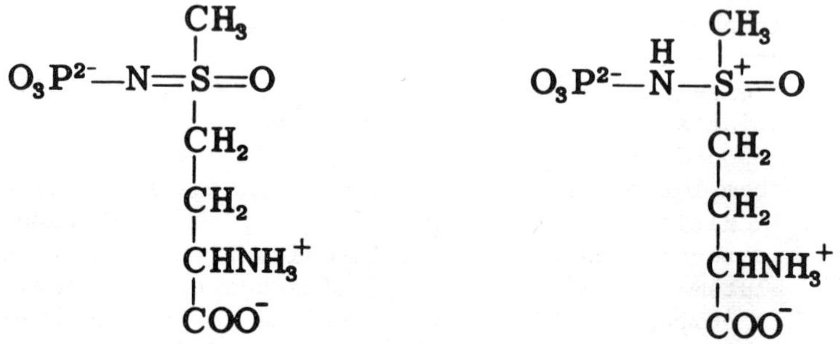

$$O_3P^{2-}\!-\!N\!=\!S\!=\!O$$

$$
\begin{array}{c}
CH_3 \\
| \\
\end{array}
$$

$$
\begin{array}{c}
CH_2 \\
| \\
CH_2 \\
| \\
CHNH_3^+ \\
| \\
COO^-
\end{array}
$$

$$O_3P^{2-}\!-\!\overset{H}{N}\!-\!\overset{+}{S}\!=\!O$$

Possible structures for methionine sulfoximine phosphate

Fig. 2

It has been known for many years that administration of methionine sulfoximine produces convulsions in a number of animal species and it seems probable that the inhibition of brain glutamine synthetase by methionine sulfoximine is related to its convulsant action (Rowe and Meister, 1970). As discussed below, only one of the four isomers of methionine sulfoximine, i.e., L-methionine-S-sulfoximine, inhibits glutamine synthetase and is phosphorylated by the enzyme. It seems notable that only this same isomer is capable of inducing convulsions. When methionine sulfoximine is administered to mice, there is prompt formation in the brain, liver, and kidney of protein bound methionine sulfoximine phosphate. The amounts of methionine sulfoximine phosphate formed in these tissues are quantitatively related to the amounts of glutamine synthetase present (Rao and Meister, 1972). Other studies have shown that methionine sulfoximine inhibits the glutamine synthetases of E. coli and of peas; inhibition by methionine sulfoximine appears to be a general property of this enzyme. As discussed below, its geometry favors binding to the active center of the enzyme.

Mapping of the Active Site

As discussed above, ovine brain glutamine synthetase exhibits a very unusual optical specificity; thus, it acts on both L- and D-glutamate to yield the corresponding isomers of glutamine. The enzyme also acts on β-glutamic acid and can therefore catalyze the amidation of all of the three possible monoaminoglutaric acids. However, it acts selectively on β-glutamic acid, amidating a specific carboxyl group to form D-β-glutamine as the product (see Meister, 1968b and c). While the enzyme can utilize both optical isomers of glutamic acid, it acts only on the L-isomer of α-methylglutamic acid. The unusual specificity of the enzyme is further illustrated by the fact that it acts upon only one of the four isomers of β-methylglutamic acid, i.e., threo-β-methyl-D-glutamic acid, and on only one of the four isomers of γ-methylglutamic acid, i.e., threo-γ-methyl-L-glutamic acid. These intriguing observations on the specificity of glutamine synthetase are now explicable. Indeed they form the basis for a series of studies on the mapping of the active site of the enzyme. It perhaps should be mentioned at this point that glutamine synthetases from mammalian tissues, plants, and bacteria exhibit similar substrate specificity and presumably therefore the conclusions discussed below are generally applicable.

Glutamine synthetase does not interact to a significant extent with monocarboxylic amino acids or with glutaric acid derivatives that lack an amino group; furthermore, aspartic acid is neither a substrate nor an inhibitor of the enzyme. It was therefore postulated that the enzyme has binding sites for both of the carboxyl groups of glutamic acid as well as for the amino group. It would appear that the distance between the carboxyl binding sites on the enzyme must be greater than the maximum possible intercarboxyl distance for aspartic acid (about 4A); this suggests that glutamic acid is in the fully extended (or almost fully extended) form when attached to the active site. (Stereophotographs of L- and D-glutamates in the extended conformation are shown in Fig. 3). However, if we assume that the enzyme site which interacts with the

85

amino group of L-glutamic acid also interacts with the amino group of D-glutamic acid, it becomes difficult to explain the enzymatic susceptibility of D-glutamic acid if the D-glutamic acid molecule is assumed to be oriented on the enzyme in the same manner as L-glutamic acid. However,

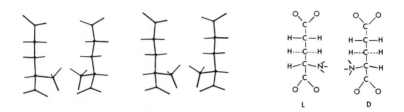

Fig. 3 stereophotographs of Dreiding models of L-glutamic acid (left) and D-glutamic acid (right).

the amino nitrogen atom of D-glutamic acid can be brought to the same relative position as that of L-glutamic acid by rotating the molecule of D-glutamic acid 69° about an axis formed by a straight line intersecting the centers of carbon atoms 1, 3, and 5 (Fig. 4). This procedure moves

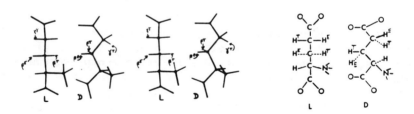

Fig. 4 Stereophotographs of Dreiding models of L-glutamic acid (left) and D-glutamic acid (right) rotated as described in the text.

the α-hydrogen atom of D-glutamic acid to a position which is essentially opposite to that occupied by the α-hydrogen atom of L-glutamic acid. If we assume that L-glutamic acid

86

attaches to the enzyme with its α-hydrogen atom directed
away from the enzyme, then substitution of the α-hydrogen
atom of L-glutamic acid by a methyl group should not interfere in binding to the enzyme. On the other hand, replacement of the α-hydrogen atom of D-glutamic acid by a methyl
group would be expected to provide considerable interference with the attachment of this substrate to the enzyme.
These considerations, which are in accord with the observed
strict stereospecificity of the enzyme toward α-methyl-L-
glutamic acid, were further extended in studies on various
β- and γ-substituted glutamic acids. Study of space fill-
ing and Dreiding models of L- and D-glutamic acid in the
conformations described above revealed that both β-hydrogen
atoms of L-glutamic acid and the erythro-β-hydrogen atom of
D-glutamic acid are on the same side of the molecule as the
α-hydrogen atom of D-glutamic acid. However, the threo-β-
hydrogen atom of D-glutamic acid lies in a position very
close to that of the α-hydrogen of L-glutamic acid. These
considerations led to the prediction that substitution by
a methyl group of either of the β-hydrogen atoms of L-glutamic acid or of the erythro-β-hydrogen atom of D-glutamic
acid would lead to a loss or marked reduction in enzymatic
susceptibility. Similarly, it was predicted that substitution of the threo-β-hydrogen atom of D-glutamic acid by a
methyl group would not lead to loss of enzymatic susceptibility. The experimental data verified these predictions.

Additional consideration of the models of L- and D-
glutamic acid in the conformations and positions postulated
at the active site of the enzyme indicated that the erythro
-γ-hydrogen atoms of both L- and D-glutamic acid occupy
about the same position in space and lie just between the
γ-carboxyl and amino groups of these molecules. It would
be expected that an erythro-γ-methyl group would provide
considerable steric hindrance. The threo-γ-hydrogen atom
of D-glutamic acid is close to the position of the α-hydrogen atom of D-glutamic acid; it would therefore be expected
that threo-γ-methyl-D-glutamic acid, like α-methyl-D-glu-
tamic acid would not be a substrate. However, the threo-
γ-hydrogen atom of L-glutamic acid is in a position close
to the α-hydrogen atom of this molecule and to that of the
threo-β-hydrogen atom of D-glutamic acid. This suggested
that threo-γ-methyl-L-glutamic acid would be a substrate
and that other γ-methylglutamic acids would not; the ex-

perimental data are in accord with these predictions.

These findings appear to explain the ability of glutamine synthetase to use α-methyl-L-glutamic acid, threo-
β-methyl-D-glutamic acid, and threo-γ-methyl-L-glutamic
acid, but not the other monomethyl substituted glutamates.
Examination of the models of the three enzymatically susceptible methyl substituted substrates in the postulated
conformations and orientations showed that the methyl
groups of these molecules are located on the same side of
the molecule, i.e., on the side opposite to that bearing
the amino group. This suggested the attractive possibility
of constructing a cyclohexane ring consisting of carbon
atoms 2,3, and 4 of the L-glutamic acid carbon chain and a
chain of 3 carbon atoms attached to carbon atoms 2 and 4 of
glutamate. The isomer of this compound that corresponds to
L-glutamic acid can exist in a form possessing a relatively
rigid 5-carbon chain identical to that of L-glutamic acid
(Fig. 5). When cis-1-amino-1,3-dicarboxycyclohexane was

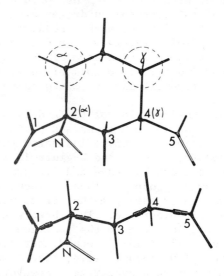

Fig. 5 Dreiding models of L-glutamic acid and cis-L-1-
amino-1,3-dicarboxycyclohexane (cis-cycloglutamate). From
Gass and Meister, 1970a.

synthesized and studied, it was found to be an excellent substrate of the enzyme.

The data reviewed above has been dealt with more fully elsewhere and a number of stereophotographs of models of the substrates and non-substrates have been published (Meister, 1968b and c; Gass and Meister, 1970b). The findings lead to the conclusion that L-glutamic acid is attached to the active site of glutamine synthetase in an extended conformation in which the carboxyl groups are as far apart (or almost so) as possible and in which the α-hydrogen atom of L-glutamic acid is directed away from the enzyme. The reasoning that leads to this conclusion starts with the postulate that the enzyme is equipped with binding sites for the two carboxyl groups and the amino group of glutamic acid. The ability of other substrates to attach to the enzyme depends upon their ability to assume a conformation and orientation on the enzyme in which their carboxyl and amino groups are in their respective proper places for attachment to the active site and in which other portions of the molecule do not provide steric hindrance sufficient to interfere with attachment. The hypothesis is supported by the finding that α-methyl-L-glutamic acid, threo-β-methyl-D-glutamic acid and threo-γ-methyl-L-glutamic acid are substrates, while the other monomethyl glutamic acid derivatives are not. The finding that the cyclic analog of glutamate, cis-1-amino-1,3-dicarboxycyclohexane is an excellant substrate affords additional strong support for the hypothesis. Other observations which can be explained by the hypothesis include the finding that only one of the four stereoisomers of methionine sulfoximine, i. e., L-methionine-S-sulfoximine inhibits the enzyme. It is also consistent with the experimental finding that both optical isomers of methionine sulfone inhibit the enzyme. The studies on inhibition of the enzyme by methionine sulfoximine led to the conclusion that this molecule attaches to both the enzyme sites for glutamate and ammonia thus serving as a bifunctional reagent.

In an extension of these studies, which were initially derived largely from consideration of molecular models of the various substrates, non-substrates and inhibitors, an attempt was made to define by means of a computer the individual points of attachment between the natural amino

acid substrate (as well as analogs and inhibitors) and the
active site of the enzyme. In this work, the three dimensional coordinates of the amino acid substrates and inhibitors were calculated and an active site was mathematically
designed. This approach has led to more quantitative explanations of the experimental findings and in addition to
several new conclusions about the mechanism of action of
the enzyme.

The calculations permitted the selection of a phosphorylation site by virtue of the coincidence in the positions of one of the γ-carboxyl oxygen atoms of L- and D-
glutamate, and also of an ammonia binding site on the enzyme derived from the calculated position of the nitrogen
atom of the hypothetical tetrahedral intermediate formed
by reaction of ammonia with γ-glutamyl phosphate. These
calculations showed that the methyl groups of methionine
sulfoximine and methionine sulfone attach to the ammonia
binding site of the enzyme. This site, probably hydrophobic, is designed for un-ionized ammonia rather than for the
ammonium ion. These calculations, described in detail
elsewhere (Gass and Meister, 1970b), also provide detailed
steric explanations for the more rapid synthesis of L-glutamine as compared to D-glutamine and for other kinetic
data. They also provide explanations for the specificity
of inhibition by methionine sulfoximine and the lack of
such specificity by methionine sulfone. The calculations
are in accord with the conclusion that methionine sulfoximine and methionine sulfone are analogs of the hypothetical
tetrahedral intermediate rather than glutamate. Thus, the
arrangement of the atoms about the sulfur atoms of the
methionine derivatives is essentially tetrahedral and
closely resembles the geometry of the tetrahedral intermediate.

The computer approach has also provided a detailed explanation for the experimental observation that β-glutamic
acid, which does not possess an asymmetric carbon atom, is
converted only to D-β-glutamine and not to the corresponding L-isomer. The computer findings about the relative
positions of phosphate and glutamate lead to additional
considerations about the reaction of glutamate with the
terminal phosphate group of ATP. The computer approach has
produced a large number of stereographs, some of which have

been published (Gass and Meister, 1970); the example given in Figure 6 shows the tetrahedral addition compound formed in the reaction of L-γ-glutamyl phosphate with ammonia.

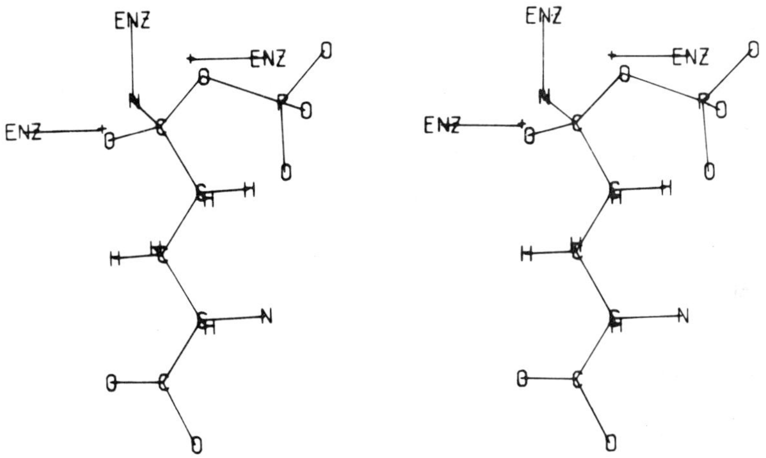

Fig. 6 Computer-drawn stereograph of the tetrahedral addition compound formed in the reaction of γ-glutamyl phosphate with NH_3. From Gass and Meister, 1970b.

The active site contains 5 points derived from the positions of L-glutamate; these are the α-carboxyl carbon atom, the α-amino nitrogen atom, the γ-carboxyl oxygen atom which is phosphorylated (ENZ, right side), the carboxyl oxygen atom which is not phosphorylated (ENZ, left side) and the nitrogen atom of the tetrahedral intermediate (ammonia binding site, central ENZ).

The accumulated data now available relating to the mechanism of action of glutamine synthetase are consistent with a general scheme which has been published previously (Krishnaswamy et al., 1962). The activated glutamic acid intermediate is represented in this proposal as enzyme bound γ-glutamyl phosphate. Another feature of this scheme

91

is that nucleotide serves as a portion of the active site required for the binding of glutamate. Thus, the experimental data indicate that the enzyme cannot bind either glutamate or glutamine in the absence of nucleotide. The data supports a sequential mechanism in which the enzyme reacts with metal-nucleotide, glutamate, and ammonia, in this order. It is of interest in this connection that although glutamate competes with methionine sulfoximine for attachment to the enzyme and therefore affords some protection against this inhibitor, addition of both ammonia and glutamate affords complete protection against inhibition. This seems to be consistent with the hypothesis that methionine sulfoximine attaches to both glutamate and ammonia binding sites of the enzyme. It is pertinent to note that the failure of ammonia to protect against inhibition in the absence of glutamate suggests that binding of ammonia to the enzyme requires prior binding of glutamate. The binding site for ammonia may be made available by a conformational or electronic change in the enzyme induced by the binding of glutamate. Detailed explanations for the γ-glutamyl transfer reaction (reaction 2) and the arsenolysis of glutamine (reaction 3) have also been proposed (Krishnaswamy et al., 1962).

The computer study of the active site of the enzyme has also led to some tentative conclusions about the mechanism of the reaction. These studies have shown, for example, that the tetrahedral intermediate cannot fit exactly into the binding site for glutamate. Thus, if one assumes that the active site is designed to afford optimum binding for the tetrahedral intermediate, L-glutamic acid would be strained on binding and this strain would tend toward the direction of the tetrahedral intermediate, thus accelerating its formation and facilitating the overall reaction. Other considerations relating to the details of the phosphorylation reaction have been discussed elsewhere (Gass and Meister, 1970).

Physical and Chemical Properties of Liver and Brain Glutamine Synthetases

The diversity of the metabolic functions served by glutamine suggests that glutamine synthetase must be subject to rigorous controls, which can regulate the supply of glutamine for various purposes. The regulatory mechanisms that control the glutamine synthetases of various cells would be expected to reflect the functions performed by glutamine in those cells. The studies of Stadtman, Holzer, and their colleagues (Stadtman et al., 1968a and b; 1970; Shapiro and Stadtman, 1970; Holzer, 1969) have shown that the bacterial glutamine synthetases (the most extensively studied being the enzyme from E. coli) are subject to multivalent regulation. Controls of this type have not been found for the ovine brain glutamine synthetase. As pointed out earlier the function of the brain enzyme is probably related to the synaptic roles of glutamate and γ-aminobutyrate. In a tissue such as the liver, however, the primary role of the synthetase is probably to provide glutamine for a wide variety of synthetic reactions, and the properties of the enzyme would be therefore expected to be tailored to meet such requirements. With such considerations in view, a comparative study of the purified liver and brain synthetases was undertaken (Tate and Meister, 1971; Tate et al., 1972).

Studies with the ovine brain glutamine synthetase have shown that the enzyme is composed of 8 apparently identical subunits with a cubelike morphological appearance (Haschemeyer, 1965, 1966). Electron microscope studies and considerations of symmetry have led to the formulation of a model for the enzyme which possesses D4 symmetry (Haschemeyer, 1968, 1970). According to this model, the octamer is formed by the isologous association of two heterologously bonded tetramers. Further, in the presence of low concentrations of urea (1-2M), solvents such as dimethylformamide, dimethyl sulfoxide, or at pH values greater than 8.1 in low ionic strength media, the native enzyme (S_{20w} = 15S) dissociated reversibly to what appeared to be a tetramer (S_{20w} = 8.6S) without extensive alteration of the tertiary structure of the enzyme (Wilk et al., 1969). Controlled treatment with N-acetylimidazole also resulted in reversible dissociation of the octamer to the tetramer (Wilk et al., 1970). More extensive treatment resulted in the formation of an apparent monomer (S_{20w} = 3.8S). The observed two-step dissociation (15S⟶8.6S⟶3.8S) is consistent with the model proposed for the enzyme, the less stable isologous bonds being cleaved first to yield a tetramer; the heterologously bonded tetramer, which has four identical bonding sets, yields only the monomer on further dissociation. Figure 7 summarizes the studies on the dissociation of the ovine brain enzyme.

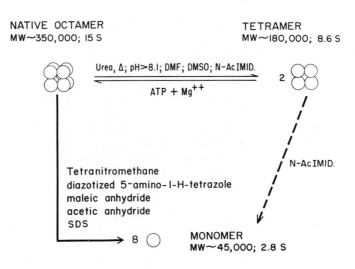

NATIVE OCTAMER
MW~350,000; 15 S

TETRAMER
MW~180,000; 8.6 S

Urea, Δ; pH>8.1; DMF; DMSO; N-AcIMID.

ATP + Mg^{++}

N-AcIMID.

Tetranitromethane
diazotized 5-amino-1-H-tetrazole
maleic anhydride
acetic anhydride
SDS

MONOMER
MW~45,000; 2.8 S

Fig. 7 Summary of studies on the dissociation of ovine brain glutamine synthetase. From Wilk et al., 1969, 1970.

The rat liver enzyme (S_{20w} = 15S) exhibits morphology in the electron microscope similar to that of the ovine brain glutamine synthetase and thus appears to have a similar octameric structure. Polyacrylamide get electrophoresis of the monomers of the liver and brain enzymes in sodium dodecyl sulfate gave monomer molecular weights of 44,000 ± 2,000 and 49,000 ± 3,000, respectively (Tate et al., 1972). Thus, the liver and brain enzymes appear to have molecular weights of 350,000 and 400,000, respectively. It is of interest that the monomers of the pea glutamine synthetase, which also appears to be octameric, and the chicken liver mitochondrial glutamine synthetase have molecular weights close to 45,000 (Tate and Meister, 1971) and

44,000 (Tate and Meister, unpublished), respectively.
Also, although the glutamine synthetases from E. coli and
B. subtilis are dodecameric (Valentine et al., 1968;
Shapiro and Ginsburg, 1968; Deuel et al., 1970), the monomers have molecular weights of approx. 50,000. Thus, the
mammalian as well as the plant and bacterial enzymes are
composed of polypeptide chains whose molecular weights fall
within the narrow range, 44,000 - 50,000.

The amino acid compositions of the rat liver and rat
brain glutamine synthetases are compared with those of the
ovine, pig, and human brain enzymes in Table I. An impressive similarity in amino acid composition is evident.
Table I also gives the amino acid compositions of pea and
E. coli glutamine synthetases. These differ from mammalian
enzymes in several respects, the most significant being
relative deficiency in half-cystine residues. In Figure
8, an attempt has been made to depict the compositions
graphically showing the departure of the values for the

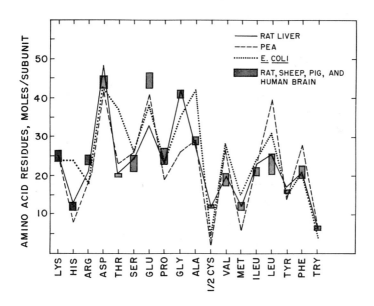

Fig. 8 Graphical representation of the amino acid compositions of glutamine synthetases from various sources.

TABLE I

Amino Acid Composition of Mammalian Liver and Brain and of Pea and E. coli Glutamine Synthetases

Amino Acid	Moles of Amino Acid/45,000 g Enzyme						M/ 50,000 g
	Rat Liver	Rat Brain	Ovine Brain	Human Brain	Pig Brain	Pea	E. coli
Lys	25	26	25	22	25	26	24
His	13	13	11	12	11	8	24
Arg	21	23	25	26	23	19	18
Asp	48	47	41	44	44	42	43
Thr	20	20	20	20	20	23	37
Ser	24	21	25	25	20	26	26
Glu	33	41	44	46	47	41	38
Pro	23	29	25	22	24	19	23
Gly	42	41	41	43	39	26	35
Ala	28	28	29	30	30	29	42
1/2 Cys	12		12			2	4
Val	20	18	18	22	18	27	28
Met	12	13	11	11	13	6	15
Ile	23	22	21	21	19	23	24
Leu	25	25	22	27	19	40	31
Tyr	17	16	16	16	16	14	15
Phe	21	23	21	20	20	28	21
Try	6		7			6	4
References	a	a	b	c	d	e	f

a. Tate et al., 1972; b. Ronzio et al., 1969; c. Wilk, S., and Meister, A., unpublished; d. Schnackerz and Jaenicke, 1966; e. Tsuda, Y., and Meister, A., unpublished; f. Woolfolk et al., 1966.

liver, pea and <u>E. coli</u> enzymes from the range of values found for the mammalian brain enzymes.

There appear to be no disulfide linkages in rat liver and ovine brain enzymes since all the half-cystine residues (11-12/subunit) reacted with 5,5'-dithiobis(2-nitrobenzoate) in the presence of 5M guanidinium hydrochloride. A conspicuous difference, however, was observed in the number of exposed sulfhydryl groups. In the native ovine brain enzyme, about 1 mole of sulfhydryl per subunit reacted rapidly with DTNB (Ronzio <u>et al</u>., 1969b); whereas in the rat liver enzyme about three sulfhydryls per subunit reacted rapidly with this reagent (Tate <u>et al</u>., 1972). In both cases, however, enzyme activity was unaffected.

Polyacrylamide gel electrophoresis in several buffer systems revealed no differences between the ovine brain and rat liver enzymes. This is in accord with close similarity in their molecular weights and amino acid compositions. However, when the two enzymes were subjected to isoelectric focussing in polyacrylamide gels, significant difference was noted in their isoelectric points. As discussed above (Wilk <u>et al</u>., 1969) the ovine brain enzyme dissociates to a tetramer at pH values above 8.1. Taking advantage of this finding, the liver and brain enzymes were mixed at pH 8.5 and then subjected to isoelectric focussing in polyacrylamide gels. Figure 9 shows the result of one such experiment demonstrating the formation of one hybrid species as predicted by considerations of subunit structure. The experiment also demonstrates that considerable homology exists between the rat liver and ovine brain enzymes. Further experiments of this type are required to study homologies between enzymes from various cell types.

The substrate specificity of rat liver glutamine synthetase resembles that of the ovine brain enzyme. Thus it acts on <u>L</u> and <u>D</u>-glutamate, only on the L-isomer of α-methylglutamate, and on other glutamate analogs such as β-glutamate, <u>threo</u>-γ-methyl-L-glutamate, and <u>cis</u>-cycloglutamate (<u>Table II</u>). The conclusions and considerations about the conformation of L-glutamate at the active site of ovine brain enzyme apply also to the rat liver enzyme, and, indeed, to the pea and E. coli enzymes as well.

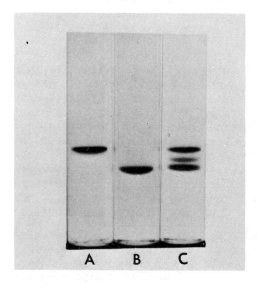

Fig. 9 Hybrid Formation Between Brain and Liver Glutamine Synthetases. Isoelectric focussing of the ovine
brain (A), the rat liver (B), and a mixture of the two
glutamine synthetases (C). The two enzymes were mixed in
equal amounts at pH 8.5 and then subjected to focussing in
4% polyacrylamide gels containing 2% carrier ampholine pH
range 3 to 10.

It was shown earlier for ovine brain glutamine synthetase (Wellner and Meister, 1966) that catalytic quantities of either ADP or ATP are required for the γ-glutamyl
transferase reaction (reaction 2), ADP being more effective
than ATP (Figure 10). Similar requirements were found for
the rat liver enzyme (Figure 10); however, ADP and ATP were
about equally effective. These studies, which were carried
out in the presence of Mn^{++}, suggest that both enzymes exhibit high affinities for nucleotides under these conditions. With rat liver enzyme, equilibrium dialysis experiments gave a value of 1.7×10^{-7}M for the dissociation
constant for the enzyme-ATP complex and the data extrapolated to a maximum of 5 moles of ATP bound per mole of
enzyme (n̄), (Figure 11). Similar studies with the ovine

TABLE II

Relative Activities of Various Substrates of Ovine Brain
and Rat Liver Glutamine Synthetases

| Amino Acid Substrate | Relative Synthetase Activities with Mg^{++} | | | |
| | With NH_3 | | With NH_2OH | |
	Brain	Liver	Brain	Liver
L-Glutamate	100	100	100	100
D-Glutamate	27	34	54	59
α-Methyl-L-glutamate	75	80	67	54
Threo-γ-Methyl-L-glutamate	27	25	63	66
Threo-γ-Hydroxy-DL-glutamate		33		18
β-Glutamate	18	19	46	28
cis-L-1-Amino-1,3-dicarboxycyclohexane (cis-cyclogluta-mate)	20	22	102	100

Ovine Brain data from Meister (1968c).

Rat Liver data from Tate et al., (1972)

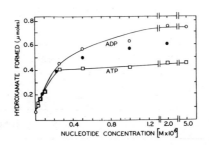

Ovine Brain (Wellner and Meister, 1966)

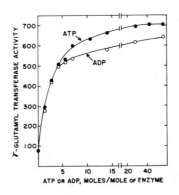

Rat Liver (Tate, Leu, and Meister, 1972)

Fig. 10 Effect of ATP and ADP on γ-glutamyl transferase activity of the brain and the liver glutamine synthetases.

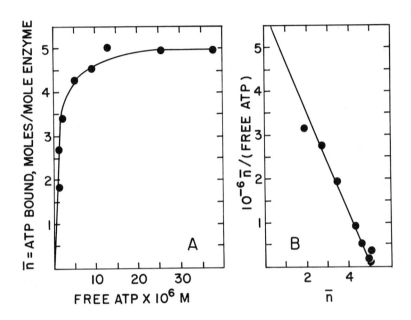

Fig. 11 Binding of ATP in presence of Mn^{++} by rat liver glutamine synthetase. B, Scatchard plot of the data in A. From Tate et al., 1972.

brain enzyme indicated that about 8 moles of ATP per mole of enzyme bound tightly and that about 4 additional moles of ATP bound less tightly (Figure 12). The data for the high affinity sites give a value of 3 x 10^{-7}M for the dissociation constant of the enzyme-ATP complex. A significant difference, therefore, exists between the two enzymes in their affinity and binding sites for ATP.

Studies cited earlier showed that the inhibition of ovine brain glutamine synthetase by the convulsant glutamate analog, L-methionine-S-sulfoximine is associated with

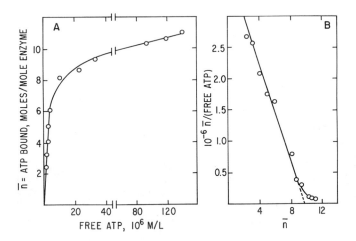

Fig. 12 Binding of ATP in presence of Mn^{++} by ovine brain glutamine synthetase. Tate and Meister, unpublished.

---

phosphorylation of this compound and its tight attachment to the active site (Ronzio and Meister, 1968; Ronzio et al., 1969a; Rowe et al., 1969). Rat liver and rat brain, as well as pea and E. coli glutamine synthetases are also inactivated by methionine sulfoximine (see Tate and Meister, 1971; Weisbrod, 1971). This substrate analog, by virtue of its tight attachment, provides a convenient probe for the study of substrate binding sites of glutamine synthetases. Earlier (recalculated data of Ronzio et al., 1969a) and more recent studies (Y. Tsuda, unpublished) had shown that between 5.7 to 6.1 moles of methionine sulfoximine bound per mole of ovine brain enzyme. Studies on the rat liver enzyme, however, showed that close to 4 moles of methionine sulfoximine are bound per mole of inactivated enzyme. Experiments on the relationship between the extent of inhibition and the amount of methionine sulfoximine bound to the enzyme, following incubation with ^{14}C inhibitor, ATP and Mn^{++}, show that essentially complete inactivation is associated with the binding of about 4 moles of inhibitor per mole of rat liver enzyme (Figure 13).

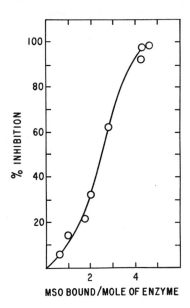

Fig. 13 Relationship between inhibition of liver glutamine synthetase and the binding of methionine sulfoximine. From Tate et al., 1972.

The sigmoidal relationship between inhibition and the
binding of methionine sulfoximine suggests the possibility
of cooperative interactions between the substrate binding
sites of different subunits.

The binding studies revealed an interesting difference between the ovine brain and rat liver glutamine synthetases, although both enzymes are octameric and display
several other similarities, the brain enzyme binds tightly
close to 8 moles each of ATP and methionine sulfoximine per
mole, whereas the liver enzyme binds only about 4 moles
each of the nucleotide and the inhibitor per mole of enzyme.
Such half-of-the sites reactivity is displayed by several
other enzymes (see Levitzki et al., 1971). These authors
and other workers have discussed several interesting
models to explain this phenomenon. One consequence of this

103

property of the rat liver enzyme is discussed below in relation to inhibition of this enzyme by several amino acids.

Regulation of Liver Glutamine Synthetase
by Various Metabolites

The rat liver glutamine synthetase was found to be significantly activated by α-ketoglutarate and less so by citrate (Tate and Meister, 1971). Figure 14 shows the ef-

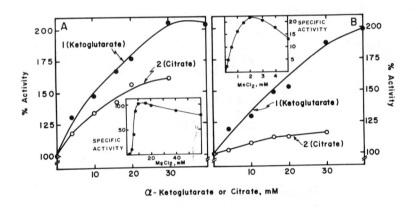

α- Ketoglutarate or Citrate, mM

Fig. 14 Effect of α-ketoglutarate and citrate on the activity of rat liver glutamine synthetase. A, 50 mM $MgCl_2$; B, 2 mM $MnCl_2$. The insets show the effect of either $MgCl_2$ or $MnCl_2$ on the activity in the absence of the effectors. From Tate and Meister, 1971.

fect of α-ketoglutarate and citrate on the Mg^{++} and Mn^{++}-dependent synthesis activity. Activation by α-ketoglutarate occurs only at saturating concentrations of L-glutamate and at Mg^{++}:ATP ratios of greater than 1. At low glutamate and Mg^{++}:ATP ratios either no effect or only slight inhibition was observed. The insets in Figure 14 show the effect of varying either $MgCl_2$ or $MnCl_2$ concentrations on the activity at 10 mM ATP and in absence of the effectors to emphasize the point that the activation by α-ketoglutarate is not due to chelation of metal ions since near-optimal metal concentrations were used. It would appear, therefore, that binding of the effector keto acid requires free metal ions; also, prior binding of substrate, L-glutamate, appears to be mandatory. Several other compounds including isocitrate, pyruvate, glyoxylate, phenylpyruvate, and succinate had no effect on the activity of rat liver and other glutamine synthetases. The ovine and rat brain synthetases are much less affected by α-ketoglutarate.

The activation of the liver enzyme by α-ketoglutarate provides a cellular mechanism of "feed-forward" activation by a precursor whereby excess α-ketoglutarate produced by the citric acid cycle and by transamination can stimulate glutamine formation. Since both of the two consecutive reactions catalyzed by glutamate dehydrogenase and glutamine synthetase utilize ammonia, activation of the synthetase by α-ketoglutarate only at high glutamate concentrations provides a mechanism by which the cell could achieve a balance between glutamate and glutamine concentrations. It will be recalled that the effect of α-ketoglutarate in increasing the glutamine formation in E. coli is mediated indirectly by stimulating deadenylylation to produce more active enzyme and by inhibiting adenylylation (Stadtman et al., 1968a and b, 1970; Shapiro and Stadtman, 1968; Shapiro, 1969).

The effect of various other metabolites on the liver and brain enzymes is shown in Table III. The amino acids studied did not affect the activities of these enzymes in presence of Mg^{++}. However, the Mn^{++}-dependent activity of the rat liver enzyme was significantly inhibited by glycine, L-alanine and L-serine. The Mn^{++}-dependent activities of the ovine and rat brain enzyme were much less affected. L-Methionine and L-cysteine exhibited smaller effects on the rat liver enzyme, whereas several other amino acids such as L-aspartate, L-asparagine, L-histidine, γ-aminobutyrate, and L-phenylalanine did not significantly affect the activities, nor did several other end-products, of glutamine metabolism such as D-glucosamine-6-phosphate, AMP, UTP, UMP, and DPN. Carbamyl phosphate, however, was a potent inhibitor of the Mn^{++}-dependent activities of rat liver and brain, and ovine brain enzymes. The mechanism of this inhibition is discussed below.

105

TABLE III

Effect of various compounds on glutamine synthetases isolated from several sources

Compound added[a]	Relative activity							
	With Mn^{++} [b]					With Mg^{++} [b]		
	Rat liver	Rat brain	Ovine brain	Hu-man brain	Pea[c]	Rat liver	Ovine brain	Pea[d]
None (control).......	100	100	100	100	100	100	100	100
α-Ketoglutarate (20)..	160	117	122	115	111	173	127	101
Glycine (20)..........	49	60	90	77	70	99	93	94
L-Alanine (20)........	52	63	85	81	47	105	95	99
D-Alanine (20)........	37	29	63	46	50	81	94	86
L-Serine (20)..........	61	57	92	85	73	114	120	100
β-Alanine (20)........	57	64	94	86	53	105	100	94
Carbamyl phosphate (10)...............	31	20	23	42	74	88	95	89
AMP (10)............	93	85	97	97		99	96	54

[a] The concentrations are given in parentheses (millimolar).

[b] Activity was determined by the γ-glutamyl hydroxamate assay method.

[c] γ-Glutamyltransferase activity was determined.

[d] The assay solutions also contained 5 mM EDTA.

From Tate, Leu, and Meister (1972)

The effect of increasing concentrations of glycine and L-alanine on the activity of the liver enzyme at saturating concentrations of glutamate and ATP is described in Figure 15A. When either L-alanine was added to mixtures containing saturating concentrations of glycine, or the reverse (curves 1A and 2A) there was no further inhibition. Thus, cumulative inhibition of the type observed with the E. coli enzyme (Woolfolk and Stadtman, 1964) is ruled out. Inhibition by glycine and L-alanine increases as glutamate concentration increases (Fig. 15B), and is independent of ammonia concentration over the range 2-20 mM. The poten-

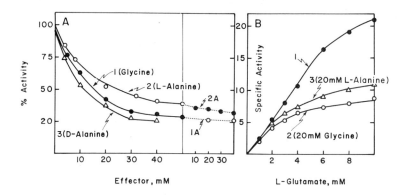

Fig. 15 Effect of glycine and alanine on rat liver glutamine synthetase activity in presence of $MnCl_2$. In curves 1A and 2A, L-alanine and glycine were added to solutions containing 50 mM glycine or 50 mM L-alanine, respectively. B, effect of L-glutamate concentration on the activity in the absence (1), and presence of 20 mM glycine (2) or 20 mM L-alanine (3). From Tate and Meister, 1971.

tiation of inhibitions by glutamate may reflect the requirement of substrate for the binding of the amino acid effectors.

It has been pointed out earlier that in the liver the utilization of the α-amino group of glutamine for the synthesis of glycine, L-alanine, and L-serine may represent a significant metabolic pathway. The reaction is catalyzed by glutamine transaminase for which glyoxylate, pyruvate, and hydroxypyruvate are amongst the best α-keto acid substrates (Cooper and Meister, 1972). Thus, the inhibition of glutamine synthetase by these amino acids may serve to regulate the utilization of glutamine for their synthesis.

The mechanism by which Mn^{++} makes the liver enzyme sensitive to the negative effectors is not yet clear. The synthesis reaction with both the liver and brain enzymes is

107

much slower in the presence of Mn^{++} than of Mg^{++}. Moreover, addition of Mn^{++} To the Mg^{++} synthesis system, even at $Mg^{++}:Mn^{++}$ ratios of 1,000, markedly inhibits the liver enzyme (Fig. 16). Similar results were obtained with the

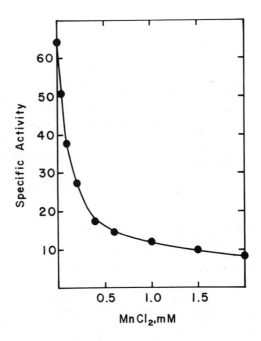

Fig. 16 Effect of $MnCl_2$ on the Mg^{++}-dependent activity of rat liver glutamine synthetase. The solutions contained 10 mM ATP and 100 mM $MgCl_2$. From Tate and Meister, 1971.

ovine brain enzyme; in this case, however, the enzyme is not affected by the negative amino acid effectors. A detailed study of the interactions of various metal ions with the brain glutamine synthetase has been published (Monder, 1965)

While glycine and L-alanine do not inhibit the liver enzyme appreciably in the presence of Mg^{++}, it was found that these amino acids inhibit substantially when inorganic phosphate is also present. Thus, as shown in Fig. 17, the presence of phosphate led to a considerable enhancement of

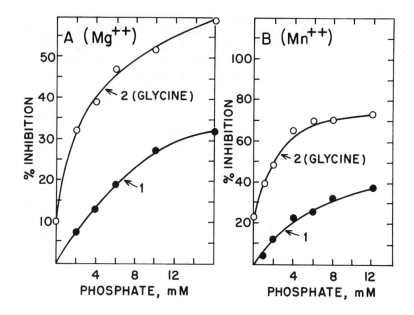

Fig. 17 Effect of phosphate and glycine on the activity of rat liver glutamine synthetase. A, inhibition by phosphate in absence (curve 1) and in presence of 30 mM glycine (curve 2); 40 mM MgCl$_2$. B, inhibition by phosphate in absence (curve 1) and presence of 10 mM glycine (curve 2); 2 mM MnCl$_2$. From Tate et al., 1972.

the inhibition by glycine in the presence of both Mg^{++} and Mn^{++}. It must be noted that phosphate itself inhibits glutamine synthetase in the presence of either metal ion. However, inhibition by both glycine and phosphate is greater than the sum of the inhibitions observed with glycine and phosphate separately. Similar results were obtained with L-alanine.

The inhibition of the liver enzyme (as well as the brain enzyme) by inorganic phosphate alone may be ascribed, at least in part, to the reversibility of glutamine synthetase (Levintow and Meister, 1954). The available evidence indicates that phosphate may bind at or close to the site that binds the terminal phosphoryl group of ATP (Tate et

109

al., 1972). It is possible that the inhibition of glutamine synthetase by inorganic phosphate is of physiological
significance. The well known activation of certain glutaminases by inorganic phosphate suggests that phosphate
may offer reciprocal control of glutaminase and glutamine
synthetase by activating the former and inhibiting the
latter enzyme. It is apparent that uncontrolled coupling
of glutaminase and glutamine synthetase represents a potentially wasteful metabolic situation, which must be normally
prevented in vivo either by compartmentalization or by control mechanism of some sort.

As discussed earlier, the rat liver enzyme exhibits
half-of-the-sites reactivity, i.e., only about 4 moles of
ATP and 4 moles of the substrate analog, methionine sulfoximine, bind per mole of enzyme which has eight subunits;
although the available data indicate that the subunits of
rat liver enzyme have the same molecular weight, there is
insufficient evidence at hand to conclude that the subunits
are identical. Thus, half of the subunits may function in
the synthesis of glutamine; the other 4 subunits may play
a role in the regulation of the enzyme by providing binding
sites for the negative amino acid effectors. That the
sites for L-glutamate and for the amino acid effectors may
bear a close resemblance is apparent from the fact that
glycine, both L- and D-alanine, and β-alanine inhibit the
liver enzyme (Table III). The steric resemblance between
these compounds and those that can bind to the glutamate
site (e.g. L- and D-glutamate, and β-glutamate) is noteworthy, and raises the possibility that half of the sites
have been modified through evolution to sites which allow
the binding of the negative amino acid effectors. Poten-
tiation of the glycine and alanine inhibition by inorganic
phosphate may reflect complementation at the "allosteric"
sites which may have retained the binding sites for the
terminal phosphoryl group of ATP.

Mechanism of Inhibition of Glutamine
Synthetase by Carbamyl Phosphate

The inhibition of liver and brain glutamine synthetases by carbamyl phosphate can provide the cell with a
means for regulating the supply of glutamine for pyrimidine

biosynthesis. Glutamine-dependent carbamyl phosphate synthetases specific for pyrimidine biosynthesis are present
in liver and other mammalian tissues (Hager and Jones,
1967; Nakanishi et al., 1968; Tatibana and Ito, 1969).

Carbamyl phosphate inhibits not only the liver and
brain glutamine synthetases, but also the enzymes from
other sources such as pea, E. coli, and chicken liver mitochondrial glutamine synthetase, especially in the presence
of Mn^{++}. While carbamyl phosphate is known to decompose to
yield inorganic phosphate and cyanate at values of pH from
about 6 to 9 (Allen and Jones, 1964), carbamyl phosphate is
considerably more inhibitory than equivalent concentrations
of inorganic phosphate. Thus, with 5 mM phosphate the rat
liver enzyme was inhibited 22%, while 52% inhibition was
observed with 5 mM carbamyl phosphate of the Mn^{++}-dependent
synthesis reaction. No inhibition of the enzyme was found
with 5 mM cyanate. The structural similarity between carbamyl phosphate, γ-glutamyl phosphate and the tetrahedral
intermediate γ-glutaminyl phosphate (postulated to be
formed by reaction of ammonia with γ-glutamyl phosphate
[see above]) as shown in Fig. 18, led us to consider the

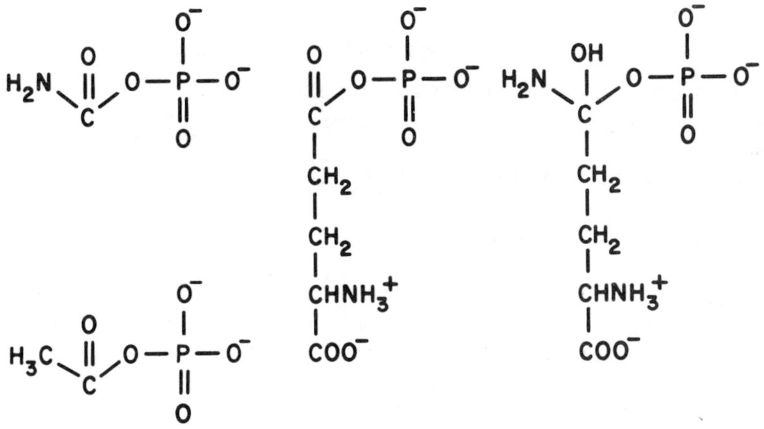

Fig. 18 Structural resemblances between carbamyl phosphate, acetyl phosphate, γ-glutamyl phosphate, and γ-glutaminyl phosphate.

111

possibility that carbamyl phosphate might bind to the active center of the enzyme at the acyl phosphate binding site. Support for this possibility came from the findings that rat liver glutamine synthetase catalyzes the synthesis of ATP from carbamyl phosphate and ADP in presence of either Mg^{++} or Mn^{++} (Table IV); the rate of ATP formation

TABLE IV

Synthesis of ATP from carbamyl phosphate (CP) and ADP catalyzed by rat liver glutamine synthetase

Experiment No.	Components of reaction mixture	ATP formed	
		Mg^{++}	Mn^{++}
		nmoles	
1	ADP + CP	28.2	6.63
2	ADP + CP + L-glutamate (5 mM)	7.8	2.65
3	ADP + CP + ATP (5 mM)	10.4	
4	ADP + CP + NH₄Cl (10 mM)	27.4	6.20
5	ADP + CP + L-alanine (20 mM)	28.8	3.90
6	ADP + CP	0	0

Experiment 6 was performed with enzyme inactivated by preliminary incubation with methionine sulfoximine, ATP, and Mg^{++}. From Tate, Leu and Meister (1972).

was equivalent to about 2% of the rate of glutamine synthesis. Addition of glutamate and ATP decreased the formation of ATP from ADP and carbamyl phosphate. L-Alanine, which inhibits liver enzyme in presence of Mn^{++}, affected ATP synthesis from ADP and carbamyl phosphate in the presence of Mg^{++}. When enzyme inactivated by prior treatment with methionine sulfoximine and ATP was used, no synthesis of ATP occurred. (Experiment 6)

The effect of varying carbamyl phosphate concentration

on ATP synthesis in presence of Mg^{++} is shown in <u>Fig. 19</u>;

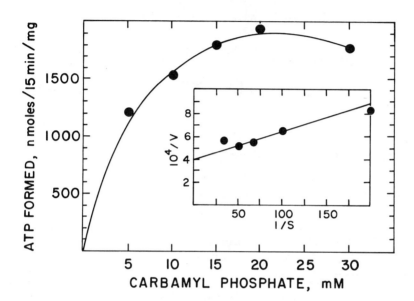

Fig. 19 Effect of carbamyl phosphate concentration on ATP synthesis from ADP and carbamyl phosphate by rat liver glutamine synthetase. From Tate <u>et al</u>., 1972

the data gave an apparent Km value for carbamyl phosphate 6.3 mM. The corresponding value for the Mn^{++} system was too low to be determined accurately by this method, but it was estimated that it is less than 0.5 mM or about 10% of the Km value for glutamate. The greater affinity of the enzyme for carbamyl phosphate in the presence of Mn^{++} than in the presence of Mg^{++} seems to explain the more substantial inhibition of the enzyme by carbamyl phosphate in presence of Mn^{++}.

Although the rat liver enzyme does not catalyze the synthesis of carbamyl phosphate from bicarbonate, ammonia (or glutamine) and ATP, it can, as expected from the series of reactions shown in Fig. 20, catalyze the synthesis of glutamine from L-glutamate, carbamyl phosphate and ADP.

$$\text{Carbamyl phosphate} + \text{ADP} \xrightarrow[\text{synthetase}]{\text{glutamine}} \text{ATP} + \text{carbamate}$$

$$\text{Carbamate} + \text{H}_2\text{O} \xrightarrow{\text{spontaneous}} \text{HCO}_3^- + \text{NH}_4^+$$

$$\text{NH}_4^+ + \text{glutamate} + \text{ATP} \xrightarrow[\text{synthetase}]{\text{glutamine}} \text{ADP} + \text{glutamine} + \text{P}_i$$

Sum:

Carbamyl phosphate + glutamate

$$+ \text{H}_2\text{O} \xrightarrow[\text{synthetase}]{\text{glutamine}} \text{HCO}_3^- + \text{glutamine} + \text{P}_i$$

Fig. 20 Reactions leading to the net synthesis of glutamine from ADP, glutamate and carbamyl phosphate.

In the course of these studies, it was found that acetylphosphate can substitute for carbamyl phosphate in the ATP synthesis reaction. As indicated in Fig. 21A, the formation of ATP and ADP and acetylphosphate takes place at about 50% of the rate observed with ADP and carbamyl phosphate. Both of these reactions take place at rates which are much slower than the reversal of glutamine synthesis (curve 1, Fig. 21A). When carbamyl phosphate was added to the reverse synthesis system in presence of Mn^{++} (curve 3, Fig. 21B) the rate of ATP formation was somewhat less than that of the reverse synthesis reaction (curve 1, Fig 21B). The findings, which show that the rates of ATP synthesis from glutamine, Pi, and ADP, and from carbamyl phosphate and ADP are not additive, are in accord with the view that both reactions are catalyzed by the same enzyme site. Additional evidence in support of this conclusion is given in Fig. 22, which describes the effect of L-glutamate on the synthesis of ATP from ADP and acetylphosphate. Glutamate inhibited competitively; the Ki value for L-glutamate is 4.5 mM, a value similar to the Km value determined

114

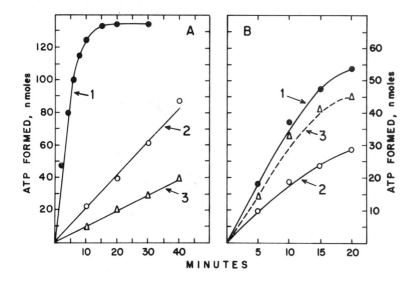

Fig. 21 Synthesis of ATP catalyzed by rat liver glutamine synthetase. A, MgCl₂. Curves 1 (A and B), reversal
of glutamine synthesis; curves 2 (A and B), phosphorylation
of ADP by carbamyl phosphate; curve 3 (A), ADP plus acetyl
phosphate; curve 3 (B), ATP synthesis in mixtures containing glutamine, P_i, ADP and carbamyl phosphate. From Tate
et al., 1972.

for glutamate in the synthesis reaction (5 mM). A satis-
factory experiment of this type with carbamyl phosphate
could not be done because formation of ammonia led to ATP
utilization for glutamine synthesis.

The finding that acetyl phosphate can also be used by
the rat liver enzyme for ATP synthesis would seem to be
consistent with the structural analogies shown in Fig. 18.
It is possible that the methyl group of acetyl phosphate
and the amino group of carbamyl phosphate occupy the same
enzyme site. This would be analogous to the postulate
that the methyl group of methionine sulfoximine attaches to

115

the ammonia binding site of the enzyme (Gass and Meister, 1970b).

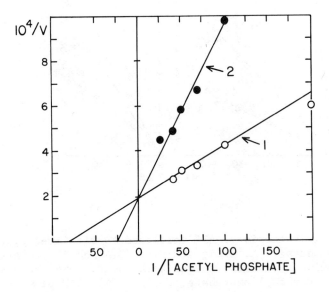

Fig. 22 Effect of glutamate on the synthesis of ATP from acetyl phosphate and ADP catalyzed by rat liver glutamine synthetase. Curve 2, 10 mM glutamate present. From Tate et al., 1972.

As indicated in Table V, the ability to catalyze ATP synthesis from ADP and carbamyl phosphate is not only a property of rat liver glutamine synthetase. The ovine and rat brain, as well as the E. coli enzymes can also catalyze this reaction. Thus, it would appear that carbamyl phosphate inhibits glutamine synthetases by interacting with the acyl phosphate binding site at the active center of these enzymes.

TABLE V

Synthesis of ATP from carbamyl phosphate and ADP catalyzed by glutamine synthetases from various sources

Source of purified glutamine synthetase	Formation of ATP	
	Mg^{++}	Mn^{++}
	nmoles/mg enzyme	
Rat Liver......................	1655	464
Rat brain.....................	1270	452
Ovine brain..................	1570	430
E. coli		325

From Tate, Leu, and Meister (1972)

Effect of Anions on Liver and Brain Glutamine Synthetases

Recent studies (Tate and Meister, unpublished) reveal that certain anions, most notably bicarbonate and chloride, significantly activate the rat liver and ovine brain synthetases when non-saturating levels of L-glutamate are used (i.e., 2-10 mM). Table VI shows the effects of various salts on the NH_3- and NH_2OH-dependent synthesis activities of the liver and brain enzymes. The assay solutions contained 10 mM each of ATP, $MgSO_4$ (brain) or magnesium acetate (liver), and L-glutamate and either 5 mM NH_4Cl or 15 mM NH_2OH. The values in parentheses are for [^{14}C] glutamine synthesis. Under these conditions a 2- to 3-fold stimulation of activity was observed. Further studies showed that bicarbonate (as well as chloride) acts by increasing the affinity of the enzyme for L-glutamate. Bicarbonate has no effect on the apparent Km value for ATP. The significance of these activations, which requires further study, may possibly be related to the regulation of intracellular hydrogen ion concentration.

117

TABLE VI

Effect of Various Salts on the Activities of Rat Liver
and Ovine Brain Glutamine Synthetases

Compound Added (mM)		Relative Activity, %			
		Rat Liver		Ovine Brain	
		NH_3	NH_2OH	NH_3	NH_2OH
None		100	100	100 (100)	100
KCl	(100)	156	167	130	137
K acetate	(100)	107	107	120	125
NaCl	(100)	159	160	138 (150)	141
Na acetate	(100)	113	109	127 (132)	128
$NaNO_3$	(100)	47	50	27	44
Na_2SO_4	(100)	46	67	75	82
$NaHCO_3$	(50)	346	159	241 (266)	176
Choline Chloride	(100)				146

DISCUSSION

This review, wnich deals primarily with two mammalian glutamine synthetases, illustrates many similarities between the enzymes and some significant differences. The latter seem to reflect mainly differences in the particular functions of the organs involved, although some species differences may also exist. The octameric structure proposed for the ovine brain enzyme (Haschemeyer, 1965, 1966, 1968) is supported by subsequent studies in this laboratory (Wilk et al., 1969, 1970). The hybridization studies between the brain enzyme and the liver enzyme and
    the electron microscope studies suggest that the liver enzyme also possesses the same type of octameric structure. It is of interest that the pea enzyme also possesses such a structure. In this respect the mammalian and plant enzymes differ from the bacterial enzymes (E. coli and B. subtilis), which are composed of 12 subunits. The brain and the liver enzymes, as well as the pea enzyme

are composed of monomeric units whose molecular weights are
close to 45,000; the subunit molecular weight of the E.
coli and B. subtilis enzymes, however, are about 50,000.
As documented in Table I, the rat liver and the several
brain enzymes exhibit impressive similarities in their
amino acid compositions. Indeed when one considers the
amino acid compositions of all of the enzymes, some interesting similarities become evident, e.g., lysine, aspartate, serine, glutamate, isoleucine, tyrosine. The pea and
the E. coli enzymes show several differences, the most impressive being their relative deficiency in half cystine
residues as compared to the mammalian enzymes.

Additional similarities between the liver and the
ovine brain enzymes include the reactions catalyzed and
substrate specificity. Indeed, the arguments and rationale
used in arriving at a model for the conformation of L-glutamate at the active site and the relation of glutamate
binding sites to the ammonia and ATP binding sites (discussed in detail by Meister, 1968b and c; Gass and Meister,
1970b), seem to apply not only to the ovine brain enzyme
but also to the liver, pea, and the E. coli enzymes.
Furthermore, several lines of evidence have been presented
and discussed in support of the intermediate formation of
an acyl phosphate in glutamine synthesis. These include
the formation of pyrrolidone carboxylate from glutamate in
absence of ammonia; the observation that the binding of
glutamate is associated with cleavage of ATP to ADP; the
ability to use β-aminoglutaryl phosphate for the synthesis
of ATP; phosphorylation of methionine sulfoximine; the formation of cycloglutamyl phosphate from cycloglutamate and
ATP; and the synthesis of ATP from carbamyl phosphate and
ADP.

All of the glutamine synthetases studied have been
found to be inhibited irreversibly by L-methionine-S-sulfoximine. The mechanism has been studied in detail with
the ovine brain enzyme and involves the phosphorylation of
methionine sulfoximine by ATP in presence of metal ions
(Mg^{++}, Mn^{++}) and the tight attachment of methionine sulfoximine phosphate to the active center. The formation of
methionine sulfoximine phosphate has been demonstrated
with the rat liver and E. coli enzymes; presumably, therefore, similar mechanisms are involved in the inhibition of

119

these glutamine synthetases. An interesting difference, however, exists in the number of methionine sulfoximine binding sites in the liver and brain enzymes. The brain enzyme binds eight moles of ATP per mole of enzyme tightly and an additional four moles with lower affinity; complete inhibition of this enzyme is associated with the binding of close to one mole of methionine sulfoximine phosphate per subunit. In the rat liver enzyme, however, only four moles of methionine sulfoximine phosphate are bound per mole of enzyme (eight subunits). This curious half-of-the-sites reactivity is considered in relation to the allosteric inhibition of the liver enzyme by glycine, alanine, and serine.

The activation of the rat liver enzyme by α-ketoglutarate seems to be of significance in relation to the control of the utilization of ammonia for the synthesis of glutamate and glutamine. The brain enzymes are much less susceptible to activation by α-ketoglutarate. Furthermore, the liver enzyme is inhibited markedly by glycine, alanine, and serine, particularly in the presence of Mn^{++}. The ovine brain enzyme is not significantly affected by these compounds and the rat brain enzyme is inhibited to a lesser degree than is the liver enzyme. Inorganic phosphate potentiates the inhibition of the rat liver enzyme by glycine and alanine in the presence of either Mg^{++} or Mn^{++}. The inhibition of the liver glutamine synthetase by glycine, alanine, and serine may serve to regulate the utilization of the α-amino group of glutamine for the synthesis of these amino acids via the liver glutamine transaminase. Inhibition by phosphate may impose reciprocal controls on the activities of glutamine synthetase and glutaminase, thus preventing a potentially wasteful cycle.

The liver, brain, pea, and the E. coli enzymes are inhibited by carbamyl phosphate, markedly so in the presence of Mn^{++}. Evidence has been presented that carbamyl phosphate inhibits the glutamine synthetase by virtue of its ability to bind at the active center, presumably at the acyl phosphate binding sites. The inhibition, however, may control the utilization of glutamine for pyrimidine synthesis. Fig. 23 shows the interrelationships and regulations involved in the metabolism of glutamate and glutamine.

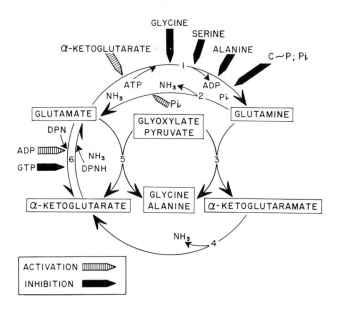

Fig. 23 Metabolic relationships between glutamine and other compounds in liver. (1) glutamine synthetase; (2) glutaminase; (3) glutamine transaminase; (4) ω-amidase; (5) glutamate transaminases; (6) glutamate dehydrogenase.

In addition to the regulatory mechanisms discussed here, other types of regulation have been observed for mammalian glutamine synthetases. Thus, steroid hormones induce synthesis of glutamine synthetase in several mammalian cells (Wu, 1964; Wu and Morris, 1970; Reif-Lehrer and Amos, 1968; Moscona et al., 1968; Reif-Lehrer and Chader, 1969; Weissman and Ben-Or, 1970). In chick embryo retina grown in tissue culture, the enzyme appears to be induced by cyclic AMP. In human cervical carcinoma cells (Demars, 1958) and in mouse fibroblast cells grown in tissue culture (Paul and Fottrell, 1963; Barnes et al., 1971; Stamatiadou, 1972) glutamine represses the levels of glutamine synthetase.

121

FOOTNOTES

[1/] This list is probably incomplete; additional pathways utilizing the amide nitrogen of glutamine will probably be found. One wonders, for example, about the origin
of the amide groups of oxytocin, vasopressin, and vitamin
B_{12}.

[2/] The glutamine transaminases, which catalyze the
conversion of a variety of α-keto acids to the corresponding amino acids and at the same time (in the presence of
ω-amidase) convert glutamine to α-ketoglutarate and ammonia (Meister, 1962a), are discussed by Cooper and Meister
later in this symposium. Recent studies (Cooper and
Meister, 1972 and unpublished) have led to the isolation
and characterization of glutamine transaminase from rat
liver and also from rat kidney. The liver and kidney
transaminases exhibit markedly different specificities and
probably serve different metabolic functions in these organs.

[3/] Krebs (1935) was the first to study the synthesis
of glutamine in tissue slices and to show that this reaction requires energy. Elliott (1948, 1951, 1953), Speck
(1947, 1949) and Leuthardt and Bujard (1947) independently
established that ATP was required for glutamine synthesis.
The synthesis of glutamine was shown to be freely reversible and from the experimentally determined equilibrium
constant it was possible to calculate a value for the
standard free energy of hydrolysis of ATP (Levintow and
Meister, 1954). The first glutamine synthetase to be isolated in essentially homogeneous form was obtained from
ovine brain (Pamiljans et al., 1962).

REFERENCES

Allen, C.M., and Jones, M.E. (1964), Biochemistry, 3, 1238.

Barnes, P.R., Youngberg, D., and Kitos, P.A. (1971), J.
    Cell Physiol., 77, 135.

Boyer, P.D., Koeppe, O.J., and Luchsinger, W.W. (1956), J.
    Am. Chem. Soc., 78, 356.

Cooper, A.J.L., and Meister, A. (1972), Biochemistry, 11, 661.

De Mars, R. (1958), Biochim. Biophys. Acta, 27, 435.

Deuel, T.F., Ginsburg, A., Yeh, J., Shelton, E., and Stadtman, E.R. (1970), J. Biol. Chem., 215, 5195.

Elliott, W.H. (1948), Nature, 161, 128.

Elliott, W.H. (1951), Biochem. J., 49, 106.

Elliott, W.H. (1953), J. Biol. Chem., 201, 661.

Gass, J.D., and Meister, A. (1970a), Biochemistry, 9, 842.

Gass, J.D., and Meister, A. (1970b), Biochemistry, 9, 1380.

Hager, S.E., and Jones, M.E. (1967), J. Biol. Chem., 242, 5674.

Haschemeyer, R.H. (1965), 150th National Meeting, American Chemical Society (Atlantic City, abstract 68).

Haschemeyer, R.H. (1966), 152nd National Meeting, American Chemical Society (New York, abstract 46).

Haschemeyer, R.H. (1968), Trans. N.Y. Acad. Sci., 30, 875.

Haschemeyer, R.H. (1970), Advan. Enzymol., 33, 71.

Holzer, H. (1969), Advan. Enzymol., 32, 297.

Khedouri, E., Wellner, V.P., and Meister, A. (1964), Biochemistry, 3, 824.

Kowalski, A., Wyttenbach, C., Langer, L., and Koshland, D.E., Jr. (1956), J. Biol. Chem., 219, 719.

Krebs, H.A. (1935), Biochem. J., 29, 1951.

Krishnaswamy, P.R., Pamiljans, V., and Meister, A. (1962), J. Biol. Chem., 237, 2932.

Leuthardt, F., and Bujard, E. (1947), Helv. Med. Acta, 14, 274.

Levintow, L., and Meister, A. (1953), J. Am. Chem. Soc., 75, 3039.

Levintow, L., and Meister, A. (1954), J. Biol. Chem., 209, 265.

Levitzki, A., Stallcup, W.B., and Koshland, D.E., Jr. (1971), Biochemistry, 10, 3371.

Meister, A. (1962a), Enzymes, 6, 193.

Meister, A. (1962b), Enzymes, 6, 443.

Meister, A. (1968a), Fed. Proc., 27, 100.

Meister, A. (1968b), Advan. Enzymol., 31, 183.

Meister, A. (1968c) Harvey Lect., 63, 139.

Monder, C. (1965), Biochemistry, 4, 2677.

Moscona, A.A. Moscona, M.H., and Saenz, N. (1968), Proc. Nat. Acad. Sci. USA, 61, 160.

Nakanishi, S., Ito, K., and Tatibana, M. (1968), Biochem. Biophys, Res. Commun., 33, 774.

Nishimura, J.S., Dodd, E.A., and Meister, A. (1963), J. Biol. Chem., 238, PC 1179.

Nishimura, J.S., Dodd, E.A., and Meister, A. (1964), J. Biol. Chem., 239, 2553.

Pamiljans, V., Krishnaswamy, P.R., Dumville, G., and Meister, A. (1962), Biochemistry, 1, 153.

Paul, J., and Fottrell, P.F. (1963), Biochim. Biophys. Acta, 67, 334.

Rao, S.L.N., and Meister, A. (1972), Biochemistry, 11, 1123.

Reif-Lehrer, L., and Amos, H. (1968), Biochem. J., 106, 425.

Reif-Lehrer, L., and Chader, G.J. (1969), Biochim Biophys. Acta, 192, 310.

Ronzio, R.A., and Meister, A. (1968), Proc. Nat. Acad. Sci. USA, 59, 164.

Ronzio, R.A., Rowe, W.B., and Meister, A. (1969a), Biochemistry, 8, 1066.

Ronzio, R.A., Rowe, W.B., Wilk, S., and Meister, A. (1969b), Biochemistry, 8, 2670.

Rowe, W.B., Ronzio, R.A., and Meister, A. (1969), Biochemistry, 8, 2674.

Rowe, W.B., and Meister, A. (1970), Proc. Nat. Acad. Sci. USA, 66, 500.

Schnackerz, K., and Jaenicke, L. (1966), J. Physiol. Chem., 347, 127.

Shapiro, B.M. (1969), Biochemistry, 8, 659.

Shapiro, B.M., and Ginsburg, A. (1968), Biochemistry, 7, 2153.

Shapiro, B.M., and Stadtman, E.R. (1968), Biochem. Biophys. Res. Commun., 30, 32.

Shapiro, B.M., and Stadtman, E.R. (1970), Ann. Rev. Microbiol., 24, 501.

Speck, J.F. (1947), J. Biol. Chem., 168, 403.

Speck, J.F. (1949), J. Biol. Chem., 179, 1387, 1405.

Stamatiadou, M.N. (1972), Biochem. Biophys. Res. Commun., 47, 485.

Stadtman, E.R., Shapiro, B.M., Ginsburg, A., Kingdon, H.S., and Denton, M.D. (1968a), Brookhaven Symposia in

Biology, 21, 378.

Stadtman, E.R., Shapiro, B.M., Kingdon, H.S., Woolfolk, C.A., and Hubbard, J.S. (1968b), Advan. Enzyme Regul., 6, 257.

Stadtman, E.R., Ginsburg, A., Ciardi, J.E., Yeh, J., Hennig, S.B., and Shapiro, B.M. (1970), Advan Enzyme Regul., 8, 99.

Tate, S.S., and Meister, A. (1971), Proc. Nat. Acad. Sci. USA, 68, 781.

Tate, S.S., Leu, Fang-Yun, and Meister, A. (1972), J. Biol. Chem., 247, 5312.

Tatibana, M., and Ito, K. (1969), J. Biol. Chem., 244 5403.

Tsuda, Y., Stephani, R.A., and Meister, A. (1971), Biochemistry, 10, 3186.

Valentine, R.C., Shapiro, B.M., and Stadtman, E.R. (1968), Biochemistry, 7, 2143.

Weisbrod, R. (1971), Doctoral dissertation, Cornell University Medical College.

Weissman, H., and Ben-Or, S. (1970), Biochem. Biophys. Res. Commun., 41, 260.

Wellner, V.P., and Meister, A. (1966), Biochemistry, 5, 872.

Wilk, S., Meister, A., and Haschemeyer, R.H. (1969), Biochemistry, 8, 3168.

Wilk, S., Meister, A., and Haschemeyer, R.H. (1970), Biochemistry, 9, 2039.

Woolfolk, C.A., and Stadtman, E.R. (1964), Biochem. Biophys. Res. Commun., 17, 313.

Woolfolk, C.A., Shapiro, B., and Stadtman, E.R. (1966), Arch. Biochem. Biophys., 116, 177.

Wu, C. (1964), Arch. Biochem. Biophys., 106, 394.

Wu, C., and Morris, H.P. (1970), Cancer Res., 30, 2675.

ACKNOWLEDGEMENT

The authors are indebted to the Public Health Service, National Institutes of Health for grant support of many of the studies reviewed in this paper.

REGULATION OF GLUTAMINE SYNTHETASE FROM RAT LIVER AND RAT KIDNEY[*]

Thomas F. Deuel, Alfred Lerner
and Diane Albrycht

Department of Medicine, Section of Hematology,
University of Chicago, and the Argonne Cancer
Research Hospital,[‡] Chicago, Ill.

Abstract

Rat liver glutamine synthetase has been purified to apparent homogeneity. Histidine and the reaction product, glutamine, are inhibitors of catalytic activity. Both histidine and glutamine are more effective when the substrate glutamate is low in concentration; neither inhibits glutamine synthetase when Mg^{2+} (50 mM) replaces Mn^{2+} in the reaction. As reported by Tate and Meister (1971), alanine and glycine also inhibit the enzyme, whereas α-ketoglutarate and citrate result in apparent activation. The activation is very dependent on divalent cation concentration, and occurs only when Mn^{2+} or Mg^{2+} are present in excess; α-ketoglutarate and citrate are inhibitory when divalent cations are limiting in concentration. Divalent cations reverse the effects of α-ketoglutarate and citrate, and other compounds chelating divalent cations mimic the effect of both. Chelation of divalent cations thus may be significant in the apparent activation of glutamine synthetase by α-ketoglutarate and citrate. High concentrations of divalent cations lower the apparent K_m for glutamate, and convert the hyperbolic substrate saturation curve (glutamate) to one with inhibition at high concentrations of substrate. Rat kidney glutamine synthetase has also been purified; initial kinetic data reveal no differences

[*]These studies were supported in part by Grant
CA 13980 from the National Cancer Institute, United States
Public Health Service.

[‡]Operated by the University of Chicago for the United
States Atomic Energy Commission.

between rat kidney and rat liver glutamine synthetases. These studies indicate that divalent cations are of major importance in the expression of glutamine synthetase, and indicate a possible significant role of glutamine in the regulation of its own synthesis.

Introduction

Glutamine is a compound of central importance to both microorganisms and higher animals. Glutamine synthetase, the enzyme responsible for the synthesis of glutamine, has been the subject of considerable interest, and particular attention has been directed toward the control of this enzyme. Studies of glutamine synthetase have shown a close correlation between the overall cellular needs for nitrogen and the regulation of this enzyme (Stadtman et al., 1968; Holzer, 1969). (A) Repression or derepression of enzyme synthesis, (B) feedback inhibition by multiple end products of glutamine metabolism, (C) modulation of catalytic activity in response to divalent cation specificity, (D) energy-dependent enzymatic adenylylation of preformed glutamine synthetase, changing the response of the enzyme to specific divalent cations as well as feedback inhibitors, and (E) metabolite regulation of enzymes responsible for adenylylation and deadenylylation of native glutamine synthetase have been identified as important mechanisms available in Escherichia coli for the fine regulation of glutamine synthetase (Stadtman et al., 1968; Holzer, 1969). Glutamine synthetase from Bacillus licheniformis (Hubbard and Stadtman, 1967a,b) and Bacillus subtilis (Deuel and Stadtman, 1970; Deuel and Turner, 1972; Deuel and Prusiner, 1972) is likewise subject to stringent regulation, although these organisms appear to lack the enzymes present in E. coli which are required for adenylylation (or de-adenylylation) of preformed native glutamine synthetase. In microorganisms, glutamine may be regarded as a major determinant controlling its own synthesis, either through the ability of glutamine to modify the activity of the enzymes responsible for adenylylation or deadenylylation of native E. coli glutamine synthetase (Stadtman et al., 1968), or, as in B. subtilis, through the action of glutamine to inhibit directly glutamine synthetase (or modify the response of the enzyme to other inhibitors)(Deuel and Prusiner, 1972).

Glutamine synthetase in animal tissues also has been investigated. The enzyme obtained from sheep brain in particular has been studied in great detail (Meister, 1968), and was found to be little influenced by inhibitors (Tate and Meister, 1971). However, data have been presented to show that pig brain glutamine synthetase is inhibited by specific amino sugars, amino acids, and nucleotides (Schnacherz and Jaenicke, 1966). Recently, rat liver glutamine synthetase has been purified to apparent homogeneity and shown to be inhibited by alanine, glycine, serine, carbamyl phosphate, and methionine sulfoximine, while α-ketoglutarate and citrate resulted in apparent activation of this enzyme (Tate and Meister, 1971). Chinese hamster glutamine synthetase has been purified also; this enzyme is activated by α-ketoglutarate, and is inhibited by ADP, phosphate, CTP, alanine, and serine (Tiemeier and Milman, 1972).

In addition to in vitro regulation of catalytic activity, multiple studies using animal tissues or cells have shown variation in apparent levels of glutamine synthetase activity in response to different stimuli (Raina and Rosen, 1968; Moscona and Piddington, 1967; Reif-Lehrer and Amos, 1968; Wu, 1964). Low activity is also found in tumor compared with normal tissue (Wu and Morris, 1970). These results emphasize the need for further investigations correlating enzyme turnover rates with other control mechanisms in understanding overall cellular regulation, as was emphasized by Schimke (1969).

In the present studies, an attempt is made to examine factors important for the overall control of glutamine synthetase in mammalian tissues. Rat liver and rat kidney glutamine synthetase have been purified; results of initial studies on the regulation of these enzymes are presented.

Materials and Methods

Glutamine synthetase has been isolated from liver and kidney acetone powder obtained from adult male Sprague-Dawley rats weighing between 250 and 300 grams. Extracts of liver acetone powder were treated with pH 5.0 precipitation, streptomycin precipitation, ammonium sulfate precipitation, differential heat inactivation, and hydroxyapatite chromatography. Details of the purification scheme will be published elsewhere (Deuel et al., 1972a). The enzyme (specific activity = 150 μM γ-glutamylhydroxamate formed/min/mg protein) is 271-fold pure, and 22% of the original material is recovered. Rat kidney glutamine synthetase is purified from acetone powder extracts by precipitation at pH 4.5, hydroxyapatite chromatography, and DEAE chromatography. The final recovery is 20%, and the product purified 300-fold; the specific activity varies between 180-220 μM γ-glutamylhydroxamate/min/mg protein.

Amino acids were obtained from Schwarz-Mann, Orangeburg, N.Y., and the nucleotides were purchased from P.L. Biochemicals, Milwaukee, Wisc. Assays were performed using either Mn^{2+} (2 mM) or Mg^{2+} (50 mM) in a standard biosynthetic assay modified from the procedure of Boyer et al. (1959). Routine assays (pH 7.2) contained divalent cation as noted, 50 mM imidazole chloride, 50 mM NH_4Cl (or 125 mM NH_2OH), 50 mM L-glutamate, and 10 mM ATP, unless otherwise stated. Assays (0.4 ml) were performed at 37° for 15 minutes. Enzyme activity during the initial stages of purification was followed by a modification of the γ-glutamyl transferase assay as described by Levintow (1954). All values reported are the result of duplicate determinations.

Results

With acetone powder prepared from rat liver as the starting material, glutamine synthetase has been purified 271-fold. The resulting protein is judged to be over 95% pure using standard acrylamide gel electrophoresis, acrylamide gel electrophoresis in urea, and acrylamide gel electrophoresis in sodium dodecyl sulfate. (The purification scheme is outlined under "Methods.") Final purification is achieved by elution of the enzyme from a column of hydroxyapatite, as illustrated in Fig. 1. The large bulk of protein elutes from the column with 0.1 M phosphate, at pH 7.0, whereas elution of rat liver glutamine synthetase requires higher concentrations of phosphate. Comparison of fractions across the activity peak demonstrates essentially no differences in the specific activity, suggesting a high degree of purity. The end product, when analyzed by gel filtration, has two peaks of activity. The first peak emerges with the void volume of the column; when pooled,

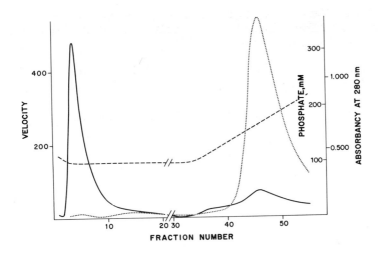

Fig. 1.   Elution pattern of rat liver glutamine synthetase activity from a column (1.5 x 10 cm) of hydroxyapatite, equilibrated with 50 mM imidazole chloride (pH
7.0), 0.5 mM 2-mercaptoethanol, 0.1 mM EDTA, and 0.1 M
phosphate.   The sample (25 ml, 6 mg/ml) is dialyzed against
the same buffer and applied to the column.   After washing
with several column volumes of buffer, elution is begun
with a linear gradient varying phosphate from 0.1 to 0.3 M.
2 ml fractions were collected.   The activity (dotted line)
is determined as noted in "Materials and Methods," and the
protein estimated by absorbancy at 280 nm (solid line).
Phosphate concentration (dashed line) is estimated by con-
ductivity measurements.

and subjected again to gel filtration on the same column,
the two peaks are again found, in a pattern similar to that
in the initial gel filtration experiment.   When the second
peak is pooled and passed over the column for a second
time, the second peak does not reaggregate; no protein or
activity could be detected in the void volume of the
eluate.   The protein from the second peak has a sedimentation coefficient (20°, 7 mg/ml) of 15 S.   The subunit
molecular weight is 45,000, as estimated from gel electrophoresis experiments in sodium dodecyl sulfate.   This value
agrees with that of Tate and Meister (1971), who also found
the molecular weight for the non-dissociated enzyme to be

364,000. The enzyme is sensitive to the sulfhydryl reagents parachloromercuribenzoate and iodoacetamide, and is
inhibited by heavy metals.

When the purified enzyme is studied in the standard
Mn^{2+} (2 mM) biosynthetic assay system, it has a pH optimum of 6.0, but retains over 80% activity at pH 7.2. The
apparent K_m for glutamate is 5.0 mM, and for ammonia, 0.25
mM. Figure 2 shows the results of increasing Mn^{2+} (2A)
and Mg^{2+} (2B) concentrations on the catalytic activity of
rat liver glutamine synthetase at 10 mM ATP. The results

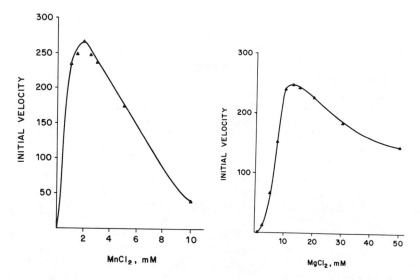

Fig. 2. Effect of increasing $MnCl_2$ (A) and $MgCl_2$ (B)
on the initial velocity of rat liver glutamine synthetase.
The standard biosynthetic assay is used, except that 125 mM
NH_2OH is substituted for NH_4Cl, and divalent cations added
at the final concentrations indicated on the abscissa.

resemble closely those reported by Tate and Meister (1971),
and indicate that the enzyme responds differently to specific divalent cations. Maximum catalytic activity results
when Mn^{2+} is present in 2 mM concentration, and when Mg^{2+}
is present at 12 mM. Concentrations of divalent cation
either below or in excess of the optimal value result in a
significant reduction of catalytic activity. Cobalt may

replace Mn^{2+} and Mg^{2+} in the biosynthetic reaction, with a specific activity intermediate between the two.

In other studies (Deuel et al., 1972a), substrate saturation curves for glutamate were compared at 2 and 8 mM Mn^{2+}; the apparent K_m for glutamate decreases from 5.0 mM to 0.33 mM as Mn^{2+} is increased from 2 to 8 mM. Likewise, the apparent K_m for glutamate falls from 7.5 mM to 1.7 mM when the Mg^{2+} concentration is raised from 10 to 50 mM. Substrate inhibition by glutamate is demonstrated at the higher concentrations of either divalent cation, whereas the glutamate substrate saturation curve is hyperbolic over the range studied (0-50 mM) at the lower concentration of either Mn^{2+} (2 mM) or Mg^{2+} (10 mM). No differences in apparent K_m for hydroxylamine are demonstrated at the different levels of divalent cation. These results thus illustrate again the importance of divalent cations in the overall activity of rat liver glutamine synthetase.

It has been reported previously that α-ketoglutarate and citrate produce activation of rat liver glutamine synthetase (Tate and Meister, 1971). Figures 3 and 4 indicate that the response of the enzyme to both compounds varies markedly with the level of divalent cation present during the assay. Figure 3 shows that when Mn^{2+} is present in a less than optimal concentration (0.5 mM) during assay, both α-ketoglutarate and citrate inhibit catalytic activity. In contrast, when Mn^{2+} is present at excess and thus inhibitory levels (5 mM), both compounds activate the enzyme to a significant degree. Minimal activation of rat liver glutamine synthetase is shown in the middle panel of Figure 3 with α-ketoglutarate and citrate at low concentrations; citrate is shown to be inhibitory at higher concentrations. Similar findings are shown in Figure 4, which demonstrates that both α-ketoglutarate and citrate activate rat liver glutamine synthetase when Mg^{2+} is in excess (50 mM), but that they act as inhibitors of the enzyme when Mg^{2+} is slightly below the levels required for optimum activity (10 mM). When divalent cation is in excess (5 mM Mn^{2+}, 50 mM Mg^{2+}), glutamine synthetase activity is inhibited (see Fig. 2). Compounds binding the divalent cation thus would result in apparent activation of the enzyme, which would be reversed by the further addition of divalent cations. In other experiments, addition of

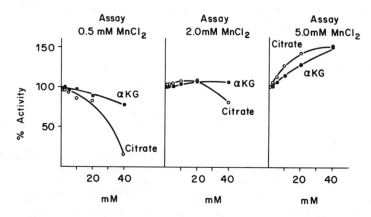

Fig. 3. Effects of increasing α-ketoglutarate and citrate concentrations on the Mn^{2+} biosynthetic activity of rat liver glutamine synthetase. Mn^{2+} is studied at each of the three concentrations indicated, and NH_2OH (125 mM) is substituted for NH_4Cl. Results are expressed as percentage activity in the presence of α-ketoglutarate (solid circles) or citrate (open circles) at the concentrations indicated, and compared with a control assayed in the absence of either compound. Twelve μg protein are used in each assay.

---

appropriate levels of divalent cation completely reverses the apparent activation of rat liver glutamine synthetase by α-ketoglutamate and citrate. EDTA, oxalate, malonate, and malate also have been shown to act as activators at 50 mM Mg^{2+}; the effectiveness of each parallels the relative affinity of the ligand for divalent cation. Therefore, chelation of divalent cation can account for the apparent activation of rat liver glutamine synthetase by citrate and α-ketoglutarate, although other effects are not ruled out (Deuel et al., 1972a).

The response of rat liver glutamine synthetase to various inhibitors is shown in Figures 5, 6, and 7. L-alanine (5A) and glycine (5B) significantly inhibit the catalytic activity studied in the Mn^{2+} biosynthetic assay system, as was previously demonstrated by Tate and Meister (1971). In both cases, the inhibition is increased when the assay is saturating for all of the respective substrates. The enhanced inhibition by either alanine or

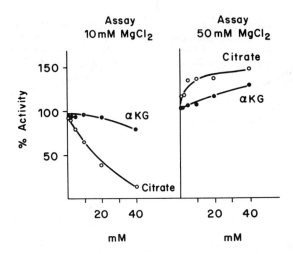

Fig. 4. Effect of increasing concentrations of α-ketoglutarate and citrate on the Mg^{2+} biosynthetic activity of rat liver glutamine synthetase. Mg^{2+} is used at the two concentrations indicated, and NH_2OH (125 mM) substituted for NH_4Cl. Results are expressed as percentage activity in the presence of either α-ketoglutarate (solid circles) or citrate (open circles) compared with a control assayed in the absence of either compound. Three μg protein are used per assay.

glycine in the presence of the substrates may thus indicate that substrate binding is required for inhibition by either of these agents.

In Figure 6, L-histidine is shown also to inhibit rat liver glutamine synthetase. Figure 6A demonstrates that histidine is a more effective inhibitor when glutamine is limiting in concentration during assay, and less effective when MnATP concentration is low. The double reciprocal plot in Figure 6B shows that histidine behaves kinetically as a purely competitive inhibitor with respect to L-glutamate (Deuel et al., 1972a).

As seen in Figure 7, L-glutamine also acts as an inhibitor of rat liver glutamine synthetase. Glutamine is a more potent inhibitor when either ammonia or glutamate is limiting in concentration during assay, and is less potent

137

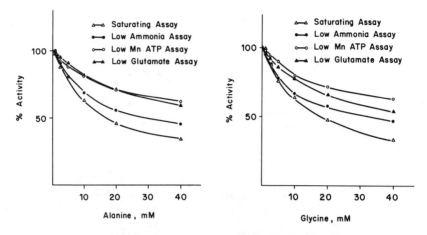

Fig. 5. Effects of increasing alanine (A) and glycine
(B) concentrations on the catalytic activity of rat liver
glutamine synthetase. The standard Mn^{2+} biosynthetic assay
is used (saturating assay, open triangles), as well as
assays limiting in each of the substrates (0.5 mM NH_4Cl,
solid circles, 5 mM glutamate, solid triangles, and 0.5
mM ATP, 0.1 mM $MnCl_2$, open circles, respectively). Percentage activity in the presence of each effector relative
to a control without effector is plotted against the
respective effector concentration. Six μg protein are used
in the saturating assays, and 12 μg in the limiting assays.

when MnATP is limiting in concentration (7A). In a double
reciprocal plot (Fig. 7B), glutamine is shown to be a mixed
type of inhibitor with respect to L-glutamate. The product of the reaction, glutamine, thus is an effective
inhibitor of its own synthesis, particularly when the substrates ammonia and L-glutamate are low in concentration.
When Mg^{2+} (50 mM) is substituted for Mn^{2+}, neither histidine nor glutamine are active as inhibitors of rat liver
glutamine synthetase. Both D-alanine and B-alanine are
effective inhibitors of the enzyme; D-histidine also
inhibits activity significantly. Histamine, histidinol,
histidine-methyl-ester, and glycyl-L-histidine (studied at
20 mM) do not significantly influence the catalytic activity (Deuel et al., 1972a).

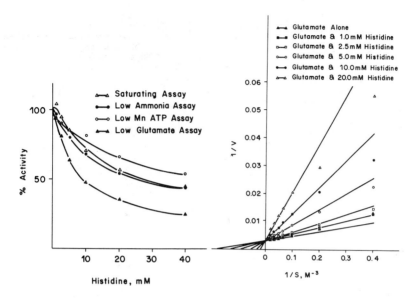

Fig. 6. Effect of histidine on rat liver glutamine synthetase. (A) Effects of increasing concentrations of histidine on the catalytic activity in the saturating assay, and with each of the three substrates limiting. Details are as in Figure 5. (B) Lineweaver-Burk plot of (velocity)$^{-1}$ versus (substrate)$^{-1}$ for glutamate in the presence of different concentrations of histidine, as shown. The standard biosynthetic assay is used with glutamate concentrations varied as indicated. Six μg protein are used per assay.

Rat kidney glutamine synthetase has also been purified to apparent homogeneity (see Methods). Comparison of this enzyme with the glutamine synthetase from rat liver shows that the preparation from kidney has slightly higher specific activity than liver (180-220 compared to 150 mM γ-glutamylhydroxamate/mg/min); in other respects, however, the enzymes are indistinguishable. Table 1 summarizes several kinetic properties of the two enzymes that have been determined thus far. The apparent K_m values for glutamate and ammonia are identical (studied at 2 mM Mn^{2+}), and the responsiveness of the enzyme to the amino acid inhibitors alanine and histidine, as well as to the product glutamine, is not significantly different. The pH

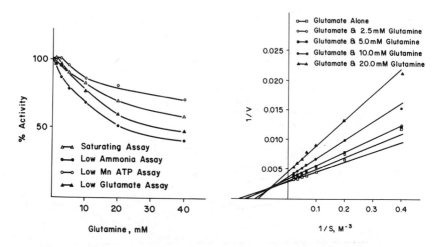

Fig. 7. Effect of the reaction product glutamine on rat liver glutamine synthetase. (A) Effects of increasing concentrations of glutamine on the catalytic activity in the saturating assay, and with each of the three substrates limiting in concentration. Details are as in Figure 5. (B) Lineweaver-Burk plot of (velocity)$^{-1}$ versus (substrate)$^{-1}$ for glutamate in the presence of glutamine at the concentrations shown. The standard Mn^{2+} biosynthetic assay is used, except that glutamate concentration is varied as indicated. Six μg protein are used per assay.

optima for the enzymes are likewise similar, and the subunit molecular weight (as estimated from gel electrophoresis in sodium dodecyl sulfate) is 44,000, in close agreement with that determined for the liver enzyme. Both enzymes are inhibited by the product, phosphate. Antibodies against purified rat liver glutamine synthetase have been prepared. No antigenic differences between rat liver and rat kidney glutamine synthetase are seen when studied by double diffusion techniques.

Discussion

The regulation of rat liver glutamine synthetase is complex. The in vitro kinetic properties of the enzyme presented here, and by Tate and Meister (1971), emphasize the importance of specific divalent cations in determining

TABLE 1

Comparison of Glutamine Synthetase Purified
from Rat Liver and Rat Kidney

Property		Rat Liver	Rat Kidney
Apparent K_m (mM)			
L-Glutamate		5.0	6.0
Ammonia		0.30	0.25
Inhibition (percent)			
L-alanine	5 mM	17	16
	20 mM	41	43
L-histidine	5 mM	14	15
	20 mM	43	45
L-glutamine	5 mM	12	15
	20 mM	34	42

Purified enzyme from kidney and liver (6 µg protein/assay) is assayed as described in "Materials and Methods," using the Mn^{2+} biosynthetic assay. K_m data are derived from Lineweaver-Burk plots and expressed in mM. Inhibition studies are evaluated at two effector concentrations (5 and 20 mM) and expressed as percent inhibition compared to a control activity in the absence of effectors.

the catalytic activity as well as the responsiveness of the enzyme to feedback inhibitors. Also, compounds with high affinity for divalent cations may activate or inhibit glutamine synthetase activity in vitro by virtue of chelating divalent cations and thus varying the levels available to the enzyme. In these studies, citrate and α-ketoglutarate are activators of rat liver glutamine synthetase when divalent cations are in excess of optimum concentration, and are inhibitors when either divalent cation is present in limiting concentrations. These compounds may thus influence catalytic activity by competing for divalent cations. Additional support for a role of chelation by α-ketoglutarate and citrate is found in experiments in which divalent cations were added to assays with α-ketoglutarate or citrate. The effects of these compounds are reversed entirely by the addition of appropriate levels of divalent cations. Other chelators of divalent cations have also been tried; these compounds, when tested in the Mg^{2+} assay system, mimic the effects of α-ketoglutarate and citrate. Thus, the effects of these compounds may be explained in terms of chelation of divalent cations. Divalent cations also may alter the binding sites on the enzyme for the carboxylic acids, or divalent cation-carboxylic acid complexes may form which behave differently from the free carboxylic acids. The present evidence is insufficient to exclude these possibilities as explanations for the marked difference seen with α-ketoglutarate and citrate at different levels of divalent cations. The studies also do not preclude a site on the enzyme for α-ketoglutarate or citrate; binding studies are needed to establish this point. These studies, along with the changes in the glutamate substrate saturation curve discussed under "Results" at different levels of divalent cations, emphasize the need for further research directed at elucidating the role of divalent cations in cellular regulation.

Rat liver glutamine synthetase is also subject to regulation by feedback inhibition. As previously demonstrated by Tate and Meister (1971), alanine, glycine, and serine are strong inhibitors of glutamine synthetase at appropriate levels. The present studies indicate that the product of the reaction, glutamine, and histidine also are potent inhibitors of rat liver glutamine synthetase, and that they are particularly effective inhibitors when glutamate concentrations are low. Significantly, in those cases studied, feedback inhibition is not seen when Mg^{2+} (50 mM) replaces Mn^{2+} (2 mM) in the biosynthetic assay. Divalent cations therefore also play an important role in the responsiveness of rat liver glutamine synthetase to feedback inhibitors.

Purified rat kidney glutamine synthetase appears to be similar to rat liver glutamine synthetase in the physical and kinetic parameters studied thus far. The overall regulation of this enzyme may, however, be very different in the kidney. In addition to the parameters presented in Table 1, both enzymes are strongly inhibited by phosphate, which may be of considerable importance in that glutaminase I is activated by phosphate (Katunuma et al., 1966, 1968). A futile cycle involving glutamine synthesis and degradation may thus be avoided.

The finding that glutamine directly inhibits its own synthesis, and is more effective as an inhibitor when glutamate and ammonia are present in low concentrations, suggests that intracellular glutamine levels may be important in the overall regulation of glutamine synthetase. This conclusion is further strengthened by reports that glutamine synthetase activity is sharply reduced by the addition of glutamine to growing animal cells (DeMars, 1958; Paul and Fottrell, 1963; Barnes et al., 1971; and Slamatiadou, 1972). Studies are in progress to evaluate enzyme levels as they are altered in rat liver and kidney in response to various physiological stimuli.

References

Barnes, P. R., Youngberg, D., and Kitos, P. A. (1971). J. Cell. Physiol. 77, 135.

Boyer, P. D., Mills, R. C., and Fromm, H. J. (1959). Arch. Biochem. Biophys. 81, 249.

DeMars, R. (1958). Biochim. Biophys. Acta 27, 435.

Deuel, T. F., Lerner, A., and Albrycht, D. (1972a). in preparation.

Deuel, T. F., Lerner, A., and Albrycht, D. (1972b). Biochem. Biophys. Res. Commun. in press.

Deuel, T. F., and Prusiner, S. (1972). in preparation.

Deuel, T. F., and Stadtman, E. R. (1970). J. Biol. Chem. 245, 5206.

Deuel, T. F., and Turner, D. C. (1972). J. Biol. Chem. 247, 3039.

Holzer, H. (1969). Advan. Enzymol. 32, 297.

Hubbard, J. S., and Stadtman, E. R. (1967a). J. Bacteriol. 94, 1007.

Hubbard, J. S., and Stadtman, E. R. (1967b). J. Bacteriol. 94, 1016.

Katunuma, N., Tomino, I., and Nishino, H. (1966). Biochem. Biophys. Res. Commun. 22, 321.

Katunuma, N., Tomino, I., and Sanada, Y. (1968). Biochem. Biophys. Res. Commun. 32, 426.

Lerner, A., and Deuel, T. F. (1972). in preparation.

Levintow, L. (1954). J. Nat. Cancer Inst. 15, 347.

Meister, A. (1968). Advan. Enzymol. 31, 183.

Moscona, A. A., and Piddington, R. (1967). Science 158, 496.

Paul, J., and Fottrell, P. F. (1963). Biochim. Biophys. Acta 67, 334.

143

Raina, P. N., and Rosen, F. (1968). Biochim. Biophys. Acta 165, 470.

Reif-Lehrer, L., and Amos, H. (1968). Biochem. J. 106, 425.

Schimke, R. T. (1969). "Current Topics in Regulation," Vol. I. Academic Press, New York.

Schnacherz, K., and Jaenicke, L. (1966). Hoppe-Seylers Z. Physiol. Chem. 347, 127.

Stadtman, E. R., Shapiro, B. M., Ginsburg, A., Kingdon, H. S., and Denton, M. D. (1968). Brookhaven Symp. Biol. 21, BNL-50116(C-53), 378.

Stamatiadou, M. N. (1972). Biochem. Biophys. Res. Commun. 47, 485.

Tate, S. S., and Meister, A. (1971). Proc. Nat. Acad. Sci. 68, 781.

Tiemeier, D. C., and Milman, F. (1972). J. Biol. Chem. 247, 2272.

Wu, C. (1964). Arch. Biochem. Biophys. 106, 394.

Wu, C., and Morris, H. P. (1970). Cancer Res. 30, 2675.

REGULATION OF GLUTAMINE SYNTHETASE
IN CHINESE HAMSTER CELLS

David C. Tiemeier[†], David Smotkin, and Gregory Milman[*]

Department of Biochemistry
University of California
Berkeley, California 94720

Abstract

The specific activity of glutamine synthetase is a sensitive function of the concentration of glutamine in the media of cultured Chinese hamster cells. Enzyme activity is 8-10 fold higher in cells grown in the absence of glutamine compared to cells grown in the presence of glutamine, although glutamine does not affect the in vitro activity of the purified enzyme. The induction of activity resulting from glutamine removal requires protein synthesis and is complete within one cell cycle (20 hours). When glutamine is added back to cells with induced levels of glutamine synthetase growing in media lacking glutamine, enzyme activity is rapidly repressed to the basal level with a half-time of 12 min. Repression does not require protein synthesis. Both induced and repressed cellular activities are identical to glutamine synthetase purified from Chinese hamster liver. RNA synthesis does not appear to be required for either induction or repression implying that regulation is a posttranscriptional event. However, the results of incubation in the presence of actinomycin D are quite complicated and suggest that a small RNA molecule is required for the glutamine mediated repression of glutamine synthetase. Glutamine synthetase is also induced in the presence of glutamine

[†]Recipient of a predoctoral fellowship from the National Science Foundation.
[*]This research was supported by the United States Public Health Service Grant CA12308 from the National Cancer Institute and Cancer Research Funds of the University of California.

145

by dibutyryl cyclic AMP, dexamethasone, and insulin, although the time course is much slower than the increase accompanying the addition of actinomycin or the removal of glutamine. The regulatory mechanisms may be elucidated by the analysis of Chinese hamster cell glutamine auxotroph mutants which have different levels of glutamine synthetase activity.

Introduction

Glutamine synthetase (L-glutamate:ammonia ligase(ADP) EC 6.3.1.2.) catalyzes the synthesis of glutamine from glutamate and ammonia.

$$\text{glutamate} + NH_3 + ATP \longrightarrow \text{glutamine} + ADP + P_i$$

The enzyme plays a central role in the metabolism of nitrogen because the amide group of glutamine is required for the synthesis of several amino acids and nucleotides. The complex mechanisms which control glutamine synthetase in bacteria have been studied in detail. Regulation occurs by enzymatic adenylylation of tyrosine residues in the enzyme's primary structure, and by cumulative feedback inhibition by the various metabolites which incorporate the amide group of glutamine. In contrast to bacterial systems, the mechanisms by which glutamine synthetase is regulated in mammals are poorly understood.

Many workers have described glutamine regulation of glutamyltransferase activity in animal cells in tissue culture. Glutamyltransferase, measured by capacity to form γ-glutamylhydroxamate, is usually assumed to be synonymous with glutamine synthetase (Levintow et al., 1955; Kirk and Moscona, 1963) although the two activities respond differently to activators, inhibitors, and divalent cations (Meister, 1962). DeMars (1958) reported that glutamyltransferase activity in HeLa cells was elevated fifteen-fold when cells were grown in media lacking glutamine, and he noted a slow decrease in enzyme activity when glutamine was added back to the growth media. Paul and Fottrell (1963) demonstrated an eight-fold stimulation of glutamyltransferase activity in mouse L cells within 48 hours after removal of glutamine from growth media. Both protein and RNA synthesis were required for increase in enzyme activity. Glutamyltransferase activity returned to the initial level within 16 to 24 hours after addition of glutamine. The glutamine mediated decrease in enzyme activity was attributed to oxidation of an unstable enzyme. Barnes et al. (1971) observed in L cells a stimulation of glutamyltransferase activity by removal of glutamine and also reported a twelve-fold induction of glutamyltransferase by cortisol in the presence and in the absence of glutamine. Kulka et al. (1972) reported clonal variation in the ability of cells to survive in glutamine free media. Survival approximately correlated with ability to accumulate glutamyltransferase in glutamine-free media. With one clone, glutamine removal led to a fifteen-fold induction of activity in 3 days. Readdition of glutamine resulted in a repression of activity with a half time of five hours. The glutamyltransferase activity of cells growing in both the presence and absence of glutamine was increased two-fold by dexamethasone.

The regulation of glutamine synthetase, measured by γ-glutamyltransferase activity has been studied in chick embryonic retina both in vivo and in tissue culture explants. Glutamine synthetase activity increases markedly during late retinal development and maturation (Piddington and Moscona, 1965). Treatment with hydrocortisone produces precocious induction of enzyme activity in embryonic tissue in vivo and stimulation of enzyme synthesis in retinal cells in organ culture (Moscona and Piddington, 1966; Piddington and Moscona, 1967). Enzyme induction is inhibited 60% by glutamine and 80% by γ-aminobutyric acid (Kirk and Moscona, 1963). Induction by hydrocortisone requires both RNA and protein synthesis and is also affected by cell-cell interactions (Kirk, 1965; Moscona and Kirk, 1965; Moscona et al. 1968; Alescio et al., 1970; Reif-Lehrer, 1971). Induction is highest in intact tissue, intermediate in aggregated tissue culture cells, and lowest in tissue culture cells in monolayers (Morris and Moscona, 1970). Sensitivity to hydrocortisone induction appears to be a function of the time after dispersion into monolayers; inducibility is lost after twenty-four hours in culture. The chick retinal studies suggest that cellular organization may be required to produce high levels of glutamine synthetase activity.

We are studying the regulation of glutamine synthetase in Chinese hamster cells. We began by investigating the properties of enzyme purified from Chinese hamster liver. Unlike the bacterial enzyme, the purified hamster enzyme exhibits no clear control by cellular metabolites. Therefore,

we examined the effect of various compounds on enzyme activity in living cells. As described below, glutamine synthetase activity in tissue culture cells is regulated by
glutamine, steroids, and cyclic AMP.

Enzyme Characterization

Glutamine synthetase may be isolated from the liver,
kidney, brain, and testes of Chinese hamsters. As previously described (Tiemeier and Milman, 1972a), the Chinese hamster liver enzyme has been purified 150-fold to apparent
homogeneity by 30-55% ammonium sulfate precipitation, G150
Sephadex chromatography, DEAE sephadex chromatography, and
sedimentation in a 20-40% glycerol gradient. Table 1 presents a comparison of basic biochemical data for glutamine
synthetase purified from Chinese hamster liver, rat liver
(Tate and Meister, 1971), and sheep brain (Pamiljans et al.,
1962; Ronzio et al., 1969). In size, subunit composition,
substrate Michaelis constants, and sensitivity to cellular
metabolites these three mammalian enzymes are very similar.

The high concentrations of α-ketoglutarate (αKG), AMP,
P_i, CTP, CDP, serine, or alanine required to affect the
Chinese hamster liver glutamine synthetase leave the physiological significance of these effectors in doubt. It is
important to note that glutamine has no effect on the in
vitro assay of purified glutamine synthetase since studies
described below indicate that glutamine plays an important
role in the in vivo regulation of the enzyme.

Regulation in Cell Culture

The Chinese hamster cells (V79) used in our studies are
descendents of the V line originally isolated by Ford and
Yerganion (1958) from Chinese hamster lung. They are grown
without antibiotics in modified versions of Ham's F12 media
containing glutamine(MCG) and media lacking glutamine(MLG).
The growth of Chinese hamster cells and the preparation of
cell extracts have been described (Tiemeier and Milman,
1972b).

We have previously described a sensitive radioisotope
assay for glutamine synthetase (Tiemeier and Milman, 1972b).
The data in Fig. 1 illustrate that glutamine formation catalyzed by cell extracts is a linear function of time for 35-
40 minutes. Thus, the quantity of glutamine produced in a

TABLE 1

Biochemical Characterization of
Mammalian Glutamine Synthetase

	Chinese hamster	Rat	Sheep
Molecular weight	335,000	352,000	392,000
Subunit weight (#)	42,000(8)	44,000(8)	49,000(8)
$s_{20,w}$(S)	13.8	15.0	15.0
Specific activity (units/mg)	9.5	9.0	12.4
K_m (mM) glutamate	3.1	5.0	2.5
ATP	1.2	–	2.3
NH_3	0.16	–	0.18
Activation (>10%) (conc., mM)	αKG (20)	αKG (20)	αKG (20)
Inhibition (>10%) (conc., mM)	ADP(3)	–	–
	AMP(10)	–	–
	dAMP(10)	–	–
	CTP(20)	CTP(4)	–
	CDP(10)	–	–
	P_i(20)	Carbamyl Phos(10)	Carbamyl Phos(10)
	–	Gly(20)	Gly(20)
	Ala(20)	Ala(20)	Ala(20)
	Ser(20)	Ser(20)	–
No effect (conc., mM)	Glu(20)	Glu	Glu
	Cyclic AMP (10)	–	–

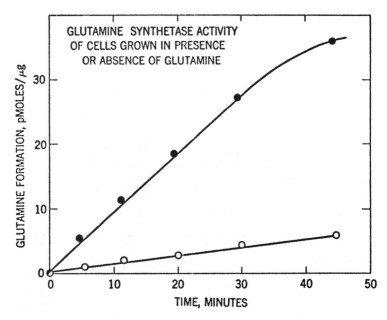

Fig. 1. Time dependence of glutamine formation by cell
extracts. Glutamine formation is determined at the indicated times for extracts of cells grown in MLG(•–•–•, 2.34 x 10^6
cells, 3.7 mg/ml protein) or MCG(o-o-o, 1.41 x 10^6 cells,
2.6 mg/ml protein).

standard assay time of 20 minutes is a direct measure of the
rate of glutamine formation. The data in Fig. 2 illustrate
that the rate of glutamine formation is also directly proportional to the protein concentration of the cell extract.
Therefore, the amount of glutamine formed per min per mg protein of cell extract is a valid measure of glutamine synthetase specific activity.

The cellular level of glutamine synthetase activity is
dependent on the glutamine concentration in the growth media.
Glutamine synthetase activity is 8-10 fold greater in extracts
of cells grown in media lacking glutamine than in extracts of
cells grown in media containing glutamine (Fig. 1). When glutamine is removed from media, the level of glutamine synthetase in cell extracts increases with time for 48 hours (Fig. 2
The specific activity of glutamine synthetase is shown as a
function of time of cell exposure to media lacking glutamine in the Fig. 2 inset. After a two hour lag, the
specific activity increases linearly for the next 18 hours

INDUCTION

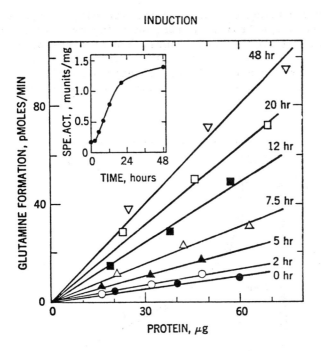

Fig. 2.   Induction of glutamine synthetase activity by glutamine removal.   Cells grown in MCG (6 days) are washed with Tricine-buffered saline (30 mM Tricine, pH 7.6, 5.3 mM KCl, 141 mM NaCl, 5.5 mM glucose, and 5 mg/liter phenol red) and then incubated for the indicated times in MLG.   Inset: Dependence of glutamine synthetase specific activity on increasing time of incubation in MLG.   Specific activities are obtained from the slopes of the lines.

or approximately one cell cycle reaching a half maximal level 12 hours after removal of glutamine.   The specific activity increases in 48 hours from 0.18 milliunits/mg to a maximal level of 1.4 milliunits/mg.   Removal of glutamine also results in a 12 hour lag in logarithmic growth after which cell division is resumed at a slower rate characteristic of growth in media lacking glutamine.

When cells having high levels of glutamine synthetase grown in the absence of glutamine are placed into media

containing glutamine, the specific activity of glutamine
synthetase drops from 1.28 milliunits/mg protein to 0.14
milliunits/mg protein in 8 hours, but half the activity is
lost within 12 minutes (Fig. 3). This rapid repression of
glutamine synthetase activity suggests a specific inactivation of the enzyme rather than normal catabolic degradation.

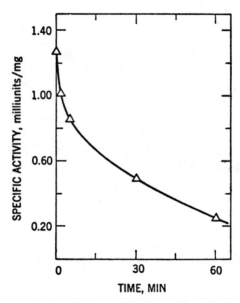

Fig. 3. Time course of repression of glutamine synthetase activity by glutamine. Cells grown in MLG (60 hours)
are then incubated for the indicated times in MCG.

The specific activity of glutamine synthetase is dependent upon the concentration of glutamine in the media.
Glutamine synthetase specific activity is decreased to half
the maximal value by 0.12 mM glutamine (Fig. 4). It is interesting to note that the K_m of the enzyme for glutamate
is 20-fold higher. Decrease in glutamine synthetase specific
activity is linearly dependent on glutamine for concentrations
less than 0.3 mM. As the glutamine concentration is increased, the specific activity of glutamine synthetase approaches
a constitutive level of 0.11 milliunits/mg protein and no
further reduction in enzyme activity is observed.

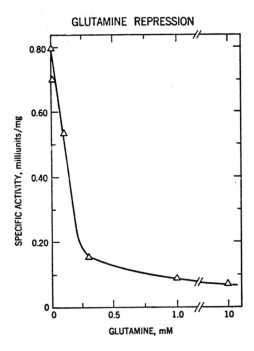

Fig. 4. Dependence of glutamine synthetase activity on glutamine concentration. Cells grown in MCG (48 hours) are incubated for 20 hours in MLG supplemented with the indicated levels of glutamine.

The number of glutamine synthetase molecules per cell may be estimated assuming that the intracellular enzyme has the same activity as purified Chinese hamster liver glutamine synthetase. A cell extract containing 1 mg protein is obtained from 10^7 cells. From the data in Table 1 and the specific activities of 0.1 milliunits/mg or 1.0 milliunits/mg for repressed and induced cell extracts respectively, we calculate that there are 2000 enzyme molecules in each repressed cell and 20,000 molecules in each induced cell.

It is reasonable that the repressed enzyme possesses residual activity even at high levels of glutamine. In addition to its metabolic role of providing glutamine, glutamine synthetase is probably involved in pH regulation, in the

153

kidney (Orloff and Burg, 1971), in ammonia detoxification in the brain (Berg and Kolenbrander, 1970), and in neurotransmission through association with the metabolism of glutamate and γ-aminobutyrate (Baxter, 1970; van den Berg, 1970).

The observed changes in glutamine synthetase activity might be explained by the existence of a metabolic inhibitor of glutamine synthetase in cells grown in the presence of glutamine or a metabolic activator in cells grown in the absence of glutamine. If an activator were present, a mixture of repressed and induced cell extracts would yield an activity greater than the sum of the two individual activities. If an inhibitor were present, a mixture would yield an activity less than the sum of the two individual activities. That mixed extracts are approximately additive (Tiemeier and Milman, 1972b) implies that intracellular metabolites are neither inhibiting nor activating the assay of the enzyme.

Our study of glutamine synthetase in cultured Chinese hamster cells is based on the radioisotope assay of glutamine synthetase developed with the enzyme purified from Chinese hamster liver. It is conceivable that in crude cell extracts the specificity of the assay might be affected by the presence of other cellular enzymes. We present the following studies to support the hypothesis that the pure liver glutamine synthetase and the enzyme activities in induced and repressed cells are identical:

1. The enzyme activity from both induced and repressed cells behaves identically with Chinese hamster liver glutamine synthetase during purification steps previously described (Tiemeier and Milman, 1972a): ammonium sulfate precipitation, G150 Sephadex chromatography and DEAE Sephadex chromatography.

2. Manganese is a strong inhibitor of Chinese hamster liver glutamine synthetase. The effect on repressed and induced activities is identical (Fig. 5).

3. The temperature inactivation profiles of the purified enzyme and the induced cell activity are identical. Furthermore, both are protected to the same extent from inactivation by incubation for 10' at 60° by preincubation in ATP (Tiemeier and Milman, 1972b). In both temperature inactivation and protection by ATP, the repressed activity behaves identically to the induced activity.

4. Induced and purified liver activities are inactivated by N-ethylmaleimide. Both are protected to the same extent by preincubation in ATP (Tiemeier and Milman, 1972b).

154

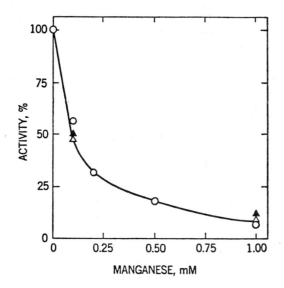

Fig. 5. Manganese inhibition of glutamine synthetase. Extracts of cells grown in the presence or absence of glutamine are partially purified as described for the Chinese hamster liver enzyme. The three activities are assayed in the presence of the indicated levels of manganese. The activity percentages are based on 2.05 units/mg for the liver glutamine synthetase (0.02 μg, Fraction IV); 16.8 milliunits/ mg for the induced cell activity (1.25, 2.50, 3.75 μg, Fraction III); and 3.2 milliunits/mg for the repressed cell enzyme (6.4, 12.9, 19.4 μg, Fraction III).

5. The data in Figs. 6 and 7 indicate that Chinese hamster cells in tissue culture synthesize a protein which fractionates identically to purified glutamine synthetase on the basis of size on a glycerol gradient, and on the basis of size and charge on polyacrylamide gel electrophoresis. [3H]leucine is added to growth media during enzyme induction (potentially a period of maximal rate of enzyme synthesis). The 3H-labeled extract prepared from the cells is fractionated on a glycerol gradient with added purified glutamine synthetase as a marker (Fig. 6). The fraction with maximal glutamine synthetase activity (fraction 17) and fractions

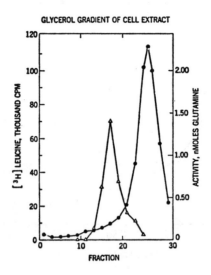

Fig. 6. Glycerol gradient of ^{3}H-labeled cell extract.
Cells grown in F12 for 4 days are washed with Tricine-buffered
saline and then incubated in 20 ml MLG containing 670 µCi
[3H]leucine (Schwarz, 0.034 Ci/mmole in media) for an additional 36 hours. A 250 µl extract from 1.39×10^7 cells
(4.0 mg/ml protein) is prepared as previously described
(Tiemeier and Milman, 1972b) and dialyzed for five hours
against 50 mM Tricine, pH 7.5, 50 mM KCl, 20 mM $MgCl_2$, 0.1 mM
EDTA, and 3.5 mM DTT. The dialyzed extract contains $2.4 \times
10^5$ cpm/µl. Ten µl of purified Chinese hamster liver glu-
tamine synthetase (Fraction IV, 3 mg/ml, 2.0 units/mg) is
added to 100 µl of the cell extract and 100 µl of the mixture
is layered onto a 4.5 ml linear 20–40% glycerol gradient prepared in 50 mM Tricine, pH 7.5, 50 mM KCl, 20 mM $MgCl_2$, 0.1 mM
EDTA, 2 mM DTT, and 1.0 mM ATP. The gradients are centrifuged
for 12 hours at 50,000 rpm in a Spinco SW 50.1 rotor. Thirty-
one 150 µl fractions are collected from the bottom.
Five µl samples of each fraction are mixed with 1.0 ml
H_2O and 10 ml scintillation fluid and counted for 10 min in
a Nuclear Chicago Unilux II at 30% counting efficiency (o-o-o)
Five µl samples are assayed for glutamine synthetase activity
(Δ-Δ-Δ) as previously described (Tiemeier and Milman, 1972b)
except that the incubation time is 30 min.

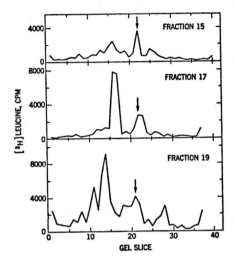

Fig. 7. Polyacrylamide gel electrophoresis of glycerol gradient fractions. Polyacrylamide gel electrophoresis is performed as described by Davis (1964) and Ornstein (1964) using 7.5 cm 4% acrylamide (Eastman) running gels (37.5 mM Tris-Cl, pH 8.9, 0.5 mg/ml TEMED (Eastman), 0.7 mg/ml ammonium persulfate) and 1.5 cm, 2.5% acrylamide stacking gels (58 mM Tris-Cl, pH 6.7, 20% sucrose, 10 μg/ml riboflavin, and 0.5 mg/ml TEMED, polymerized by 30 min exposure to white fluorescent lamp). One hundred μl samples of fractions 13, 15, 17, 19, and 21 from the glycerol gradient (Fig.6) are layered on to the polyacrylamide gels in 150 μl volumes containing 25 mM Tris-Cl, pH 7.4, 10% glycerol, 0.01 M β-mercaptoethanol, and 0.75 μg bromophenol blue. Electrophoresis at 2.5 ma per gel is performed at $0°$ until the dye is 0.5 cm from the bottom (3.5-4.0 hours). The gels are stained with 0.1% Coomassie Blue in 7% acetic acid and 45% methanol and destained for 30 min in 7% acetic acid using a Canalco quick gel electrophoretic destainer. Only a single protein band is observed corresponding to the added glutamine synthetase. The band is the darkest staining in fraction 17 (maximal glutamine synthetase activity, Fig. 6) and is barely visible in the gels for fractions 13 and 21. Each gel is cut into slices approximately 1-2 mm in width using a Canalco gel slicer. The slices are numbered from top to bottom. The

gel slices are placed in scintillation vials and solubilized
according to the method of Goodman and Matzura (1971) except
that 0.5 ml of the alkaline peroxide is used and the reactions require forty-eight hours. Scintillation fluid is
added to each vial and radioactivity is determined. The cpm
are shown as a function of gel slice. The arrow marks the
migration position of the glutamine synthetase marker.

on either side (fractions 13, 15, 19, and 21) are further
fractionated by polyacrylamide gel electrophoresis. When the
gels are stained for protein, only a single band corresponding to the marker glutamine synthetase is observed. Radioactivity in gels for fractions 15, 17, and 19 is shown in
Fig. 7. Only two major peaks of label are observed for
fraction 17. One peak corresponds to the position of marker
glutamine synthetase and is presumably enzyme synthesized
during the labeling period. The label peak corresponding to
enzyme is also observed in almost equal amounts in the gel
for fraction 15 (where it is the major peak) and in the gel
for fraction 19 (where many other labeled protein bands are
apparent). The slower migrating label peak observed in the
gel for fraction 17 is barely visible in the gel for fraction
15 and increases in intensity in the gel for fraction 19,
and therefore it is apparently not related to glutamine synthetase activity.

Regulatory Mechanisms

We have investigated the protein synthesis and RNA requirements for the glutamine-mediated induction and repression of glutamine synthetase using cycloheximide and actinomycin D as probes. Cycloheximide at 0.10 mM which blocks
95% of protein synthesis has no effect on repression but almost totally blocks induction (Tiemeier and Milman, 1972b).
These data suggest that induction results from new enzyme
synthesis, but that no new proteins are required for glutamine-mediated repression.
The action of actinomycin D on cellular glutamine synthetase levels depends on the concentration of actinomycin D,
the time of exposure to the drug, and the concentration of
glutamine in the media. The effect of actinomycin D concentration on the induction of glutamine synthetase by glutamine
removal at a fixed time of induction and exposure to the drug
is presented in Fig. 8. While low levels of actinomycin D

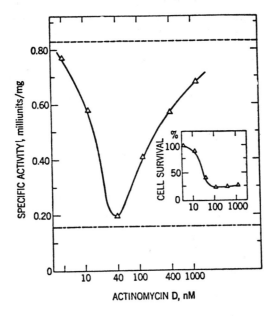

Fig. 8. Effect of actinomycin D concentration on induction. Cells growing in MCG (3 days) are incubated in MCG
plus the indicated concentrations of actinomycin D for one
hour. The cells are then incubated in MLG plus the indicated
concentrations of actinomycin D for 14 hours. The lower
dashed line represents the initial specific activity of glutamine synthetase. The upper dashed line represents the
final induced activity of glutamine synthetase with no added
actinomycin D. Cell survival at the various levels of actinomycin D is presented in the inset.

(25-100 nM) block induction, high levels of actinomycin D
(above 100 nM) allow increased glutamine synthetase activity
nearly equal to the induction in the absence of actinomycin.
The concentration curve suggests that two actinomycin D sensitive processes may be involved in the induction by glutamine removal, and prompted us to study the time course of

159

induction at fixed concentrations of actinomycin. Fig. 9
presents the data for the time course of induction by glutamine removal in the presence of high (500 nM) and low (25 nM)
actinomycin D concentrations. Both levels of actinomycin D
initially enhance the induction but after 10 hours, enzyme

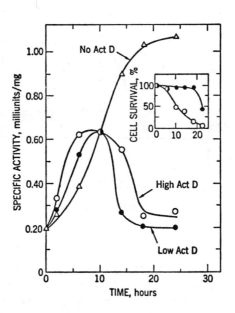

Fig. 9. Time course of actinomycin D effect on induction.
Cells growing in MCG (3 days) are incubated in MCG + 500 nM
actinomycin D (0.6 µg/ml) o-o-o, or MCG + 25 nM actinomycin
D (0.03 µg/ml) ●-●-●, or MCG (no actinomycin D) for one hour.
The cells are then incubated in MLG plus the respective levels
of actinomycin D. At the indicated times, glutamine synthetase specific activity is determined. Cell survival is presented in the inset.

activity reaches a plateau and within 18 hours returns to the
basal level. Since the high actinomycin D appears to maintain
the elevated level of glutamine synthetase longer than low
actinomycin D, the concentration curve in Fig. 8 is strongly

dependent on the time at which enzyme levels are measured as well as on the concentration of actinomycin employed.

Actinomycin D at 120 nM fails to block repression when added with media containing glutamine to cells grown in media lacking glutamine, suggesting that RNA synthesis is not required for repression. As a control, we investigated the effect of actinomycin D on cells in media containing glutamine. Unexpectedly, we found (Fig. 10) that a high level of actinomycin D (500 nM) overcomes glutamine repression and produces an induced level of enzyme activity. High levels of actinomycin D in the presence of glutamine may produce a

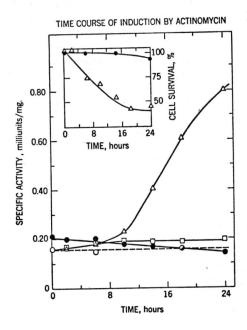

TIME COURSE OF INDUCTION BY ACTINOMYCIN

Fig. 10. Cells grown in MCG (4 days) are incubated in MCG + 500 nM actinomycin D (0.6 μg/ml) Δ-Δ-Δ, MCG + 25 nM actinomycin D (0.03 μg/ml) □-□-□ , MCG + 36 μM cycloheximide (10 μg/ml) ●-●-●, or MCG with no addition in controls o-o-o. Cell survival is presented in the inset: for plates incubated in 500 nM actinomycin D Δ-Δ-Δ and in 36 μM cycloheximide or 25 nM actinomycin D ●-●-●.

similar state in the cells to that caused by the removal of glutamine. Except for the difference in lag time, the rate of increase of glutamine synthetase activity in response to high actinomycin D parallels the induction by glutamine removal. Furthermore, cells induced for glutamine synthetase by growth in media lacking glutamine show no stimulation ("super-induction") of glutamine synthetase activity by actinomycin D. The induction of glutamine synthetase by high actinomycin D suggests that either a small RNA molecule or a metabolic product resulting from it interacts with glutamine to maintain a repressed enzyme level.

We have previously shown that dibutyryl cyclic AMP and, to a lesser degree, hydrocortisone increase the specific activity of glutamine synthetase in both induced and repressed Chinese hamster cells (Tiemeier and Milman, 1972b). The data in Table 2 indicate that the repression of glutamine synthetase specific activity by glutamine can also be overcome by insulin and dexamethasone. However, the rate of increase in activity caused by these compounds is much slower

TABLE 2

Induction of Glutamine Synthetase
in the Presence of Glutamine

Inducer	Time of exposure (days)	S.A. (mU/mg)
Dibutyryl cyclic AMP (1.0 mM)	6	0.90
Insulin (0.2 units/ml)	3	0.65
Dexamethasone (0.1 mM)	3	0.86
Insulin (0.2 units/ml) + Dexamethasone (0.1 m)	3	0.95
Repressed enzyme	–	0.14
Induced enzyme	–	1.30

than that caused by glutamine removal. All compounds exhibit a 12-24 hour lag before glutamine synthetase activity begins to rise compared to the 2 hour lag which precedes the increase in glutamine synthetase activity when glutamine is removed. Furthermore, the time to reach a corresponding level of activity is longer--3 to 6 days--than the induction which occurs

when glutamine is removed. It is not clear whether induction
by glutamine removal and induction by hormones or cyclic AMP
operate by similar or distinct mechanisms.

Prospectus

In Chinese hamster cells, the glutamine regulation of
glutamine synthetase is measured by glutamine formation represents a relatively rapid method of control. Induction by
glutamine removal presumably operates at a translational
level requiring approximately one cell cycle to be expressed.
The repression of activity has a half time of only 12 minutes
although glutamine synthetase is a stable protein with a half
life of over twenty hours (Reif-Lehrer, 1971). Moreover,
repression by glutamine occurs in the absence of RNA or protein synthesis, implying that new repressor is not required
to shut off glutamine synthetase. We have previously shown
that glutamine has no effect on glutamine synthetase activity in vitro. Therefore, the repression by glutamine must
be mediated by some other component.

In procaryotes, Stadtman and coworkers (Shapiro and
Stadtman, 1970; Stadtman et al., 1970; Brown et al., 1971)
have found that covalent adenylylation controls the activity
of E. coli glutamine synthetase. The adenylylated enzyme
possesses lower specific activity, and displays different
susceptibilities to inhibition by metal ions and cellular
metabolites.

The increase of glutamine synthetase activity in the
presence of dibutyryl cyclic AMP in Chinese hamster cells is
reminiscent of the regulation of phosphorylase by a specific
phosphorylase kinase kinase which has cyclic AMP as a cofactor (Walsh et al., 1968). There are many examples of
cyclic AMP dependent protein kinases. They are found in the
liver where they phosphorylate histones (Yamamura et al.,
1970) and glycogen synthetase (Stadtman, 1970) and in the
particulate fraction of neural tissue (Maeno et al., 1971)
and might be involved in the regulation of glutamine synthetase.

By analogy with modifications observed in other systems,
we propose that during repression of glutamine synthetase,
a glutamine-activated modification factor changes the enzyme
to a less active form. The slower repressions of glutamyltransferase in hepatoma and L cells (5 1/2 and 24 hours respectively) are consistent with this model. The modification might be in the covalent structure of the enzyme, in

163

its intracellular localization, or in its state of aggregation. Specific and controlled degradation of the enzyme is
also possible but would seem unlikely.

The determination of glutamine synthetase protein as
well as activity is important for further study of enzyme
biosynthesis and regulation. We have prepared antibody to
purified Chinese hamster liver glutamine synthetase that
gives a single band with crude extracts on Ouchterlony double
diffusion plates. The enzyme–antibody complex retains glutamine synthetase activity which permits quantitative measurement of antibody-precipitated enzyme.

Chinese hamster cells are pseudodiploid and grow with a
doubling time of 8–19 hours. These characteristics may enable the selection of mutants in the various steps of glutamine synthetase biosynthesis and regulation. We have selected glutamine auxotrophs which possess different levels of
glutamine synthetase activity. The analysis of these mutants
both biochemically and immunologically may provide insight
into glutamine synthetase regulatory mechanisms.

References

Alescio, T., Moscona, M., & Moscona, A. A. (1970). Exp. Cell
Res. 61, 342–346.
Barnes, P. R., Youngberg, D., & Kitos, P. A. (1971). J. Cell
Physiol. 77, 135–144.
Baxter, C. F. (1970). Handbook of Neurochemistry (Lajtha, A.,
ed) Vol. 3, pp. 289–353, Plenum Press, New York.
Berg, C. P., & Kolenbrander, H. M. (1970). Comparative Bio-
chemistry of Nitrogen Metabolism (Campbell, J. W., ed)
pp. 795–916, Academic Press, New York.
Brown, M. S., Segal, A., & Stadtman, E. R. (1971). Proc. Nat.
Acad. Sci. U.S.A. 68, 2949–2953.
Davis, B. J. (1964). Ann. N.Y. Acad. Sci. 121, 404–427.
DeMars, R. (1958). Biochim. Biophys. Acta 27, 453–536.
Ford, D. K., & Yerganian, G. (1958). J. Nat. Cancer Inst.
21, 393–408.
Goodman, D., & Matzura, H. (1971). Anal. Biochem. 42, 481–486.
Kirk, D. L. (1965). Proc. Nat. Acad. Sci. U.S.A. 54, 1345–
1353.
Kirk, D. L., & Moscona, A. A. (1963). Develop. Biol. 8, 341–
357.
Kulka, R. G., Tomkins, G. M., & Crook, R. B. (1972). J. Cell
Biol. 54, 175–179.

Levintow, L., Meister, A., Hogeboom, G. H., & Kuff, E. L. (1955). J. Amer. Chem. Soc. 77, 5304-5308.

Maeno, H., Johnson, E. M., & Greengard, P. (1971). J. Biol. Chem. 246, 134-142.

Meister, A. (1962). The Enzymes (Boyer, P. D., Lardy, H., and Myrback, K., eds) Vol. 6, pp. 443-468, Academic Press New York.

Morris, J. E., & Moscona, A. A. (1970). Science 167, 1736-1738; (1971) Develop. Biol. 25, 420-444.

Moscona, A. A., & Kirk, D. L. (1965). Science 148, 519-521.

Moscona, A. A., & Piddington, R. (1966). Biochim. Biophys. Acta 121, 409-411.

Moscona, A. A., Moscona, M. H., & Saenz, N. (1968). Proc. Nat. Acad. Sci. U.S.A. 61, 160-167.

Orloff, J., & Burg, M. (1971). Ann. Rev. Physiol. 33, 83-130.

Ornstein, L. (1964). Ann. N.Y. Acad. Sci. 121, 321-349.

Pamilians, V., Krishnaswamy, P. R., Dumville, G., and Meister, A. (1962). Biochemistry 1, 153.

Paul, J., & Fottrell, P. (1963). Biochim. Biophys. Acta 67, 334-336.

Piddington, R., & Moscona, A. A. (1965). J. Cell Biol. 27, 247-252.

Piddington, R., & Moscona, A. A. (1965). Biochim. Biophys. Acta 141, 429-432.

Reif-Lehrer, L. (1971). J. Cell Biol. 51, 303-311.

Ronzio, R. A., Rowe, W. B., Wilk, S., & Meister, A. (1969). Biochemistry 8, 2670-2674.

Shapiro, B. M., & Stadtman, E. R. (1970). Ann. Rev. Microbiol. 24, 501-524.

Stadtman, E. R. (1970). The Enzymes (Boyer, P. D., ed) Vol. 1, pp. 397-459, Academic Press, New York.

Stadtman, E. R., Ginsburg, A., Ciardi, J. E., Yeh, J., Hennig, S. B., & Shapiro, B. M. (1970). Advan. Enzyme Regul. 8, 99-118.

Tate, S. S., & Meister, A. (1971). Proc. Nat. Acad. Sci. U.S.A. 68, 781-785.

Tiemeier, D. C., & Milman, G. (1972a). J. Biol. Chem. 247, 2272-2277.

Tiemeier, D. C., & Milman, G. (1972b). J. Biol. Chem. 247, In press.

van den Berg, C. J. (1970). Handbook of Neurochemistry (Lajtha, A., ed) Vol. 3, pp. 355-379, Plenum Press, New York.

Walsh, D. A., Perkins, J. P., & Krebs, E. G. (1968). J. Biol. Chem. 243, 3763-3774.
Yamamura, H., Takeda, M., Kumon, A., & Nishizuka, Y. (1970). Biochem. Biophys. Res. Commun. 40, 675-682.

GLUTAMATE SYNTHETASE (GOGAT); A KEY ENZYME IN THE ASSIMILATION OF AMMONIA BY PROKARYOTIC ORGANISMS

D. W. Tempest, J. L. Meers* and C. M. Brown†

Microbiological Research Establishment, Porton, Salisbury, England, *Agricultural Division, I.C.I. Ltd, Billingham, Teeside TS23 1LB, England, and †Department of Microbiology, Medical School, University of Newcastle upon Tyne, NE1 7RU, England.

Abstract

Nitrogen-limited cultures of bacteria have been found to assimilate ammonia by a route that does not involve the participation of glutamate dehydrogenase. Initially, ammonia is metabolized via the glutamine synthetase reaction, and then the glutamine amide-nitrogen is transferred (by a hitherto unclassified oxido-reduction reaction) to the 2-position of 2-oxoglutarate. Together, glutamine synthetase and glutamine (amide): 2-oxoglutarate amidotransferase oxido-reductase (GOGAT or glutamate synthetase) effect the net synthesis of glutamate from ammonia and 2-oxoglutarate (just as does glutamate dehydrogenase) but, unlike glutamate dehydrogenase, this pathway can function efficiently at extremely low ammonia concentrations. Since its discovery in Aerobacter aerogenes, glutamate synthetase (and the "glutamine pathway" of ammonia assimilation) has been detected in a wide variety of bacteria. It appears to play an essential role in nitrogen fixation and in the utilization of nitrate by prokaryotic marine organisms.

Introduction

In order to grow in a simple salts medium in which ammonia provides the sole source of utilizable nitrogen, microorganisms must possess some primary mechanism for the synthesis of amino acid from ammonia and intermediary metabolites. It has been generally assumed that in most bacteria this requirement is met by glutamate dehydrogenase

167

(which reductively aminates 2-oxoglutarate to glutamate) or, in those organisms which lack glutamate dehydrogenase (for example, several species of Bacillus), by some analogous amino acid dehydrogenase (particularly alanine dehydrogen-ase; see Shen et al., 1959). However, two observations indicated that an alternative pathway of ammonia assimilat-ion might be present in bacteria: first, the finding of Freese et al. (1964) that mutants of Bacillus subtilis lacking both glutamate dehydrogenase and alanine dehydro-genase still would grow readily in a minimal medium; and second, our finding (Tempest et al., 1970a) that the synthesis of glutamate dehydrogenase in Aerobacter (Klebsiella) aerogenes could be almost totally repressed without affecting the ability of the organisms to assimil-ate ammonia and grow in a simple salts medium. A detailed investigation of the primary products of ammonia assimilat-ion in these glutamate dehydrogenase-repressed (ammonia-limited) A. aerogenes organisms revealed the existence of a novel pathway of glutamic acid synthesis involving a previously unknown glutamine-utilizing enzyme. In this paper we present information on the regulation and distri-bution of this "glutamine pathway" of glutamic acid syn-thesis in microorganisms, and attempt to assess its physiological significance in natural ecosystems.

Pathways of Synthesis of Glutamate in Aerobacter aerogenes

Substantial amounts of glutamate dehydrogenase could be detected in extracts of Aerobacter aerogenes organisms that had been grown in chemostat cultures with growth limited by the availability of either glucose or phosphate. But when these organisms were limited in their growth by the availability of ammonia (the sole utilizable nitrogen source), the level of glutamate dehydrogenase was greatly decreased (Table 1). Indeed, when this enzyme was assayed at near-neutral pH values (possibly similar to the H^+ concentration within growing A. aerogenes organisms) no activity could be detected, indicating that glutamic acid synthesis in these ammonia-limited organisms proceeded by some other route. Examination of ammonia-limited organisms for aspartase, alanine dehydrogenase, leucine dehydrogenase and valine dehydrogenase (supposed routes of ammonia assim-ilation in bacteria) failed to demonstrate their presence, and thus it was concluded that ammonia incorporation into amino acids, in these ammonia-limited organisms, occurred

168

by some hitherto undescribed route.

TABLE 1

Influence of Environment on the Concentration of
some Enzymes and 'Pool' Constituents in Aerobacter
aerogenes Organisms, Growing in a Chemostat

Growth Condition	'Pool' Concentrations (mM)		Enzyme Activities	
	Ammonia	Glutamate	GS*	GD
Glucose-limited	>10	3	0.1	560
Ammonia-limited	1	6	1.6	19
N(glutamate)-limited	0.7	5	2.8	<1
Phosphate-limited	>10	2	0.2	600
Phosphate-limited plus glutamate	>10	>20	0.2	10

*GS = Glutamine synthetase (arbitrary units); GD = Gluta-
mate dehydrogenase (n moles $NADPH_2$ oxidised/min/mg protein).

Since, when growing in an ammonia-limited chemostat
culture, amino acid synthesis must be limited solely by
the availability of ammonia, a clue to the pathway of its
incorporation into amino acids could be obtained by pulsing
a small amount of ammonia into the culture (thereby
disturbing the steady state) and studying the transient
changes in the intracellular concentrations of free amino
acids. When ammonia (final concentration of 10 mM) was
added rapidly to a steady state ammonia-limited culture of
A. aerogenes and samples of organisms taken after 2 min, a
25-fold increase in 'pool' free glutamine was observed
(compared with an increase of less than 2-fold for all
other amino acids)(Fig. 1). This suggested that ammonia
incorporation into amino acids, in ammonia-limited organisms,
possibly might proceed via the amination of glutamic acid,
and comparison of these N-limited organisms with glucose-
limited A. aerogenes showed that their glutamine synthetase
activity was greatly increased (Table 1; see also Wu and
Yuan, 1968; Rebello and Strauss, 1969). Since the Km for
ammonia of A. aerogenes glutamine synthetase is low (and
therefore could function efficiently in an ammonia-limited
environment) it seemed feasible that glutamine could lie on
the pathway of synthesis of amino acids from ammonia,
particularly when the intracellular concentration of ammonia
was low. But organisms would then need to possess an

169

Figure 1. Transient changes in 'pool' free amino acids of ammonia-limited <u>Aerobacter aerogenes</u> organisms following addition of a pulse of ammonia to the chemostat culture. (●) Glutamic acid; (○) glutamine; and (■) alanine; these being the principal amino acids present in the 'pool'

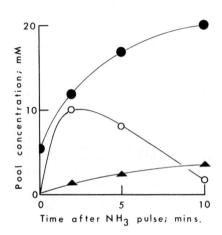

Time after NH_3 pulse; mins.

enzyme (or system of enzymes) capable of transferring the amide-nitrogen of glutamine to the α-amino position of some amino acid. And if this reaction involved a 2-oxo acid as the amino-acceptor molecule (as does glutamate dehydrogenase) then it would also require a coupled oxido-reduction step. In fact the simplest route whereby the amide-nitrogen of glutamine possibly could be incorporated into an amino acid would be by a reaction analogous to that effected by glutamate dehydrogenase, <u>viz</u>:

$$\text{Glutamine} + \text{2-oxoacid} + \begin{array}{c} NADPH_2 \\ or \\ NADH_2 \end{array} \rightarrow \text{Glutamic} + \text{Amino} + \begin{array}{c} NADP \\ or \\ NAD \end{array}$$
$$\text{acid} \quad \text{acid}$$

Incubation of cell-free extracts of ammonia-limited <u>A. aerogenes</u> with glutamine, $NADPH_2$, $NADH_2$ and either pyruvate or oxaloacetate produced no net synthesis of amino acid, although some glutamic acid (plus an equivalent amount of ammonia) was formed due to the action of glutaminase in the extracts. However, incubation of these

170

extracts with glutamine, $NADH_2$, $NADPH_2$ and 2-oxoglutarate did result in a considerable net synthesis of glutamic acid without concomitant formation of ammonia; and subsequent spectrophotometric assays (at 340 nm) showed that the active pyridine nucleotide was $NADPH_2$.

Clearly, then, net synthesis of glutamate in ammonia-limited A. aerogenes organisms can occur by a two-stage process which involves first the synthesis of glutamine and then the reductive transfer of the amide group to the 2-position of 2-oxoglutarate, thereby forming two molecules of glutamate (Fig. 2).

Figure 2.   Pathway of synthesis of glutamic acid from ammonia in Aerobacter aerogenes organisms.

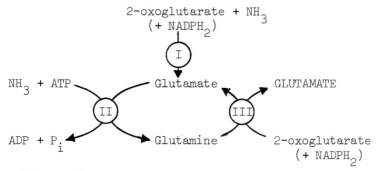

Overall reactions:
(I)   Glutamate dehydrogenase:
  NH_3 + 2-oxoglutarate + $NADPH_2$ → Glutamate + NADP + H_2O
(II) Glutamine synthetase:
  NH_3 + glutamate + ATP → Glutamine + ADP + P_i
(III) Glutamate synthetase (GOGAT):
  Glutamine + 2-oxoglutarate + $NADPH_2$ → 2 Glutamate + NADP
Sum of reactions II and III:
  NH_3 + 2-oxoglutarate + ATP + $NADPH_2$
                     → Glutamate + ADP + P_i + NADP

The relationship between this "glutamine pathway" of glutamate synthesis and that involving glutamate dehydrogenase also is clearly evident in this figure. Overall, the two reactions are almost identical except that one (ie, that functioning at low ammonia levels) involves the

171

participation of ATP. Presumably this expenditure of
energy is essential in order to assimilate ammonia under
conditions where it is present intracellularly in low
concentrations. In this connexion, it might be significant
that the "glutamine pathway" of glutamic acid synthesis is
virtually absent from glucose-limited A. aerogenes
organisms in which ammonia is present in considerable excess
of requirement (Table 1) but the supply of energy is
severely restricted.

As mentioned previously, the presence of glutamate
synthetase (GOGAT) in extracts of ammonia-limited
A. aerogenes could be detected by the formation of glutamic
acid when mixtures containing the enzyme plus glutamine
(5 mM), 2-oxoglutarate (5 mM) and NADPH$_2$ (0.25 mM) were
incubated (37°C; 50 mM Tris-HCl, pH 7.6) for varying
periods of time. However, this procedure was much too
cumbersome for routine usage, and therefore the glutamine-
dependent oxidation of NADPH$_2$ (which could be followed
spectrophotometrically at 340 nm) was made the basis of
subsequent enzyme assays.

Properties of Glutamate Synthetase (GOGAT)

As already stated, no net synthesis of amino acid
occurred when either pyruvate or oxaloacetate replaced
2-oxoglutarate in the GOGAT assay system. Similarly (with
the enzyme from A. aerogenes, at least) there appeared to
be a near-absolute requirement for glutamine and neither
asparagine, ammonia nor urea was active as an amino-donor
molecule. Almost identical results were obtained with
Klebsiella pneumoniae (Nagatani et al., 1971) and with
various marine pseudomonads (Brown et al., 1972); invariably glutamine and 2-oxoglutarate were required, but the
pyridine nucleotide specificity differed from organism to
organism (see Nagatani et al., 1971; Brown et al., 1972).
However, no organism has been found to possess both the
NAD- and NADP-linked enzymes.

Although glutamate synthetase (GOGAT) and glutamate
dehydrogenase are functionally similar (effecting synthesis of glutamate from 2-oxoglutarate by an oxida-
reduction reaction), they nevertheless are very different
enzymes. Apart from having different substrate requirements
(glutamine versus ammonia, as the amino-donor molecule) the
GOGAT-mediated reaction is virtually irreversible and
completely inhibited by the glutamine analogue, 6-diazo-5-

172

oxo-L-norleucine (DON). This analogue was found to have no
effect whatsoever on the glutamate dehydrogenase reaction,
when tested at a concentration that totally inhibited GOGAT
activity (Nagatani et al., 1971).

Using the above-mentioned spectrophotometric assay,
the influence of various parameters on the activity of
GOGAT from different organisms was investigated and
compared. Varying the concentration of either glutamine
or 2-oxoglutarate in the incubation mixture caused GOGAT
activity to change in a manner characteristic of a Michaelis
-Menten-type function. With the enzyme from ammonia-
limited A. aerogenes, the concentrations of glutamine and
2-oxoglutarate supporting half-maximal activity were 1.8
and 2.0 mM, respectively (Table 2). The corresponding Km
values of enzymes extracted from other Gram-negative
bacteria, and from Gram-positive organisms, also are shown
in this Table; invariably they were in the range of 0.1 to
2.0 mM.

TABLE 2

Apparent Km Values for Glutamine and 2-Oxoglutarate
of GOGAT Extracted From Several Different Bacteria

| Organism | Km Values for: | |
	Glutamine (mM)	2-Oxoglutarate (mM)
Aerobacter aerogenes	1.8	2.0
Pseudomonas fluorescens	0.4	0.1
Erwinia carotovora	1.6	0.1
Bacillus subtilis W23	0.2	0.04
B. subtilis var. niger	1.7	0.1
B. megaterium	0.3	0.04

Hydrogen-ion concentration also had a marked effect on
the GOGAT enzymes from both Gram-positive and Gram-negative
bacteria (Fig. 3). Invariably an optimum activity was
expressed at an alkaline pH value (in the region 7.5 - 8.5)
just as with the glutamate dehydrogenase reaction.

Of the other low molecular weight substances known to
be present intracellularly in substantial amounts in
growing bacteria, potassium and phosphate had little effect
on GOGAT activity (except at very high concentrations), but
magnesium was a potent inhibitor (Fig. 4).

Figure 3. Influence of pH on the activity of glutamate synthetase (GOGAT) extracted from (a) Gram-negative bacteria - <u>Aerobacter aerogenes</u> (O) and <u>Erwinia carotovora</u> (●), and (b) Gram-positive bacteria - <u>Bacillus subtilis</u> var. <u>niger</u> (Δ) and <u>B. subtilis</u> W23 (▲).

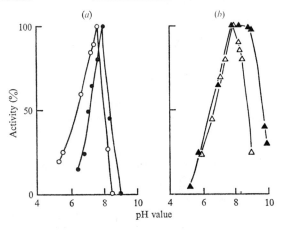

Figure 4. Influence of various concentrations of sodium chloride (O), potassium chloride (●) and magnesium chloride (Δ) on GOGAT extracted from <u>Aerobacter aerogenes</u>

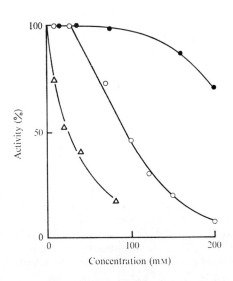

Regulation of Glutamate Synthetase Synthesis and Activity

With ammonia-limited and glucose-limited cultures of
Aerobacter aerogenes a reciprocal relationship seemingly
existed between the cellular contents of glutamate
synthetase (GOGAT) and glutamate dehydrogenase. Thus,
conditions that favoured the synthesis of GOGAT caused
suppression of glutamate dehydrogenase synthesis, and vice-
versa (Table 1). In fact, with all the bacterial species
so far examined, ammonia-limitation invariably caused
repression of glutamate dehydrogenase synthesis (in those
organisms that otherwise produced this enzyme) and
promoted synthesis of GOGAT. Some organisms (particularly
Erwinia carotovora, but also Bacillus subtilis W23 and
Bacillus megaterium; see Elmerich and Aubert, 1971, 1972)
lacked glutamate dehydrogenase but still could grow readily
in a simple salts medium in which ammonia provided the sole
source of utilizable nitrogen; in these cases GOGAT was
synthesized constitutively (Table 3).

Figure 5.   Influence of L-glutamate concentration on the
activities of GOGAT and glutamate dehydrogenase extracted
from different bacteria. (a) Glutamate dehydrogenase (▲)
and GOGAT (●) from Aerobacter aerogenes (glucose- and
ammonia-limited, respectively) and from Erwinia carotovora
(O). (b) Glutamate dehydrogenase (▲) and GOGAT (◉) from
Bacillus subtilis var. niger (glucose- and ammonia-limited,
respectively) and GOGAT from ammonia-limited Bacillus
subtilis W23 (O).

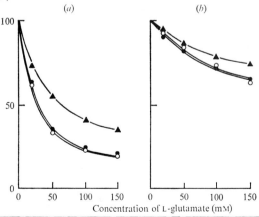

As with glutamate dehydrogenase, synthesis of GOGAT could be repressed by adding either casein hydrolysate (1%, w/v) or L-glutamate (50 mM) to the growth medium (Table 1). Inexplicably, even when exogenous glutamate provided the sole source of utilizable nitrogen, and availability of nitrogen was made the growth-limiting component of the medium, synthesis of these enzymes still was severely repressed. Under these N(glutamate)-limiting conditions the intracellular concentrations of ammonia and free glutamate were no greater than in ammonia-limited organisms (Table 1) and therefore neither of these compounds could be acting directly as repressors. Indeed, in support of this conclusion Meers et al. (1970) found that when the intracellular free glutamate concentration of A. aerogenes organisms (either ammonia-limited or glucose-limited) was increased several-fold, by adding NaCl to the growth environment, synthesis of neither GOGAT nor glutamate dehydrogenase was excessively repressed. Clearly, the nature of the mechanisms regulating synthesis of these enzymes is complex.

Although the intracellular concentration of free glutamate did not markedly affect the syntheses of GOGAT and glutamate dehydrogenase, glutamate did inhibit the activities of both of these enzymes in vitro (Fig. 5). In this respect there were obvious differences between Gram-positive and Gram-negative bacteria; enzymes isolated from the former were much less sensitive to end-product inhibition, and this correlated with them having a much greater pool content of free glutamate, than had Gram-negative organisms, when grown in a similar simple salts medium (see Tempest et al., 1970b).

The Distribution of Glutamate Synthetase (GOGAT) and
Glutamate Dehydrogenase in Bacteria Grown in Different
Media

Using the chemostat to culture organisms in environments that contained, alternately, high and low concentrations of ammonia, a study was made of the content and distribution of glutamate synthetase (GOGAT) and glutamate dehydrogenase in several species of bacteria (Aerobacter aerogenes, Erwinia carotovora, Pseudomonas fluorescens, Bacillus subtilis, B. megaterium) and yeasts (Candida utilis, Saccharomyces cerevisiae). Also various tissues (liver, brain) from rat, mouse and guinea-pig were

176

examined. It was found that whereas glutamate dehydrogenase was widely distributed, the glutamine pathway of
glutamic acid synthesis was only evident in prokaryotic
organisms; invariably, though not exclusively, in ammonia-
limited bacteria (Table 3). Yeast cells showed absolutely
no GOGAT activity, irrespective of the growth environment.
Further, their content of glutamate dehydrogenase did not
diminish when the organisms were grown in an ammonia-
limited environment. In this connexion, it might be significant that whereas bacterial glutamate dehydrogenases generally have a low affinity for ammonia (Km >5 mM; see
Sanwal and Zink, 1961; Wiame et al., 1962), those from
yeast organisms seemingly have much higher affinities (Km
< 1 mM for glutamate dehydrogenases from C. utilis and
S. cerevisiae).

Some bacteria were found that either totally lacked
glutamate dehydrogenase, or possessed only a small amount
of this enzyme when grown in the presence of excess ammonia.
In each of these cases, GOGAT, and therefore, presumably,
the glutamine pathway of glutamic acid synthesis, could be
detected (Table 3). In fact, at the time of writing it is
true to state that every prokaryotic organism which has
been examined so far, and which can grow in a simple salts
medium, has been found to possess the glutamine pathway of
ammonia assimilation. And this includes nitrogen-fixing
organisms (Nagatani et al., 1971) and several species of
blue-green algae (C.M. Brown, unpublished observation) as
well as anaerobes (Dainty, 1972), psychrophilic bacteria
(Brown et al., 1972; Brown and Stanley, 1972) and thermo-
philic bacteria (D.W. Tempest, unpublished observation).

With nitrogen-fixing organisms, ammonia is the primary
product of the fixation process and yet this metabolite is
a potent repressor of nitrogenase synthesis (Hill et al.,
1972). Clearly, then, for organisms to continue to fix
nitrogen, some mechanism must exist for the efficient
removal of ammonia as soon as it is formed. Recently,
Nagatani et al. (1971) have shown that nitrogen-fixing
organisms possess GOGAT, and the glutamine pathway of
ammonia assimilation; this, presumably, functions primarily
to mop up ammonia immediately it is formed.

Not surprisingly, mutants of nitrogen fixing organisms
that lacked GOGAT would not grow in media containing
molecular nitrogen as the sole nitrogen source (Nagatani
et al., 1971). Presumably these mutants would first
accumulate glutamine, until all the endogenous glutamate

TABLE 3

The Distribution of Glutamate Synthetase (GOGAT)
and Glutamate Dehydrogenase in Bacteria and Yeast
Organisms Growing in Different Media

Organism	Growth Conditions	Specific Activity*	
		Glutamate Synthetase	Glutamate Dehydrogenase
A. aerogenes	Glucose-limited	<1	560
	NH_3-limited	66	19
	N(glutamate)-limited	<1	<1
	Phosphate-limited	1	600
	P-limited + glutamate	1	10
E. carotovora	Glucose-limited	50	<1
	NH_3-limited	26	<1
	N(glutamate)-limited	1	<1
	Phosphate-limited	25	<1
	P-limited + glutamate	<1	<1
Ps. fluorescens	Glucose-limited	77	480
	Ammonia-limited	64	10
B. subtilis var. niger	Glucose-limited	<1	800
	Ammonia-limited	140	30
B. subtilis W23	Magnesium-limited	78	2
	Ammonia-limited	43	5
B. megaterium	Glucose-limited	750	10
	Ammonia-limited	140	10
	Batch-grown†	120	—
	Batch-grown + glutamate†	<1	—
C. utilis	Glucose-limited	<1	2050
	Ammonia-limited	<1	2100
S. cerevisiae‡	Glucose-limited	<1	720
	Ammonia-limited	<1	1040

*Values determined under optimum conditions; expressed as
nmoles $NADPH_2$ oxidised/min/mg protein.

†Data of Elmerich and Aubert (1971)

‡Thirty other species of yeast were examined; none contained a detectable glutamate synthetase.

was depleted, and then ammonia. But ammonia would switch off nitrogenase synthesis, and hence nitrogen fixation, and growth would cease. In agreement with this conclusion, it was observed that these mutant organisms were also unable to utilize many other inorganic and organic nitrogen sources - principally those which normally were metabolized via ammonia (eg, nitrate, nitrite and urea).

In marine environments many parameters may modulate the growth of bacteria (eg, low temperatures and high concentrations of salt) but since these environments often contain only low concentrations of inorganic and organic nitrogen, growth will most frequently be modulated by the supply of these metabolites. The ability of marine organisms to fix atmospheric nitrogen is probably not widespread (although some species may do so; see Kriss, 1963). However, the ability of organisms to utilize nitrate is widespread and this ion is probably the principal source of usable nitrogen in marine environments (being present in concentrations ranging from 1-500 µg/ml).

Preliminary experiments with a marine bacterium (Pseudomonas sp., strain PL-1) showed that nitrate-grown cells contained an NADP-linked glutamate synthetase (GOGAT). As with non-marine organisms, no activity was recorded when either ammonia or asparagine replaced glutamine in the assay system, nor when pyruvate or oxaloacetate replaced 2-oxoglutarate. The temperature optimum for this enzyme was $25^{\circ}C$, the pH optimum 7.6, and there was no pronounced requirement for NaCl (Brown et al., 1972). Extracts of a further eleven strains of marine bacteria (all Gram-negative pseudomonas-like organisms) possessed substantial amounts of GOGAT when the organisms had been grown on nitrate as the sole nitrogen source (Table 4). All the strains tested (except ET-4 and PL-6) contained the NADP-linked enzyme and had a high affinity for glutamine and 2-oxoglutarate (Km values in the range 1-5 mM). Further, all these organisms synthesised an active glutamine synthetase when grown on nitrate, and therefore possessed a functional "glutamine pathway" of glutamic acid synthesis.

With the exception of strains ET-7 and PL-6, all these marine organisms contained an NAD-linked glutamate dehydro-genase (Table 4). However, invariably the Km for ammonia of this enzyme was in excess of 10 mM and it seems unlikely, therefore, that it could contribute substantially to nitrog-en assimilation when these organisms were growing in a low-

179

nitrogen environment. It seems reasonable to conclude, then, that in natural marine environments bacteria assimilate nitrate (and ammonia) via the "glutamine pathway" of glutamate synthesis.

Conclusion

Glutamate occupies a central position in bacterial amino acid metabolism, acting as a reservoir of α-amino nitrogen which can be fed directly into the synthesis of other amino acids and their derivatives. Thus, control of glutamate synthesis is fundamental to the control of amino acid synthesis generally and, thereby, to the control of protein synthesis and growth. Glutamate can be formed directly from 2-oxoglutarate and ammonia by glutamate dehydrogenase and, in bacteria possessing this enzyme, no other mechanism of glutamate synthesis would seem to be necessary. However, bearing in mind that bacterial glutamate dehydrogenases seemingly require very high concentrations of ammonia to function optimally, they clearly are of little operational value when either high concentrations of ammonia preclude the operation of some fundamental physiological process (eg, nitrogenase synthesis in nitrogen-fixing organisms) or where the environment contains only low concentrations of ammonia (eg, many natural environments). Under these N-limiting conditions bacteria generally synthesise an active glutamine synthetase (see Woolfolk et al., 1966; Meister, 1968; Wu and Yuan, 1968; Pateman, 1969; Shapiro and Stadtman, 1970; Meers and Tempest, 1970) which, as pointed out by Umbarger (1969), is better equipped to scavenge traces of ammonia from the environment than is glutamate dehydrogenase. But since the glutamine synthetase reaction requires an amino acid (glutamate) to synthesize an amino acid (glutamine) it does not in itself provide a functional alternative to glutamate dehydrogenase. When coupled with the glutamate synthetase (GOGAT) reaction, however, glutamine synthetase provides not only this alternative, but a highly efficient process for the assimilation of ammonia into α-amino nitrogen (ie, glutamic acid). And the ubiquitous distribution of this "glutamine pathway" of ammonia assimilation surely must attest to its fundamental role in the growth of organisms in natural ecosystems.

The apparent absence of the "glutamine pathway" of ammonia assimilation from all the eukaryotic organisms so

TABLE 4

Enzyme Contents of Several Marine Bacteria, Grown
in Batch Culture with Nitrate as the Sole Source of
Usable Nitrogen

Strain No. of Organism[†]	Glutamate Synthetase			Glutamate Dehydrogenase		
	Coenzyme	Km[*] Glut- amine	Content[°]	Coenzyme	Km NH$_3$	Content
AT-1	NADP	1.0	56	NAD	25	50
AT-2	NADP	1.0	74	NAD	10	39
AT-13	NADP	1.6	56	NAD	22	72
PL-1	NADP	3.3	113	NAD	25	118
PL-3	NADP	2.5	59	NAD	28	74
PL-6	NAD	1.5	26	Not detected		
SW-2	NADP	1.3	131	NAD	24	144
ET-1	NADP	2.5	63	NAD	17	89
ET-4	NAD	5.0	12	NAD	30	5
ET-6	NADP	0.6	121	NAD	10	12
ET-7	NADP	5.0	161	Not detected		
ET-8	NADP	3.0	27	NAD	25	63

[†]All strains were Gram-negative pseudomonas-like organisms
(see Brown et al., 1972)
[*]Km values are mM concentrations
[°]Enzyme activities are expressed as nmoles pyridine nucleotide oxidised/min/mg protein

far examined is most surprising, and totally inexplicable
(by us, at least!). Whether or not this is connected with
the greater degree of subcellular organization possessed by
these eukaryotic organisms one cannot adduce, but it is
worth mentioning that no eukaryotic organism yet studied
has been found to possess the ability to fix atmospheric
nitrogen. The two processes (ie, nitrogen fixation and
the glutamine pathway of ammonia assimilation) are clearly
related functionally; it is not unreasonable to suppose
that they may possibly share some common evolutionary
history.

References

Brown, C.M. and Stanley, S.O. (1972). *J. appl. Chem. Biotechnol.* 22, 363.

Brown, C.M., Macdonald-Brown, D.S. and Stanley, S.O. *J. Marine Biological Assn* (1972). In press.

Dainty, R.H. (1972). *Biochem. J.* 126, 1055.

Elmerich, C. and Aubert, J-P. (1971). *Biochem. Biophys. Res. Comm.* 42, 371.

Elmerich, C. and Aubert, J-P. (1972). *Biochem. Biophys. Res. Comm.* 46, 892.

Freese, E., Park, S.W. and Cashel, M. (1964). *Proc. Nat. Acad. Sc.* 51, 1164.

Hill, S., Drozd, J.W. and Postgate, J.R. (1972). *J. Appl. Chem. Biotechnol.* 22, 541

Kriss, A.E. (1963). "*Marine Microbiology (Deep Sea)*"Trans. by J.M. Shewan and Z. Kabata. London: Oliver & Boyd.

Meers, J.L. and Tempest, D.W. (1970). *Biochem. J.* 119, 603.

Meers, J.L., Tempest, D.W. and Brown, C.M. (1970). *J. Gen. Microbiol.* 64, 187.

Meister, A. (1968). *Adv. Enzymol.* 31, 183.

Nagatani, H., Shimizu, M. and Valentine, R.C. (1971). *Arch. Mikrobiol.* 79, 164

Pateman, J.A. (1969). *Biochem. J.* 115, 769.

Rebello, J.L. and Strauss, N. (1969). *J. Bact.* 98, 683.

Sanwal, B.D. and Zink, M.W. (1961). *Arch. Biochem. Biophys.* 94, 430.

Shapiro, B.M. and Stadtman, E.R. (1970). *Ann. Rev. Microbiol.* 24, 501.

Shen, S.C., Hong, M.M. and Braunstein, A.E. (1959). *Biochim. Biophys. Acta,* 36, 290

Tempest, D.W., Meers, J.L. and Brown, C.M. (1970a). *Biochem. J.* 117, 405

Tempest, D.W., Meers, J.L. and Brown, C.M. (1970b). *J. Gen. Microbiol.* 64, 171.

Umbarger, H.E. (1969). *A. Rev. Biochem.* 38, 323.

Wiame, J.M., Pierard, A. and Ramos, F. (1962). In *Methods in Enzymology*, 5, 673. New York: Academic Press Inc.

Woolfolk, C.A., Shapiro, B. and Stadtman, E.R. (1966). *Arch. Biochem. Biophys.* 116, 177.

Wu, C. and Yuan, L.H. (1968). *J. Gen. Microbiol.* 51, 57.

GLUTAMATE SYNTHASE FROM ESCHERICHIA COLI: AN IRON-SULFIDE FLAVOPROTEIN

Richard E. Miller*

Laboratory of Biochemistry
National Heart and Lung Institute
National Institutes of Health
Bethesda, Maryland 20014

Abstract

Glutamate synthase catalyzes the NADPH-dependent con-version of 2-ketoglutarate and L-glutamine to glutamate (Tempest et al., 1970). Coupled with glutamine synthetase and transaminases, glutamate synthase provides a previously unrecognized exergonic pathway for the assimilation of ammonia to form amino acids. Levels of the enzyme are similar in E. coli grown on 4 mM or 100 mM NH_4Cl but sig-nificantly lower levels of enzyme are produced when gluta-mate replaces NH_4Cl as the source of nitrogen. Glutamate synthase purified to homogeneity from E. coli contains 7.8 moles of flavin, 38.4 moles of iron, and 30.4 moles of labile sulfide per 800,000 g of protein. The purified enzyme sediments as a single symmetrical boundary ($s_{20,w}$ = 20S) in the analytical ultracentrifuge. A molecular weight of 800,000 was determined by sedimentation equilibrium and confirmed by gel filtration. The absorption spectrum of of the enzyme has maxima at 278 nm, 380 nm and 440 nm. Dithionite reduced enzyme is approximately 60% reoxidized by 2-ketoglutarate + L-glutamine. Absortpion and electron paramagnetic reasonance spectra of NADPH reduced enzyme suggest the formation of a stable flavin semiquinone inter-mediate. The purified protein migrates as a single com-

*Present Address: Division of Metabolic Disease, Depart-ment of Medicine, University of California at San Diego, La Jolla, California 92034

ponent on polyacrylamide gel electrophoresis at pH 7.2 and
pH 8.5. Enzyme treated with sodium dodecyl sulfate, urea
or guanidine migrates as two components (mol. wt. 53,000
and 135,000) on SDS polyacrylamide gel electrophoresis,
suggesting that glutamate synthase is composed of at least
two types of dissimilar subunits. Apparent Km's for NADPH,
2-ketoglutarate and L-glutamine are 7.7 μM, 7.3 μM and
250 μM, respectively. The pH optimum is 7.6. D- and
L-aspartate, L-methionine, D-glutamate, and NADP are
among the most potent inhibitors of enzyme activity.

INTRODUCTION

Recently Tempest and coworkers (1970) identified a
previously unknown pathway for glutamate biosynthesis in
Aerobacter aerogenes. Since then existence of the pathway,
in a large number of microorganisms has been established
(Meers et al., 1970; Nagatani et al., 1971). One of the
reactions (reaction 1) of this pathway involving the reduced pyridine nucleotide dependent synthesis of glutamate
from 2-ketoglutarate and L-glutamine is catalyzed by
glutamate synthase [L-glutamate : NADP oxidoreductase
(deaminating, glutamine forming), EC 1.4.1.X]. When
coupled with the reactions catalyzed by glutamine synthetase (reaction 2) and the various transaminases (reaction
3), the

$$\text{2-ketoglutarate} + \text{L-glutamine} + \text{NADPH} \longrightarrow \text{2 glutamate} + \text{NADP} \tag{1}$$

$$\text{L-glutamate} + \text{ATP} + \text{NH}_4^+ \longrightarrow \text{L-glutamine} + \text{ADP} + \text{Pi} \tag{2}$$

$$\text{L-glutamate} + \text{RCOCOOH} \longrightarrow \text{RCHNH}_2\text{COOH} + \text{2-ketoglutarate} \tag{3}$$

$$\text{Sum: RCOCOOH} + \text{ATP} + \text{NADPH} + \text{NH}_4^+ \longrightarrow \text{RCHNH}_2\text{COOH} + \text{ADP} + \text{Pi} + \text{NAD} \tag{4}$$

glutamate synthase catalyzed reaction provides a previously
unrecognized ATP-dependent (essentially irreversible) pathway for the assimilation of ammonia to form the various
amino acids (reaction 4). Due to the high affinity of

184

glutamine synthetase for ammonia (Km $\lessgtr 0.2$ mM) (Denton and Ginsburg, 1970 ; Miller and Stadtman, 1972) this pathway has the capacity to function at levels of free ammonia far below those necessary for the production of glutamate by glutamate dehydrogenase (Km for NH_4^+ = 1.5 to 3 mM) (Miller and Stadtman, 1972). The significance of this scheme in ammonia assimilation is demonstrated by the fact that when grown under conditions of nitrogen excess, A. aerogenes contains almost no glutamate synthase activity but has high glutamate dehydrogenase levels. However, when growth is limited by the availability of ammonia the glutamate synthase activity is relatively abundant while glutamate dehydrogenase activity is low (Tempest et al., 1970; Meers et al., 1970). In Bacillus megaterium, which lacks glutamate dehydrogenase, the glutamate synthase catalyzed reaction appears to provide the major if not the sole pathway for glutamate biosynthesis (Elmerich and Aubert, 1971). In both A. aerogenes and B. megaterium glutamate synthase is repressed when glutamate is the nitrogen source.

Thus, at least in ammonia starved microorganisms, the synthesis of glutamate from 2-ketoglutarate and L-glutamine may be considered to be the first step in a highly divergent pathway leading to the biosynthesis of a large number of metabolites.

Nomenclature

To avoid confusion it seems worthwhile to devote some space to a discussion of an appropriate name for the enzyme here referred to as glutamate synthase. The enzyme has been referred to by several names. Tempest et al., (1970) named the enzyme glutamine amide -2-oxoglutarate amino-transferase (oxidoreductase, NADP). Nagatani et al.,(1970) refer to it as glutamate synthetase and Elmerich (1972) and Miller and Stadtman (1972) used the name glutamate synthase. The question of nomenclature was referred to Dr. A. Braun-stein of the International Union of Biochemistry, Commission on Enzymes. Dr. Braunstein tentatively suggested the systematic name L-glutamate: $NADP^+$ oxidoreductase (deaminating, glutamine-forming) in compliance with the nomenclature rules 12, 15, and 16 of the 1964 edition of the Report of the Commission on Enzymes. He agreed with the use of the trivial name glutamate synthase since the enzyme catalyzes the synthesis of glutamate but it does not utilize ATP.

Bacterial Growth Experiments

Unlike B. megaterium, E. coli does not lack glutamate dehydrogenase and unlike A. aerogenes the E. coli levels of glutamate synthase and glutamate dehydrogenase activities are relatively unaffected by the concentration of ammonium salt in the growth medium. E. coli grown in batch culture in media initially containing 4 mM, 20 mM or 100 mM NH_4Cl were harvested in the late log phase of growth and the levels of glutamate synthase, glutamate dehydrogenase and glutamine synthetase were measured in sonic extracts (Miller and Stadtman, 1972). As shown in Table 1, the levels of glutamate synthase and glutamate dehydrogenase were relatively unaffected by the NH_4Cl concentration. However, the level of glutamine synthetase varied inversely with the concentration of NH_4Cl; the activity in cells grown in the presence of 4 mM NH_4Cl was 3.7 times greater than that in cells from media initially containing 100 mM NH_4Cl.

The possibility that glutamate synthase is repressed by glutamate is suggested by other experiments which show that growth in glutamate as the sole source of nitrogen leads to significantly lower levels of the enzyme than are found when high NH_4Cl is the source of nitrogen (Table I). In these experiments glutamate synthase in glutamate grown cells was only 27% as great as that found in high ammonia grown cells. A similar effect was observed for glutamate dehydrogenase. In contrast, the activity of glutamine synthetase was more than six times greater in glutamate grown cells than in cells grown on high ammonia.

It is apparent from these data that glutamate synthase and glutamate dehydrogenase levels vary only slightly over a range of ammonia concentrations which produces considerable variation in the level of glutamine synthetase. In addition, the data suggest that glutamate might repress the synthesis of glutamate synthase and glutamate dehydrogenase.

TABLE I

E. COLI BACTERIAL GROWTH EXPERIMENTS

Nitrogen Source			Relative[a] Specific Activity		
NH_4Cl	L-Glutamate		Glutamate synthase	Glutamate Dehydrog- enase	Glutamine Synthetase
mM	mM				
4	--		1.13	1.45	3.72
20	--		1.08	1.38	1.52
100	--		1.00	1.00	1.00
--	20		0.27	0.33	6.47

a. Relative specific activity for each growth condition was calculated by using an arbitrary value of unity for the specific activity of the enzyme from cells grown in media initially containing 100 mM NH_4Cl.

Enzyme Purification

Glutamate synthase was purified to homogeneity from E. coli utilizing streptomycin sulfate, ammonium sulfate, acetone and heat steps, followed by agarose gel filtra- tion and DEAE cellulose chromatography (Miller and Stadman, 1972). By these procedures a 480-fold purification was achieved with a 16% yield (Table II).

The elution profile of glutamate synthase from an 8% agarose column is shown in Fig. 1. Note that the peak activity occurs at an elution to void volume ratio of 1.13 (corresponding to a molecular weight of approximately 800,000), and that the absorbance due to flavin (A_{440}) correlates well with glutamate synthase activity. It is also noteworthy that glutamate dehydrogenase activity is eluted as a nearly homogeneous protein at an elution to void volume ratio of 1.48 corresponding to a molecular weight of approximately 200,000 (Miller, 1972).

Glutamate synthase eluted from the agarose column and

187

TABLE II
PURIFICATION OF GLUTAMATE SYNTHASE FROM E. COLI

Purification Step		Total Units units*	Specific Activity units/mg	Yield %
I	Crude Extract	24,218	0.055	100
II	Streptomycin SO_4	17,461	0.041	72
III	First $(NH_4)_2SO_4$ precipitation	17,760	0.362	73
IV	42 % acetone	13,499	1.10	56
V	Second $(NH_4)_2SO_4$ precipitation	11,119	1.91	46
VI	Heat (62°)	8,571	7.19	35
VII	Third $(NH_4)_2SO_4$ precipitation	7,986	11.1	33
VIII	Agarose (8%) Gel Filtration	5,565	20.7	23
IX	DEAE Cellulose Chromatography	3,889	26.2	16

*Activity was determined by measuring the initial rate of
NADPH oxidation. Assay mixtures (1.0 ml) contained 0.16mM
NADPH, 1.0 mM 2-ketoglutarate, 2.0 mM L-glutamine,
1.0 mM EDTA, 50 mM K^+-HEPES (N-2-hydroxyethylpiperazine-
N'-2-ethansulfonic acid) (pH 7.5) and sufficient enzyme to
produce an absorbancy change at 340 nm of 0.05 to 0.35 per
minute at 37°. Units, μmoles of NADPH oxidized per minute.

chromatographed on DEAE cellulose was homogeneous according
to the following criteria: It migrated as a single protein
band on polyacrylamide gel electrophoresis at pH 7.2 and
pH 8.5 and on cellulose acetate electrophoresis at pH 8.4;
it sedimented as a single nearly symmetrical boundary in
the analytical ultracentrifuge, $s_{20,w} = 20S$; sedimentation
equilibrium data yielded nearly linear plots of the logarithm of protein concentration against the square of the
radial position in the ultracentrifuge cell.

Stability of the purified enzyme was increased in the
presence of 2-ketoglutarate and 2-mercaptoethanol (Miller
and Stadtman, 1972). Preincubation of the enzyme in the
presence of L-glutamine or L-glutamate resulted in marked
time and temperature dependent losses in activity which

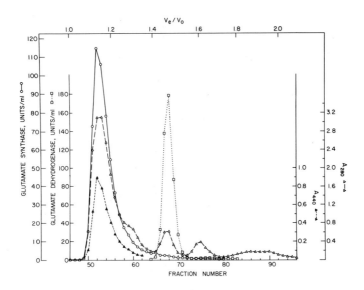

Figure 1: Elution profile of E. coli glutamate synthase from an 8% agarose column equilibrated and eluted with 10 mM imidazole-HCl (pH 7.0), 2 mM 2-ketoglutarate, 1 mM EDTA, 5 mM 2-mercaptoethanol, and 100 mM KCl. Sample: 715 mg of partially purified enzyme in 14.8 ml of the above buffer. Column: two tandem columns 4 x 112 cm. Flow rate: 1.5 ml/cm^2/hr. Fractions: 19.2 ml. Activity was determined as described in the legend to Table II. Ve/Vo, the ratio of elution volume to void volume. (From Miller, 1972)

were partially prevented by 2-ketoglutarate. Purified enzyme stored at -80° in the presence of 2-ketoglutarate and EDTA retained full activity for more than one year.

Hydrodynamic Properties of Purified Glutamate Synthase

The purified enzyme sedimented as a single, nearly symmetrical boundary in the analytical ultracentrifuge (Miller and Stadtman, 1972). Photoelectric scanner records taken at 280 nm showed slight boundary assymmetry compatible with the presence of more slowly sedimenting species. Using a partial specific volume ($\bar{v}$) of 0.73 calculated

RICHARD E. MILLER

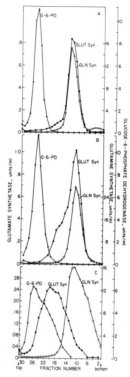

Figure 2: Sucrose density gradient sedimentation analysis
of glutamate synthase activity performed as described by
Martin and Ames (1961). Samples in 0.1 ml of buffer lay-
ered on 5 to 20% sucrose gradients were centrifuged in a
L2-65B Beckman ultracentrifuge at 60,000 rpm (4°) in an
SW 65K rotor for 3 hrs (A) or 3 1/2 hrs (B and C). Sample
and gradient were prepared in buffer containing 20 mM
K$^+$-HEPES, (pH 7.2), 2 mM 2-ketoglutarate, 1 mM EDTA, 11 mM
MgCl$_2$, 5 mM 2-mercaptoethanol and 100 mM KCl. Samples
contained: (A) 87.3 µg purified glutamate synthase, 81.3 µg
E. coli glutamine synthetase, and 20.3 µg yeast glucose-6-
phosphate dehydrogenase; (B) the components in A in the
following amounts 83 µg, 156 µg, and 18.4 µg, respectively,
plus 625 µg of partially purified glutamine synthetase
adenylylating enzyme: (C) 4.54 mg protein from a dialyzed
30 to 45% (NH$_4$)$_2$SO$_4$ fraction of crude extract, 156 µg
E. coli glutamine synthase, and 18.4 µg yeast glucose-6-
phosphate dehydrogenase. (From Miller and Stadtman, 1972)

from amino acid analysis (Miller and Stadtman, 1972), $s_{20,w}$ of 20S was calculated for a 0.57 mg/ml enzyme solution.

The sedimentation coefficient of purified glutamate synthase and that for the enzyme activity in an $(NH_4)_2SO_4$ fraction of crude extract were determined using sucrose density gradient centrifugation. E. coli glutamine synthetase ($s_{20,w}$ = 20.3S) and yeast glucose-6-phosphate dehydrogenase ($s_{20,w}$ = 6.1S) were used as markers. The sedimentation coefficient of the purified enzyme observed in the analytical ultracentrifuge (20S) was confirmed (Fig. 2a). There was a small but significant more slowly sedimenting component (13S) in the purified enzyme. This component was not increased by preincubation in 10% sucrose for 48 hours. It did increase somewhat when total protein was increased (Fig. 2b). Enzyme activity in the $(NH_4)_2SO_4$ fraction of crude extract had a sedimentation coefficient of approximately 13S (11 to 15S) (Fig 2c) suggesting that the purified enzyme behaves as a polymer, perhaps a dimer or tetramer of the active species.

The molecular weight of purified glutamate synthase was estimated using sedimentation equilibrium techniques (Miller and Stadtman, 1972). Plots of the logarithm of recorder pen deflection (c) against the square of the distance from the center of rotation (r^2) were linear for scans taken at 440 nm. There was some concavity near the meniscus and the slope of the line was somewhat less (approximately 6%), for data from scans taken at shorter wavelength (Fig. 3). The explanation for this remains undetermined, however, baseline uncertainty at shorter wavelength, protein dissociation, the presence of an impurity or a combination of these may contribute. There was no significant dependence of apparent molecular weight upon protein concentration over the range of 0.40 mg/ml to 1.38 mg/ml. Using a partial specific volume of 0.73, a molecular weight of 800,000, was calculated. This value is in good agreement with the molecular weight estimated from the ratio of elution to void volume (1.13) calculated from the elution profile of glutamate synthase activity during 8% agarose gel filtration (Fig. 1).

191

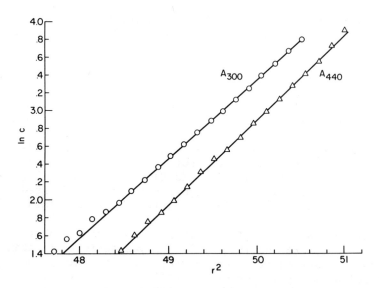

Figure 3: Sedimentation equilibrium analysis of purified glutamate synthetase using a Spinco Model E ultracentrifuge equipped with a photoelectric scanner. The logarithm of the recorder pen deflection (c) is plotted against the square of the distance from the center of rotation (r^2), evaluated after 72 hrs at 4400 rpm. The initial protein concentration was 1.30 mg/ml of buffer containing 10 mM imidazole-HCl (pH 7.0), 2 mM 2-ketoglutarate, 1 mM EDTA, 5 mM 2-mercaptoethanol and 100 mM KCl. The rotor temperature was maintained at 4°. Scans were taken at 300 nm (o–o) and 440 nm (△–△). (From Miller and Stadtman, 1972)

Absorption Spectrum

The absorption spectra of oxidized and non-enzymically reduced forms of purified glutamate synthase are shown in Fig. 4 (Miller and Stadtman, 1972). The spectrum of the native (oxidized) enzyme exhibits maxima at 278 nm, 380 nm and 440 nm and a broad low amplitude shoulder centered at 480 nm. Reduction of the enzyme with dithionite results in elimination of the absorption maximum at 440 nm. The spectral characteristics suggested that the enzyme is an iron containing flavoprotein.

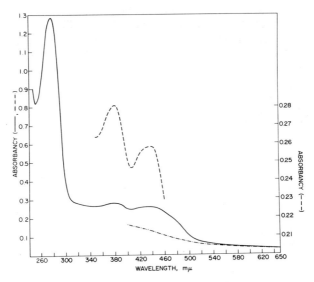

Figure 4: Absorption spectrum of purified glutamate synthase in the oxidized state (———, - - - - -) and after reduction with an excess of dithionite (·⁻·⁻··-). Sample: glutamate synthase, 1 mg/ml of 20 mM K⁺-HEPES (pH 7.2), 2 mM 2-ketoglutarate, 1 mM EDTA in a 1 cm pathlength curvette. Reference: dialysate. (From Miller and Stadtman, 1972)

Flavin, Iron and Labile Sulfide

Thin layer chromatography of the prosthetic group released from boiled glutamate synthase separated two flavin components with mobilities corresponding to those of authentic FAD and FMN. Quantitative estimation of these flavin nucleotides was performed by the fluorometric procedure of Burch et al. (1948). From the fluorescence data it was calculated that the enzyme contains at least 6.3 moles of total flavin nucleotide per 800,000 g protein, and that the ratio of FAD to FMN is between 1.2 and 1.5. In addition, total flavin content was calculated from the difference in absorbance at 450 nm between oxidized and dithionite reduced holoenzyme. This calculation neglects any spectral contribution from the non-heme iron chromophore and therefore gives an upper limit for enzyme flavin content. The value so obtained was 9.2 moles of

193

TABLE III
PROPERTIES OF GLUTAMATE SYNTHASE

Sedimentation coefficient	$s_{20,w}$ = 20S
Molecular Weight	800,000 (for $\bar{v}$ = 0.73)
Flavin Content	7.8 moles/800,000 g
FAD : FMN Ratio	1.5
Iron Content	38.4 moles/800,000 g
Labile Sulfide Content	30.4 moles/800,000 g

flavin per 800,000 g of protein.

Glutamate synthase iron content determined by atomic absorption spectroscopy and labile sulfide content determined chemically (Fogo and Popowsky, 1949) were 38.4 and 30.4 equivalents per mole, respectively. Molybdenum was not detectable. Some of the properties of the enzyme are summarized in Table III.

Role of Flavin in the Glutamate Synthase Catalyzed Reaction

Evidence for the involvement of flavin in the glutamate synthase catalyzed reaction was obtained from studies of the capacity of nonenzymically reduced flavoprotein to serve as an electron donor in glutamate synthesis. Substrate quantities of purified enzyme were reduced with dithionite in an anaerobic curvette until a marked decrease in absorbancy at 440 nm was observed (Fig. 5). Addition of 2-ketoglutarate and L-glutamine cause a rapid increase in absorbance at 440 nm indicating partial (64%) reoxidation of the reduced enzyme. Further addition of substrate failed to cause any greater oxidation of the flavin but upon the introduction of air the original spectrum was observed. There was no reoxidation of the reduced enzyme by either substrate alone (not shown in Fig. 5) and the enzyme was fully active at the termination of the experiment. The reoxidation of the reduced flavoprotein by substrates only to the extent of 64% correlates with the FAD content (55 to 60%) of the flavin chromophore suggesting that the enzyme bound FMN might be catalytically inactive. When the natural electron donor, NADPH, was used instead of dithionite, the enzyme bound flavin was reduced only to the extent of 60% again suggesting that FAD rather than FMN might be the active cofactor for glutamate synthesis. It is noteworthy that this interpretation neglects

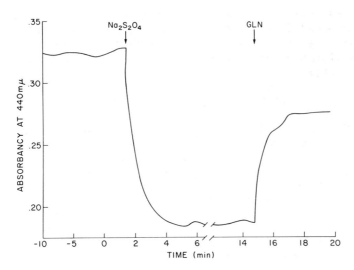

Figure 5: Reoxidation of non-enzymically reduced glutamate
synthase by substrates. 0.85 ml of purified enzyme
(1.24 mg/ml) in 20 mM K$^+$-HEPES, (pH 7.2), 2 mM 2-ketogluta-
rate, 1 mM EDTA, in an anaerobic 1 cm pathlength curvette,
was monitored continuously at 440 nm in a thermostated
(30°) spectrophotometer. Between 0 and 1.4 minutes 1 μl
aliquots of neutralized 100 mM Na$_2$S$_2$O$_4$ were sequentially
added and the A$_{440}$ was immediately recorded. At t =
1.4 min a total of 0.5 μmoles of Na$_2$S$_2$O$_4$ had been added.
At t = 14.8 minutes 1 μmole of L-glutamine was added to
the curvette and recording was resumed. (From Miller and
Stadtman, 1972)

the spectral contribution from the non-heme iron chromophore. The addition of NADH caused no reduction of the
flavoprotein

These experiments suggest that the glutamate synthase
catalyzed reaction is the sum of two partial reactions:

$$H^+ + NADPH + E \cdot Flavin \longrightarrow E \cdot flavin \cdot H_2 + NAD^+$$

$$E \cdot flavin \cdot H_2 + 2\text{-ketoglutarate} + L\text{-glutamine} \longrightarrow$$
$$2 \text{ glutamate} + E \cdot flavin$$

Sum: $NADPH + H^+ + 2\text{-ketoglutarate} + L\text{-glutamine} \longrightarrow$
$$2 \text{ glutamate} + NAD^+$$

195

The first step involves the NADPH dependent reduction of enzyme bound flavin, perhaps involving iron and sulfide. The second step involves the conversion of 2-ketoglutarate and L-glutamine to glutamate with regeneration of oxidized flavoprotein.

Flavin Semiquinone Intermediate

Partial reoxidation of dithionite-reduced enzyme by substrates and partial reduction of enzyme bound flavin by NADPH may reflect stable semiquinone formation rather than selective oxidation or reduction of FAD. Preliminary absorption and electron paramagnetic resonance (EPR) spectra performed on NADPH reduced enzyme support this possibility (Miller, 1972). Figure 6 shows absorption spectra of glutamate synthase under anaerobic conditions prior to any treatment, and also after addition of a 5-fold excess of NADPH over enzyme bound flavin, and after the addition of 17-fold excess of dithionite over flavin in addition to the NADPH. As shown in Fig. 6, there is incomplete reduction of enzyme bound flavin by NADPH as compared with dithionite. In addition, reduction of the flavoprotein with NADPH results in the development of a broad spectral absorption band centered at 630 nm which disappears upon complete reduction of the enzyme with dithionite. A long wavelength absorption band nearly identical to this was observed by Rajagopalan et al. (1962) upon reduction of dihydroorotate dehydrogenase and its development was correlated with EPR spectral evidence for the development of a flavin semiquinone.

Electron paramagnetic resonance spectra of glutamate synthase reduced with a 70-fold excess of NADPH over enzyme bound flavin demonstrate the appearance of an organic free radical signal (at g = 2.00) which is not present in the native (fully oxidized) enzyme (Fig. 7). Quantitation of the signal was performed using the organic free radical 2,2-diphenylpicrylhydrazyl as a standard and it was demonstrated that about 50% of the enzyme flavin was in the semiquinone oxidation state. These absorption and EPR spectral data suggest that the flavin prosthetic group of glutamate synthase might shuttle between the fully oxidized and the semiquinone states during catalysis as has been postulated for dihydroorotate dehydrogenase (Rajagopalan et al., 1962).

196

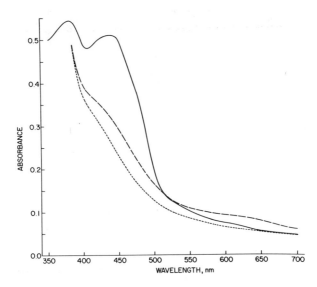

Figure 6: Absorption spectrum of purified glutamate synthase in the oxidized state (———) and after sequential
addition of NADPH (-----) followed by $Na_2S_2O_4$ (·····).
Sample: 0.8 ml of purified glutamate synthase (1.91 mg/
ml) in 20 mM of K^+-HEPES (pH 7.25), 2 mM 2-ketoglutarate,
1 mM EDTA in an anaerobic curvette. (———), spectrum
prior to additions; (-----), spectrum after the addition of
120 nmoles (10 μl) of NADPH; (·····), spectrum after the
addition of NADPH followed by the addition of 400 nmoles
(4 μl) of neutralized $Na_2S_2O_4$. (From Miller, 1972)

It is noteworthy that NADPH reduction of the enzyme
did not result in the development of an iron +2 EPR signal
(in the region of g = 1.94). Such a signal has been
observed upon reduction of other non-heme iron flavo-
proteins and its development has been taken as evidence
for the participation of iron in catalysis (Rajagopalan
et al., 1962). The failure to observe the Fe^{+2} signal
in glutamate synthase may simply reflect the conditions
used for EPR spectra. At another enzyme concentration or
in another buffer the signal may be demonstrable. The
possibility that non-heme iron is not involved in glutamate synthase catalysis must also be considered.

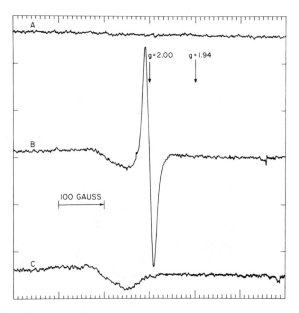

Figure 7: Electron paramagnetic resonance spectra of glutamate synthase. Instrument: Varian E-3 EPR spectrometer. Settings: time constant: 1.0 sec, scan time: 16 min, modulation amplitude 4G, modulation frequency: 100 KHz, receiver gain: $2x10^5$, temperature: liquid N_2, microwave power: 50mW, microwave frequency: 9.148. The sample (0.42 ml) contained 14.42 mg/ml purified glutamate synthase (0.14 mM bound flavin) 9.52 mM NADPH, 20 mM K-HEPES, pH 7.3, 2 mM 2-ketoglutarate, and 1 mM EDTA. Five minutes after mixing enzyme, buffer, and NADPH at 4° under nitrogen, sample tubes were sealed then frozen in liquid nitrogen. Spectra: A, enzyme omitted; B, no omissions; C, NADPH omitted. (From Miller, 1972)

Subunit Composition

Enzyme disaggregation studies suggest that glutamate synthase is composed of at least two types of dissimlar subunits. Purified glutamate synthase treated with 1% sodium dodecyl sulfate (SDS), 8 M urea, or 6 M guanidine-HCl migrates as two components (molecular weights 53,000 and 135,000) on polyacrylamide gel electrophoresis performed in the presence of 0.1% SDS. Similarly, enzyme preincubated in 8 M urea migrates as two components on disc

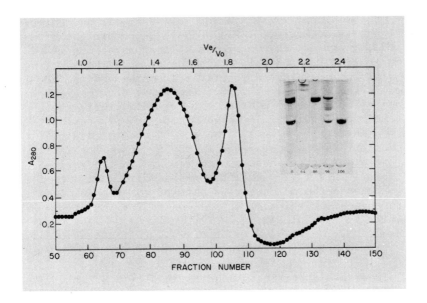

Figure 8: Elution profile of sodium dodecyl sulfate (SDS) disaggregated glutamate synthase from a 4% agarose column (91.5 x 1.6 cm). Sample: 33 mg of SDS disaggregated glutamate synthase in 1.4 ml after prior reduction with 2-mercaptoethanol and alkylation with iodoacetate. Buffer: 10 mM NaPO$_4$ (pH 7.1) 0.2% SDS. Flow rate: 4 ml/cm^2/hr. Fractions: 1.2 ml. Insert: SDS acrylamide gels performed on the original sample (o) and on selected fractions from the column. Gel number corresponds to fraction number.

gel electrophoresis in the presence of 8 M urea.

From spectrophotometric scans of Coomassie blue stained SDS polyacrylamide gels (assuming a proportionally between protein content and staining intensity) it can be calculated that the ratio between protein content in the high molecular weight band and that in the low molecular band is 2.58. This ratio is in good agreement with the ratio of the respective molecular weights (2.55) and it suggests that the two dissimilar subunits occur in the native protein in equimolar amounts.

The two types of dissimilar subunits have been separated using agarose gel filtration and their non-iden-

199

tity has been confirmed by amino acid analysis and N-terminal analysis (Miller, 1972). The elution profile of SDS disaggregated enzyme from a 4% agarose column is shown in Fig. 8. SDS polyacrylamide gels of the starting material and of selected fractions from the column are shown in the insert to Fig. 8. The first elution peak accounts for about 8% of the total protein eluted and may represent an impurity in the enzyme or non-dissaggregated enzyme (insert to Fig. 8). The second and third proteins eluted correlate with the two types of dissimilar subunits and the areas under these two curves are in the ratio of 2.9 to 1. This ratio is in reasonable agreement with the ratio between the molecular weights of the two species again suggesting a 1 to 1 stiochemetry for the dissimilar subunits in the holoenzyme.

Catalytic Parameters

Substrate specificity studies show that <u>E. coli</u> glutamate synthase requires L-glutamine, 2-ketoglutarate and NADPH for catalytic activity (Miller and Stadtman, 1972). In the absence of both 2-ketoglutarate and L-glutamine or in the presence of either of these substrates alone the purified enzyme had negligible NADPH oxidase activity. NADH did not support activity. Neither NH_4Cl, L-asparagine or D-glutamine substituted for L-glutamine nor did oxalacetate or pyruvate substitute of 2-ketoglutarate. When oxalacetate and L-asparagine were substituted for 2-ketoglutarate and L-glutamine, respectively, in the presence of either NADPH or NADH there was negligible activity.

Incubation of purified glutamate synthase with D- or L-glutamate and NADP resulted in no NADPH generation, suggesting that the glutamate synthase catalyzed reaction is essentially irreversible.

The activity pH profile of the purified enzyme was relatively broad with an optimum (pH 7.6) between pH 7.4 and 7.8. Substrate saturation kinetics were hyperbolic. The Km's for NADPH, 2-ketoglutarate, and L-glutamine were 7.7, 7.3 and 250 μM, respectively. It is noteworthy that due to the low Km's of glutamate synthase for substrates it has been possible to develope a rapid, convenient

200

assays for quantitation of as little as 20 nanomoles of
L-glutamine or 2-ketoglutarate (Miller, 1972).

The Km of E. coli glutamate synthase for 2-ketogluta-
rate (7.3 μM) and L-glutamine (250 μM) differ markedly
from those reported for the enzyme in A. aerogenes crude
extracts (Meers et al., 1970), but are comparable to those
reported for the partially purified enzyme from
B. megaterium (Elmerich, 1972). The pH optimum of
E. coli glutamate synthase is identical to that reported
for the enzyme in crude extracts of A. aerogenes (Meers
et al., 1970) and in the same range as that reported for
partially purified enzyme from B. megaterium (Elmerich,
1972). Whether these kinetic parameters reflect the
physiological milieu in the organism studied or simply the
assay conditions used in their determination remains to be
determined.

Inhibitors

The susceptability of glutamate synthase to feed
back inhibition by certain amino acids supports the hypothesis that the enzyme is of major importance in amino acid
biosynthesis. Among the most potent inhibitors of enzyme
activity are D- and L-aspartate, D-glutamate, and L-methionine. Less potent amino acid inhibitors in order of de-
creasing potency are L-cysteine, L-serine, glycine,
L-homoserine, L-glutamate, L-asparagine, L-alanine,
L-histidine and L-tryptophan. Amino acids producing no
significant inhibition of enzyme activity include
D-methionine, D-alanine, N-acetyl-glycine, L-valine,
L-leucine, L-isoleucine, DL-lysine, L-threonine, L-proline,
L-arginine, L-citrulline, and L-ornithine.

The mechanism of inhibition of five amino acids has
been studied in detail (Miller, 1972). Data from L-methionine studies are shown in Fig. 9. L-alanine, L-glutamate,
L-serine, L-methionine, and glycine are all classically
competitive with L-glutamine. L-glutamate is also compet-
tive with 2-ketoglutarate while there is no significant
inhibition by the other four amino acids when the 2-keto-
glutarate concentration is near the Km for that substrate
and NADPH and L-glutamine are at saturating levels.

NADP is a potent inhibitor of glutamate synthase while

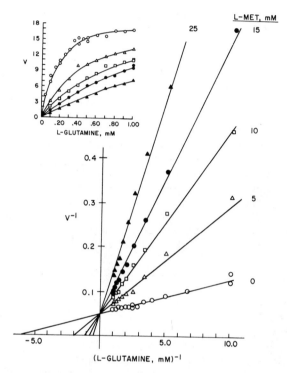

Figure 9: Inhibition of purified glutamate synthase by
L-methionine. Assay mixutres (1.0 ml) contained 50 mM
K^+-HEPES, 1.0 mM 2-ketoglutarate, 1.0 mM EDTA, 0.16 mM
NADPH and the indicated L-glutamine and L-methionine con-
centrations. Assays were performed at 30°. V, μmoles
NADPH oxidized per minute per mg protein.

NAD has no effect upon activity. Neither nucleoside
triphosphates nor tricarboxylic acid cycle intermediates
have significant effects upon catalytic activity.

The anions Cl^-, Br^-, and NO_3^- are modest inhibitors
of glutamate synthase whereas acetate, citrate and $SO_4^=$
have no significant effect upon activity. Sodium, potassium and ammonium ions are without effect. In contrast,
divalent cations including Ba^{2+}, Mg^{2+}, Ca^{2+}, Co^{2+}, Cd^{2+},
and Ni^{2+} produced significant inhibition of the enzyme.

The reason for inhibition by certain amino acids and

not by others is not known but may reflect assay conditions rather than a physiological phenomenon. It is noteworthy that inhibition of glutamate synthase by alanine, serine, glycine, histidine and tryptophan seems redundant since these amino acids also inhibit glutamine synthetase (Shapiro and Stadtman, 1970). Aspartate, methionine, asparagine and homoserine, on the other hand, are not reported to have an effect upon glutamine synthetase activity.

Conclusions

The physiological role of E. coli glutamate synthase in ammonia assimilation is probably similar to that proposed by Tempest et al. (1970) for the enzyme in A. aerogenes. Under conditions of severe ammonia starvation E. coli glutamate dehydorgenase (Km for NH_4^+ 1.5 to 3.0 mM) would be relatively inactive while glutamate synthase coupled with glutamine synthetase (Km for NH_4^+ < 0.2 mM) might provide the major if not the only route for ammonia assimilation. The data suggest that glutamate synthase activity might be regulated in vivo by end product feedback inhibition, anions and divalent cations, and repression of enzyme synthesis by glutamate.

The available data suggest that glutamate synthase might be composed of eight subunits, four of molecular weight 135,000 and four of mouecular weight 53,000. In addition, each molecule probably contains 32 iron atoms, 32 labile sulfide atoms and 8 noncovalently bound flavin molecules. That the purified enzyme (mol. wt. 800,000) might be composed of four catalytically active dimers is suggested by the fact that the enzyme activity in an ammonium sulfate fraction of crude extract sediments more slowly than purified enzyme (Fig. 3). Such dimers might be composed of one of each of the two types of non-identical subunits, eight iron atoms, eight labile sulfide atoms and two flavin molecules. No clue regarding the arrangement of flavin, iron and sulfide has yet been obtained, nor does the arrangement of components of the enzyme proposed satisfactorily accomodate unequal numbers of FAD and FMN molecules. The possibility that the ratio between FAD and FMN is unity in the native protein cannot be excluded since alterations in flavin composition of the

RICHARD E. MILLER

enzyme may occur during purification.

Acknowledgements

I am indebted to Dr. E R. Stadtman in whose laboratory this work was done for his guidance, encouragement, and support and for his valuable criticism of this manuscript.

References

Burch, H.B., Bessey, O.A. and Lowry, O. H. (1948) J. Biol. Chem. 175, 457.

Denton, M.D. and Ginsburg, A. (1970) Biochem. 9, 617.

Elmerich, C. (1972) European J. Biochem. 27: 216.

Elmerich, C. and Aubert, J. (1971) Biochem. Biophys. Res. Commun. 42, 371.

Fogo, J. K. and Popowsky, M. (1949) Anal. Chem. 21, 732.

Martin, R. G. and Ames, B. N. (1961) J. Biol. Chem. 236, 1372.

Meers, J. L., Tempest, D. W., and Brown, C. M. (1970) J. Gen. Microbiol. 64, 187.

Miller, R. E. (1972) in preparation.

Miller, R. E. and Stadtman, E. R. (1972) J. Biol. Chem. In press.

Nagatani, H., Shimizu, M., and Valentine, R. C. (1971) Arch. Mikrobiol. 79, 164.

Rajagopalan, K. V., Aleman, V., Heinen, W., Palmer, G., and Beinert, H. (1962) Biochem. Biophys. Res. Commun. 8, 220.

Shaprio, B. M. and Stadtman, E. R. (1970) Ann. Rev. Microbiol. 24, 501.

Tempest, D. W., Meers, J. L. and Brown, E. M. (1970) Biochem. J. 117, 405.

GLUTAMINE TRANSAMINASES FROM LIVER AND KIDNEY

A.J.L. Cooper and Alton Meister

Department of Biochemistry
Cornell University Medical College
New York, N.Y. 10021

ABSTRACT

Glutamine transaminase from rat liver has been purified to apparent homogeneity by the criteria of ultracentrifugation ($S_{20,w}$ 6.25 S), and polyacrylamide gel electrophoresis. The enzyme, which has a molecular weight of about 110,000 and consists of two subunits (mol. wt. $\sim$ 54,000), exhibits absorbance maxima at 278 and 415 nm. The enzyme is stabilized by addition of α-keto acids and by 2-mercaptoethanol. The specificity of the enzyme was examined in studies carried out with 27 α-keto acids and 40 amino acids. The most active α-keto acid substrates are α-ketoglutaramate, α-keto-γ-methiolbutyrate, β-mercaptopyruvate, glyoxylate, pyruvate, and β-hydroxypyruvate. The best amino acid substrates are glutamine, glutamic acid-γ-ethyl ester, γ-glutamylmethylamide, methionine, and ethionine. The equilibrium constants of several glutamine-α-keto acid transamination reactions were determined.

Liver glutamine transaminase also interacts with a number of γ-glutamylhydrazones of α-keto acids, e.g., the γ-glutamyl hydrazone of glyoxylate, to yield the corresponding L-amino acids and tetrahydro-6-pyridazinone-4-hydroxy-4-carboxylic acid. On acidification, the latter compound undergoes dehydration to yield 1,4,5,6-tetrahydro-6-pyridazinone-3-carboxylic acid.

Liver glutamine transaminase also utilizes several glutamine analogs including S-carbamyl-L-cysteine, O-carbamyl-L-serine, and L-albizziin. These compounds, when incubated with the enzyme and an α-keto acid, are converted to the corresponding α-keto acids, which undergo spontaneous cyclization to yield analogs of 5-hydroxy-2-pyrrolidone-5-carboxylate (the cyclic form of α-ketoglutaramate). Two of these compounds (4-hydroxy-2-thiazolidone-4-carboxylate and 4-hydroxy-2-imidazolidone-4-carboxylate) undergo facile dehydration in acid solution.

Glutamine transaminase has also been purified from rat kidney. This enzyme differs substantially from the liver enzyme with respect to substrate specificity. It is much more specific toward the amino donor; thus, γ-glutamylmethylamide, α-amino-γ-cyanobutyric acid and glutamine analogs in which the γ-carbon is replaced by a hetero atom,are not substrates. Only phenylalanine, tyrosine and methionine can replace glutamine. The kidney enzyme, which does not interact with the γ-glutamylhydrazones of α-keto acids, exhibits a broad specificity toward α-keto acids, but the specificity is different from that shown by the liver transaminase. Among the most active α-keto acids are phenylpyruvate, p-hydroxyphenylpyruvate, α-keto-γ-methiolbutyrate and α-ketocaproate; in contrast to the liver glutamine transaminase, glyoxylate is a poor substrate.

INTRODUCTION

Glutamine transaminase catalyzes the reversible transfer of the α-amino group of glutamine to an α-keto acid to form the corresponding L-amino acid and α-ketoglutaramate. The latter compound exists in solution in equilibrium with the corresponding cyclic ketolactam form (5-hydroxy-2-pyrrolidone-5-carboxylate). α-Ketoglutaramate is hydrolyzed to α-ketoglutarate and ammonia by α-keto acid ω-amidase, an enzyme present in rat liver and in other tissues. The combined action of glutamine transaminase and the ω-amidase explains the previously observed activation of glutaminase by α-keto acids ("glutaminase II") (Greenstein and Carter, 1946). Although early studies on "glutaminase II" were complicated by the use of enzyme preparations that contained several enzyme activities, the reaction sequence shown in Figure 1 was established (1) by experiments with ^{15}N-amide-labeled glutamine which showed that the ammonia formed is derived from the amide group of glutamine, and (2) by studies with γ-methyl-

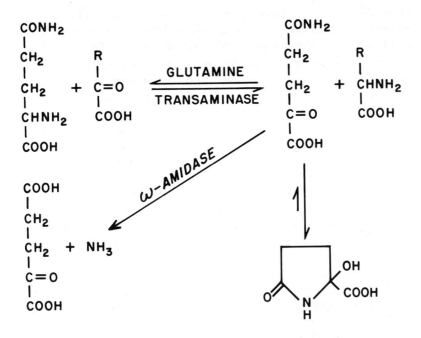

Fig. 1 Reactions catalyzed by glutamine transaminases and
ω -amidase.

glutamine, whose α-keto analog could be demonstrated as a product of the reaction since this α-keto acid is not a substrate of the
amidase (Meister and Tice, 1950; Meister, 1954). Early studies
on the specificity of the enzyme were not only complicated by the
presence of the ω-amidase, but also by the presence of other
transaminases such as glutamate-alanine transaminase, and glutamate-aspartate transaminase. Later, Braunstein and T'ing Sen
(1960) succeeded in obtaining an 80-fold purification of the
enzyme, which did not exhibit ω-amidase activity, but which did
exhibit glutamate-alanine transaminase. They raised the question
as to whether glutamate-alanine transaminase and glutamine transaminase might be identical. Yoshida (1967) subsequently succeeded
in purifying the liver enzyme about 200-fold, and obtained spectral
evidence for the presence of pyridoxal 5'-phosphate. This preparation did not exhibit glutamate-alanine transaminase activity. In
1971 we obtained a 900-fold purification of the enzyme, which
was free of ω-amidase, glutamate-aspartate transaminase and
glutamate-alanine transaminase activities. Although the reactions
catalyzed by glutamine transaminases are, like other transamination
reactions freely reversible and exhibit equilibrium constants not
far from unity, the rapid non-enzymatic cyclization of α-ketoglutaramate to 5-hydroxy-2-pyrrolidone-5-carboxylate effectively
drives the reaction in the direction of glutamine utilization.
Furthermore, under physiological conditions in the presence of α-
keto acid ω-amidase, the open-chain form of α-ketoglutaramate
produced initially by the transamination reaction undergoes enzyme
catalyzed hydrolysis rather than cyclization. The transamination
of glutamine is therefore essentially irreversible under physiological conditions and the action of this enzyme therefore leads to the
production of ammonia and to the formation of new amino acids
from the corresponding α-keto acids. Although it can be demonstrated that the enzyme can synthesize glutamine from its α-keto
analog, this reaction does not, in the absence of evidence for a
pathway for the synthesis of α-ketoglutaramate, offer an alternative route to glutamine.

[For literature references to the work of Greenstein and
co-workers on "glutaminase II", and the elucidation of this
system by Meister and co-workers, see Cooper and Meister (1972a)].

Physical Properties of Rat Liver Glutamine Transaminase

In the course of this work a number of methods were used
for the assay of glutamine transaminase activity; these involved the
determination of a variety of α-keto acids and α-amino acids. The
detailed experimental procedures employed are fully described
elsewhere (Cooper and Meister, 1972a).

Rat liver glutamine transaminase was isolated from the
soluble portion of rat liver homogenate. The enzyme was stabilized
by the addition of sodium pyruvate and 2-mercaptoethanol. The
isolation procedure involves a step in which the enzyme was
brought to a temperature of 63-65 degrees for twenty minutes in
the presence of 0.01 M sodium pyruvate. After cooling and centrifugation, the supernatant solution was fractionated with ammonium
sulfate and then chromatographed successively on columns of DE-
52 and hydroxylapatite. The details of the purification procedure
have been given (Cooper and Meister, 1972a).

The purified liver glutamine transaminase does not exhibit
glutamate-aspartate transaminase, glutamate-alanine transaminase,
or ω-amidase activities. The enzyme moves as a single band on
polyacrylamide gel electrophoresis, and it sediments as an apparently homogeneous component in the analytical ultracentrifuge; it
exhibited a sedimentation coefficient ($S_{20,w}$) of 6.25 S ($24°$; 5
mM potassium phosphate [pH 7.2]; protein, 1 mg per ml). An estimate of the molecular weight by gel filtration gave a value of
about 110,000; polyacrylamide gel electrophoresis in sodium
dodecyl sulfate gave a single protein component which exhibited
an estimated molecular weight of 54,000. The data thus indicate
that the isolated enzyme has a molecular weight of 110,000 and
that it probably consists of 2 subunits of molecular weight about
54,000.

The enzyme exhibits absorbance maxima at 278 and 415
nm (Fig. 2). The absorbance at 415 nm is consistent with the
presence of pyridoxal 5'-phosphate; however, conclusive evidence
is not yet at hand. The enzyme was not inactivated by dialysis

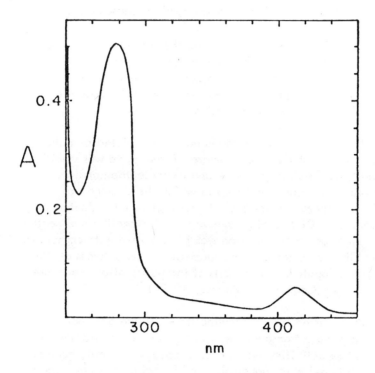

Fig. 2 Spectrum of purified rat liver glutamine transaminase (enzyme concentration, 0.79 mg/ml in potassium phosphate buffer; pH 7.2; 25°).

against solutions of 10 mM glutamine or alanine, or against a solution containing 20 mM cysteine and 4 M urea. At no time during the purification did addition of pyridoxal 5'-phosphate enhance the activity. However, the enzyme was inactivated by high levels of carbonyl reagents such as 100 mM isonicotinic acid hydrazide and by reduction with sodium borohydride. Prior incubation of the enzyme with glutamine prevents this inactivation. We could not demonstrate the presence of pyridoxal 5'-phosphate using the Wada and Snell (1963) phenylhydrazine method. However when we used 2,4-dinitrophenylhydrazine in place of phenylhydrazine a value of one mole per mole of enzyme subunit was obtained. These findings suggest that glutamine transaminase contains pyridoxal 5'-phosphate which is bound in an unusually tight form.

It should be noted from Table I that one of the best α-keto acid substrates is α-keto-γ-methiolbutyrate; however, considerable substrate inhibition occurs at concentrations greater than 5 mM. This phenomenon probably prevented the previous recognition of this α-keto acid as a good substrate for glutamine transaminase. We have also shown that the equilibrium constants are not far from unity for transamination between glutamine and the following α-keto acids: pyruvate, glyoxylate, and α-keto-γ-methiolbutyrate. The value of 0.3% for the concentration of the open chain form for α-ketoglutaramate at pH 7.0 (determined by Hersh, 1971) was used in this calculation.

Catalytic Properties of Rat Liver Glutamine Transaminase

Studies on the specificity of the liver enzyme with respect to α-keto acids are summarized in Table I. Although glyoxylate, pyruvate, β-hydroxypyruvate, and α-ketosuccinamate are active substrates, β-mercaptopyruvate, α-keto-γ-methiolbutyrate and several related α-keto acids exhibit very substantial activity. Of the amino acids, glutamine is the most active substrate; considerable activity, however, was observed with several glutamine analogs, methionine and several amino acids related to methionine (Table II). A summary of the data obtained on the specificity of glutamine transaminase is given in Table III. It is interesting that although the α-keto analogs of cysteine, glycine, asparagine, serine, alanine, α-aminobutyrate, and norleucine are fairly good substrates, the amino acids corresponding to these α-keto acids are much less active. The apparent K_m values for alanine and glycine are considerably greater than those for the other amino and α-keto acids studied. The data seem to show that α-amino and α-keto acid substrates possessing side chains of the type, $-CH_2CH_2X$, where $X = -CONH_2$, $-CONHCH_3$, $-CN$, $-COOR$, $-S(O)CH_3$, $-SCH_3$, and $-SC_2H_5$, effectively

213

TABLE I

Relative Rates of Transamination Between Glutamine and Various α–Keto Acids by Rat Liver Glutamine Transaminase*

α–Keto Acid	Rel. Rates	α–Keto Acid	Rel. Rates
Glyoxylate	[100]	S–Methyl–β–mercapto–pyruvate	80
Pyruvate	28		
α–Ketobutyrate	24	S–Methyl–β–mercapto–pyruvate (5 mM)	60
α–Keto–n–caproate	24		
α–Ketoisocaproate	3	α–Keto–γ–ethiol–butyrate	66
(D– and L–)–α–Keto–β–methylvaleric	0.2		
α–Ketoisovalerate	0	α–Keto–γ–ethiol–butyrate (5 mM)	250
Trimethylpyruvate	0	α–Ketoglutarate γ–ethyl ester	77
β–Hydroxypyruvate	50	β–Sulfopyruvate	6
α–Keto–γ–hydroxy–butyrate	24	Mesoxalate	16
α–Keto–β–hydroxy–butyrate	0	Oxaloacetate	4
α–Keto–γ–methiol–butyrate	50	α–Ketoglutarate	3
		Phenylpyruvate	17
α–Keto–γ–methiol–butyrate (5 mM)	240	p–Hydroxyphenyl–pyruvate (2 mM)	0
β–Mercaptopyruvate	195	α–Keto–δ–carbamido–valerate	10
		α–Keto–δ–guanidino–valerate	6
		α–Ketosuccinamate	62

*The reaction mixtures contained 20 mM L-glutamine, 20 mM α–Keto acid (except where noted) and 0.93 units of enzyme in 0.1 ml 50 mM Tris–HCl buffer, pH 8.4. After incubation at 37° for 3 to 60 minutes, the rate of transamination was determined (from Cooper and Meister, 1972a).

TABLE II

Relative Rates of Transamination with Various Amino Acids and
Rat Liver Glutamine Transaminase[1]

Amino Acid	Relative Rates		
	Gly*	Pyr**	KGAM†
L-Glutamine	[100]	28	134
L-γ-Glutamylmethylamide	63	17	
L-Glutamic γ-methyl ester	57		20
L-γ-cyano-α-aminobutyric acid[2]	85		
L-Albizziin[2]	87		
O-Carbamyl-L-serine[2]	42		
S-Carbamyl-L-cysteine[2]	50		
L-Glutamic γ-benzyl ester			6
L-Methionine	20	3	18
L-Ethionine	30		11
L-Methionine (SR)-sulfoxide	36	5	17
L-Methionine sulfone	10	2	5
L-Methionine (SR)-sulfoximine	12	3	
L-Methionine (SR)-sulfoximine phosphate	6		
L-Homoserine	10		14
DL-Homocysteine (40 mM)	8		
L-Phenylalanine	8	1	4
L-2-Amino-4-oxo-5-chloro-pentanoic acid	4		
L-Asparagine	2		5
L-Cysteine	2		6
L-Alanine	2		2
L-Serine	2		4
Glycine	0.05	0.05	1
L-Norleucine	1.0		
L-α-Aminobutyric acid	1.0		
S-Methyl-L-cysteine	10		

Footnotes to Table II

*Gly = Glyoxylate; **Pyr = Pyruvate; [+]KGAM = α-Ketoglutara-mate.

1/
    The reaction mixtures contained the sodium salt of the α-keto acid (pyruvate, 20 mM; glyoxylate, 20 mM: α-ketogluta-ramate, 90 mM), 20 mM amino acid (except as indicated) and 0.93 unit of enzyme in 0.1 ml Tris-HCl buffer, pH 8.4. After incubation at 37° the initial rates were determined (from Cooper and Meister, 1972a).

2/
    Data from Cooper and Meister, 1972b.

---

interact with the enzyme. However, a number of compounds that do not possess such a side chain are nevertheless quite active, but only when an α-keto group is present in the molecule. We may tentatively interpret these findings to indicate that the binding of substrates to the enzyme may involve the carboxyl group as well as the moiety attached to carbon atom 4 of these molecules. However such a moiety is clearly not required for the binding of certain α-keto acid substrates (see compounds 13-18, Table III). The fact that certain α-keto acids bind more effectively than the corresponding α-amino acids may be due to a more favorable geometry, i.e., planar rather than tetrahedral about the α-carbon atom, or possibly to the increased acidity of the α-keto acids.

    We have recently discovered that purified rat liver glutamine transaminase catalyzes the internal oxidation-reduction and hydrolysis of a number of γ-glutamylhydrazones of α-keto acids to yield the corresponding amino acid and 3-hydroxy-3-carboxytetrahydro-6-pyridazinone. This compound is irreversibly converted to the known compound 1,4,5,6-tetrahydro-6-pyri-dazinone-3-carboxylic acid in strong acid (Fig. 3). The enzyma-tic reaction does not appear to involve the intermediate formation

TABLE III

Summary of Data on the Specificity of Rat Liver Glutamine Transaminase

R = $^-$OOCCO$^-$ or R = $^-$OOCCHN$^+$H$_3$$^-$	Relative Rates		Approx Km Value (mM)	
	Amino Acid + Glyoxylate	Glutamine + α-Keto Acid	Amino Acid	Keto Acid
RCH$_2$CH$_2$CONH$_2$	(100)	54	2	0.2
RCH$_2$NHCONH$_2$	87			
RCH$_2$CH$_2$CN	85			
RCH$_2$CH$_2$CONHCH$_3$	63			
RCH$_2$CH$_2$COOCH$_3$[or OC$_2$H$_5$]	57	[31]	4	
RCH$_2$SCONH$_2$	50			
RCH$_2$OCONH$_2$	42			
RCH$_2$CH$_2$SOCH$_3$	36		2	
RCH$_2$CH$_2$SCH$_2$CH$_3$	30	(100)		
RCH$_2$CH$_2$SCH$_3$	20	96	2	3
RCH$_2$CH$_2$OH	10	10		
RCH$_2$SCH$_3$	10	32		
RCH$_2$SH	2	78		5
RH	0.05	40	>1000	8
RCH$_2$CONH$_2$	2	25		
RCH$_2$OH	2	20		
RCH$_3$	2	11	>1000	11
RCH$_2$CH$_3$	<1	10		
RCH$_2$CH$_2$CH$_2$CH$_3$	<1	10		

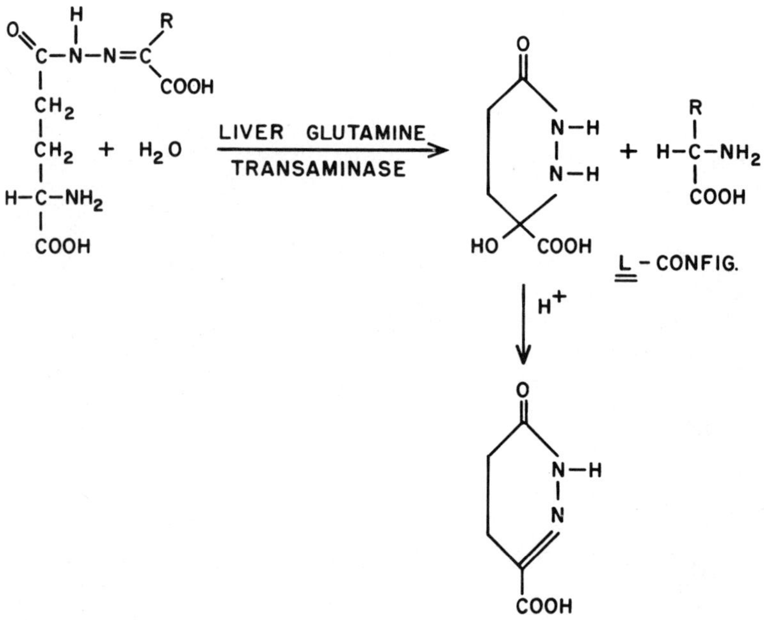

Fig. 3 Action of rat liver glutamine transaminase on the γ-glutamylhydrazones of α-keto acids.

of a free α-keto acid; thus, it is not inhibited by high levels of lactate dehydrogenase and DPNH. In general, the specificity and rate are similar to the analogous glutamine-α-keto acid reaction. Thus, free glyoxylate is a good substrate with glutamine, as is also the γ-glutamylhydrazone of glyoxylate, whereas α-ketoglutarate is a poor substrate both free and as the γ-glutamylhydrazone.

When the effect of substrate concentration on activity was investigated, a linear double reciprocal plot was obtained indicating that the system behaved as a single substrate system. The apparent Km value for the γ-glutamylhydrazone of glyoxylate was relatively high (27 mM) as compared to 2 mM for glutamine (in the presence of 20 mM glyoxylate). The reaction was competitively inhibited by free glyoxylate and by L-glutamine, but not by L-γ-glutamylhydrazide. A scheme for the reaction involving pyridoxal 5'-phosphate at the active site is shown in Figure 4.

As far as we know, this reaction is specific for liver glutamine transaminase. It is not catalyzed by the kidney glutamine transaminase or by glutamate-alanine transaminase or glutamate-aspartate transaminase from pig heart.

Another series of reactions catalyzed by the liver enzyme is summarized in Figure 5. In these studies the γ-methylene group of glutamine was replaced by another group, X. Thus, L-albizziin (α-amino-β-ureidopropionic acid, X = NH), S-carbamyl-L-cysteine (X = S) and O-carbamyl-L-serine (X = O) have been found to be good substrates. The α-keto acids produced in each case were substrates for the ω -amidase, but in the absence of this enzyme these compounds cyclized to yield five-membered heterocyclic rings analogous to the keto-lactam form of α-ketoglutaramate, whose structure was shown by Otani and Meister (1957). These cyclic compounds were also prepared by the oxidation of the parent amino acid with L-amino acid oxidase in the presence of catalase. The oxidation product of O-carbamyl-L-serine (2-oxazolidone-4-hydroxy-4-carboxylic acid) was prepared as the barium salt in a fashion similar to that employed by Meister (1953) for the preparation of the barium salt of α-ketoglutaramic acid. The oxidation products of L-albizziin and S-carbamyl-L-cysteine, although initially in the keto-lactam form, underwent dehydration during isolation to yield 2-imidazolinone-

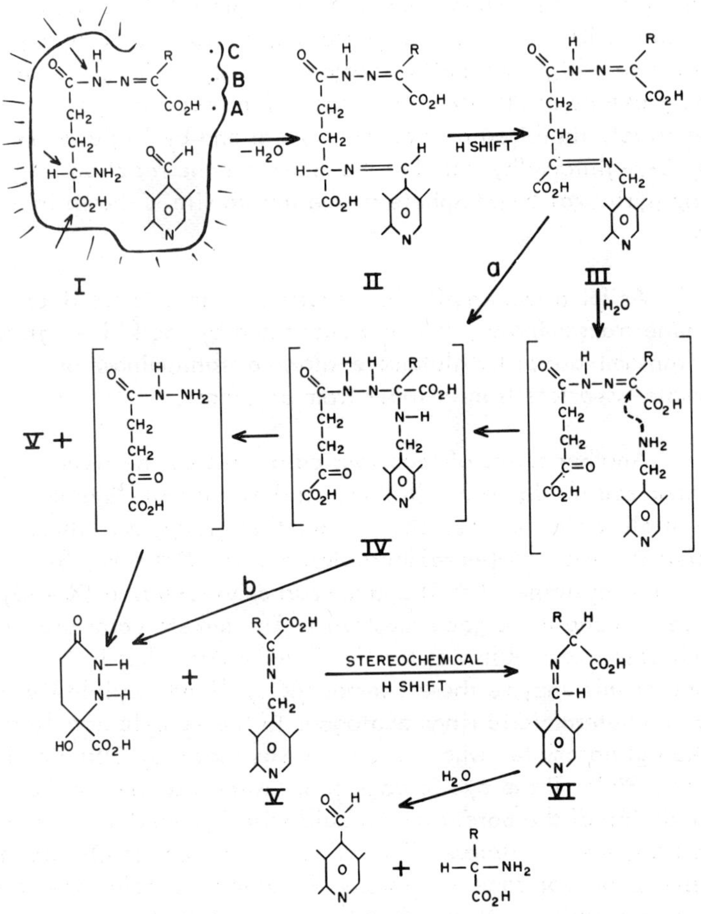

Figure 4.

Footnote to Fig. 4: A postulated mechanism for the reaction of rat liver glutamine transaminase with γ-glutamylhydrazones of α-keto acids. The hydrazone initially binds as substituted amide derivative of glutamine forming a ketimine with pyridoxal 5'-phosphate (structure II). The heavy arrows represent possible points of contact on the enzyme surface. The conversion of III to IV and of IV to V may involve a single concerted reaction or a two step reaction. Structure V is a normal ketimine intermediate which undergoes a stereochemical hydrogen shift to structure VI, which hydrolyzes to the L-amino acid pyridoxal 5'-phosphate. (For convenience the pyridoxal 5'-phosphate is shown free rather than bound to the enzyme as an aldimine).

-4-carboxylic acid and 2-thiazolinone-4-carboxylic acid, respectively. This apparently irreversible dehydration is readily catalyzed by dilute acid.

We may now summarize the specificity of liver glutamine transaminase toward the amino donor. Several classes of amino acids are substrates:

Class 1 – Amino acids which have a methyl or methylene group added to the basic L-glutamine structure e.g., γ-methyl and γ-methylene-glutamine; and glutamic acid-γ-methylamide (Meister, 1954).

Class 2 – Glutamines in which the γ-CH_2-group is replaced by -NH-, -O- or -S- (Cooper and Meister, 1972b).

Class 3 – Glutamines in which the amide group is replaced by another uncharged group such as -CN, -COOR, $-SCH_3$ or $SOCH_3$ (Cooper and Meister, 1972a).

Class 4 – γ-Glutamylhydrazones of α-keto acids (Cooper and Meister, 1972b).

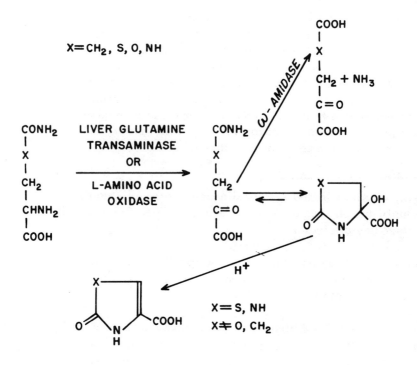

Fig. 5 Action of rat liver glutamine transaminase on various glutamine analogs in which the γ-carbon is replaced by a hetero atom. The five-membered keto lactam ring is obtained in each case, but when X = NH- or -S- the final product is a dehydrated compound.

The α-keto acid substrates are those corresponding to
class 3 and in addition lower molecular weight α-keto acids such
as glyoxylate, pyruvate, β-mercaptopyruvate and β-hydroxypyruvate. α-Keto acids with polar side groups or with branches
at the β position are inactive or poor substrates (Cooper and
Meister, 1972a).

Properties of Rat Kidney Glutamine Transaminases

We have recently isolated glutamine transaminase from
rat kidney; the properties of this enzyme are currently under
investigation. This enzyme came to light while we were examining some apparently conflicting reports. For a long time it was
thought that highest levels of glutamine transaminase activity
were present in the liver, and that smaller amounts were present
in kidney. However, in 1971 Kupchick and Knox, using phenylpyruvate as the α-keto acid acceptor, reported that kidney
exhibited a six-fold higher specific activity than did liver. When
we measured the activities with our standard glutamine-glyoxylate
reaction we found that kidney was only about 10% as active as
liver. This apparent contradiction was resolved by purification
and detailed study of the specificity and properties of kidney
glutamine transaminase. The enzyme was purified by heat treatment, ammonium sulfate precipitation, Sephadex G-150
chromatography and affinity chromatography. Like the liver
enzyme it exhibits a molecular weight of about 110,000, consists
of two subunits, and contains tightly bound pyridoxal 5'-phos-
phate. However, it has different physical properties; thus,
it is precipitated by 42% ammonium sulfate at pH 5.0, whereas
the liver enzyme is not. It also exhibits different mobilities in
two polyacrylamide gel electrophoresis systems.

The kidney enzyme exhibits a narrow specificity toward
the amino donor as compared to the liver enzyme. Thus, of all
the amino acids tested only phenylalanine, tyrosine and methionine can replace glutamine. The γ-glutamylhydrazones and the
glutamine analogs in which the γ-carbon is replaced by a hetero
atom, are not substrates. (Using the γ-glutamylhydrazone of
glyoxylate as substrate, we screened a number of tissues for

"liver-type" enzymatic activity. Liver was by far the most active, but some activity was found in kidney, brain, and heart). The specificity of the kidney glutamine transaminase toward α-keto acids is quite broad, although markedly different from that of the liver enzyme. The best α-keto acid substrates are phenylpyruvate and α-keto-γ-methiolbutyrate. Glyoxylate is a poor substrate, but increasing the chain length of the α-keto acid increases the rate of transamination; thus, α-keto-n-caproate is a good substrate. Curiously, α-ketoisovalerate is a moderately good substrate, but D- and L-α-ketomethylvalerate are not substrates. It thus appears that α-keto acids with large non-polar side groups such as phenylpyruvate, α-keto-γ-methiolbutyrate and α-ketocapro-ate are good substrates. Small molecules, such as glyoxylate and α-keto acids with polar side groups, such as α-ketoglutarate are poor substrates. Again, it is notable that methionine and its α-keto acid analog are substrates for the kidney enzyme as well as for the liver enzyme.

CONCLUSION

Two distinct glutamine transaminases, one from rat liver and one from rat kidney have been purified to apparent homo-geneity. They both exhibit molecular weights of about 110,000 and apparently contain tightly bound pyridoxal 5'-phosphate; however, they exhibit different physical and catalytic properties. The liver enzyme has a broad specificity toward both the α-keto acid and amino acid; thus, it can catalyze transamination between a number of glutamine analogs and α-keto acids. The liver enzyme can also catalyze the non classical internal oxidation-reduction and hydrolysis of a number of γ-glutamylhydrazones of α-keto acids. The kidney enzyme is much more specific toward the amino donor. Only phenylalanine, tyrosine and methionine can replace glutamine. Both enzymes are effective in catalyzing the reversible transamination of glutamine and methionine. Although both transaminases can synthesize glutamine from its α-keto analog this pathway does not seem to offer an alternative route to glutamine, since there is no evidence for a system that can catalyze the synthesis of α-ketoglutaramate. Perhaps, the relatively high activity of methionine and its α-keto analog represents quantitatively a more important pathway for the metabolism of this compound than has been generally supposed.

The observation of Tate and Meister (1971), that rat liver glutamine synthetase is inhibited by L-serine, L-alanine and glycine suggests that a regulatory mechanism exists in which products produced by glutamine transaminase can inhibit the synthesis of glutamine. Since the glutamine molecule offers a means of carrying ammonia in a non toxic form, its release may be controlled by glutaminase and by the combined action of glutamine transaminase and ω-amidase. This would produce α-ketoglutarate which could enter the citric acid cycle and at the same time provide a physiological mechanism for the amination of a number of α-keto acids so as to provide the corresponding amino acids for protein synthesis. Without such a mechanism the carbon chains of the α-keto acids would be lost by degradative processes. The possibility that glutamine transaminase and ω-amidase are physically linked in the cell (also considered by Hersh, 1971) requires investigation.

ACKNOWLEDGEMENT

This research was supported in part by a grant from the Public Health Service, National Institutes of Health.

REFERENCES

Braunstein, A.E., and T'ing Sen, H. (1960), Biokhimiya 25, 758.

Cooper, A.J.L., and Meister, A. (1972a), Biochemistry, 11, 661.

Cooper, A.J.L., and Meister, A. (1972b). Manuscripts in preparation.

Greenstein, J.P., and Carter, C.E. (1946), J. Biol. Chem., 165, 741.

Hersh, L.B. (1971), Biochemistry 10, 2884.

Kupchick, H.Z., and Knox, W.E. (1970), Arch. Biochem. Biophys., 136, 178.

Meister, A. (1953), J. Biol. Chem., 200, 571.

Meister, A. (1954), J. Biol. Chem., 210, 17.

Meister, A., and Tice, S.V. (1950), J. Biol. Chem., 187, 173.

Otani, T.T., and Meister, A. (1957), J. Biol. Chem., 224, 137.

Tate, S.S., and Meister, A. (1971), Proc. Natl. Acad. Sci. U.S., 68, 781.

Wada, H., and Snell, E.E. (1961), J. Biol. Chem., 236, 2089.

Yoshida, T. (1967), Vitamins (Japan), 35, 227.

REGULATORY MECHANISMS OF GLUTAMINE CATABOLISM

N. KATUNUMA, T. KATSUNUMA, T. TOWATARI
and I. TOMINO*

Department of Enzyme Chemistry, Institute for Enzyme
Research, School of Medicine, Tokushima University,
Tokushima, Japan.

Introduction

Inorganic ammonia is formed during the catabolism of
amino acids and it is strongly toxic to living cells. In
higher animals there is a mechanism which maintains the
intracellular ammonia level within a certain range by
synthesis of glutamine from ammonia; this is the most important intermediate in detoxication.

The elucidation of glutamine metabolism and its regulation in different organs is not only important to account
for organ specific function in nitrogen metabolism but also
to understand the phylogenic aspects of nitrogen excretion.

Glutaminases, glutamine-ketoacid transaminases and
various glutamine amidotransferases are considered to be the
most important enzymes involved in glutamine catabolism.
The different metabolic roles of the enzymes and the regulation of glutamine catabolism in various organs will be
discussed in this report.

I Glutaminase isozymes: Purification and properties

Distribution of glutaminase isozymes in rat organs

The authors found that glutaminase exists in the form
of two isozymes in various organs (Katunuma, N., et al.,
1967, Katunuma, N., et al., 1968). One of these isozymes
is phosphate dependent glutaminase (PD) and requires phosphate to show activity. The other isozyme is a phosphate
independent glutaminase (PI) and is activated by maleate.
Some phosphate independent glutaminases are activated N-
acetyl amino acids. Table 1. indicates the distribution of
glutaminase isozymes of various rat organs. It is possible
to classify the various organs into three groups: (a) Kidney,
liver, brain, bone marrow, spleen and lung have both phosphate dependent glutaminase and phosphate independent
glutaminase activities, (b) intestine, heart and testicle

---

* Present address: Department of Nutrition , Jikei Univer-
              sity of Medicine, Tokyo, Japan.

227

have only the activity of phosphate dependent glutaminase
and, (c) blood plasma and blood cells do not have any
glutaminase activity.

TABLE 1
Distribution of Glutaminase Activities in Rat Organs

	PI Maleate 0.02 M (S.A.)	PD K_2HPO_4 0.1 M (S.A.)
Bone marrow	0.62	0.98
Spleen	0.58	1.08
Lung	0.77	2.74
Intestinal mucosa	0	1.30
Intestinal muscle	0	0.36
Heart	0	0.28
Testis	0	0.50
Blood plasma	0	0
Blood cell	0	0
Kidney	2.5 - 5.0	4 - 15
Liver	0.3 - 1.0	0.3 - 1.2
Brain	0.45	7

S.A.= specific activity (units/mg)

In mammals, glutamine is the most important intermediate
in detoxification of free ammonia liberated from amino acids
in various organs with the exception of liver and kidney.
Glutamine which is produced in various organs is released
into blood, and transported to liver and kidney. The
concentration of glutamine in blood is about 0.4 mM. It is
reasonable that no glutaminase is detected in blood; otherwise, the breakdown of glutamine would result in a high
concentration of blood ammonia.

Phosphate independent glutaminases in different organs

Phosphate independent glutaminase, which was first
demonstrated in kidney, has also been found in other organs
in rat (Katunuma, N., et a., 1967, Katunuma, N., et al.,

1968, Katunuma, N., et al., 1966). In this section we
report on the purification and properties of liver phosphate independent glutaminase, kidney phosphate independent
glutaminase and brain phosphate independent glutaminase.
Mechanisms of enzyme activation at the molecular level and
the different metabolic roles of organ specific glutaminase
isozymes will also be discussed.

Purification of kidney, liver and brain phosphate independent glutaminase

Purification of kidney phosphate independent
glutaminase:

TABLE 2

Purification of Phosphate Independent Glutaminase
in Kidney

Procedure	T. Act. (units)	Sp. Act. (units/mg)	Purity
Crude extract	1,080,000	2.7	1
Ammonium sulfate fraction (25-50% sat'n)	966,456	5.97	2.2
65,000 g precipitation	556,110	10.09	3.7
Bromelain treatment	540,000	26.6	9.85
Ammonium sulfate fraction (70-100% sat'n)	270,135	450	166.6
DEAE cellulose column	170,622	947.9	351.0
Calcium phosphate gel column	135,000	2,000	740.7
Crystallization		2,090	774.0

We have previously reported the purification method for
phosphate independent glutaminase. The preparation was
judged to be a homogeneous protein by ultracentrifugation
analysis (U.C.A.) and by cellulose acetate electrophoresis
(Katunuma, N., et al., 1967, Katunuma, N., et al., 1968).
An outline of the purification of the enzyme is given in
Table 2. The precipitate of last ammonium sulfate fractionation was dissolved in the small amount of 0.01 M potassium phosphate buffer (pH 7.5). The protein concentration was

about 40 mg/ml. Solid ammonium sulfate was gradually added until just a slight turbidity appeared with stirring. Isinglass shaped crystals appeared after a few days at 4° (Fig. 1.). Specific activity of the first crystalline preparation was approximately 2,100 units/mg.

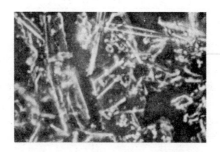

Fig. 1   Crystal of kidney phosphate independent glutaminase

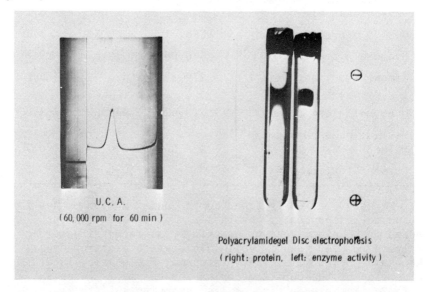

Fig. 2   Homogenity of phosphate independent glutaminase in rat kidney

TABLE 3

Amino Acid Composition of Rat Kidney Phosphate Independent Glutaminase

Amino acid	Residue molecule		
	24 hour	48 hour	
Tryptophan	(9.3)		9
Lysine	41.7	41.0	41
Histidine	18.5	16.1	17
Ammonia	94.5	102.0	98
Arginine	40.2	40.3	40
Aspartic acid	73.5	73.7	74
Threonine	53.2	51.4	53
Serine	61.3	56.7	61
Glutamic acid	76.3	76.2	76
Proline	44.0	44.2	44
Glycine	70	70	70
Alanine	77.1	79.0	78
Half-cystine	(7.2)		7
Valine	63.1	69.8	70
Methionine	17.5	17.1	17
Iso leucine	42.2	46.6	47
Leucine	59.9	61.4	61
Tyrosine	28.8	28.6	29
Phenyl alanine	32.3	32.6	32
Hexose-amine			56

amino acid 826   MW ≒ 98, 829

Purification of liver phosphate independent glutaminase: As shown in Table 4, we purified liver phosphate independent glutaminase by the following steps: acetone powder extraction, acid treatment, high speed centrifugation, ammonium sulfate fractionation, and DEAE-cellulose column chromatography. A 470-fold purification was achieved by this method.

TABLE 4

Purification of Phosphate Independent Glutaminase in Liver

Procedure	T. Act. (units)	Sp. Act. (units/mg)	Purity
Homogenate	14, 700	0.4	1
Acetone powder extraction	7, 000	2.0	5
Acid treatment (pH 5.5 p.p.t.)	8, 080	6.1	15
AmSO$_4$ fraction (25 - 45 % sat' n)			
Centrifugation (146, 000 xg sup)			
AmSO$_4$ fraction (0 - 33 % sat' n)	1, 239	35.0	88
DEAE cellulose column	670	180.0	470

Purification of brain phosphate independent glutaminase: The enzyme is very unstable in comparison with phosphate independent glutaminases of other organs, and it is very difficult to solublize. As shown in Table 5, we purified brain phosphate independent glutaminase by the following steps: freezing and thawing extraction, deoxycholate treatment, high speed centrifugation and citrate fractionation. About a 10-fold increase in purity of the enzyme was achieved.

Purity of these enzyme preparations: The crystallized enzyme of kidney phosphate independent glutaminase sedimented as a homogeneous protein in the analytical ultracentrifuge. Disc electrophoresis in polyacrylamide gel showed a single protein band that corresponded to the enzyme activity. The partially purified enzyme of liver phosphate independent glutaminase sedimented as two components in the analytical ultracentrifuge, and was observed to be 2 or 3 bands on electrophoresis. These enzyme preparations were free from glutamine transaminase and phosphate dependent glutaminase activities and were used for the experiments reported here.

TABLE 5

Purification of Phosphate Independent Glutaminase in Brain

Procedure	T. Act. (units)	Sp. Act. (units/mg)	Purity
Homogenate	37,000	3.3	1
Centrifugation (2,000 x g-17,000 x g p.p.t.)			
Freezing and thawing (17,000 x g p.p.t.)	21,000	7.4	2.2
Deoxycholate treatment			
Citrate fraction (14.4-24.3 g/dl)	6,800	20.4	6.2
Centrifugation (100,000 x g sup)			
Citrate fraction (11.4-24.3 g/dl)	3,880	32.0	9.7

Some properties of the three isozymes

Molecular weight: The sedimentation constant (S_{20w}) of the kidney enzyme is 4.1. The molecular weight was found to be about 95,000 by gel-filtration and to be about 90,000 by the Archibald method of sedimentation analysis. The molecular weights of the liver and brain enzymes were observed to be 140,000-250,000 by sucrose density gradient centrifugation.

pH Optima for three isozymes: The pH optima of these isozymes differ widely from tissue to tissue. The phosphate independent glutaminases of liver and brain have the same pH optimum of 8.6. The pH optimum of kidney enzyme is 7.5 with and without activator (20 mM maleate).

Effect of heat treatment, PCMB and product: The activity of glutaminase in the kidney is resistant to heat at 50° for 2 minutes and 5.0×10^{-4}M PCMB treatment, while the activities in the liver and brain are extremely labile. Complete loss of activity was observed with the liver and brain preparations under the conditions described above.

The effects of glutamate added on the phosphate independent glutaminases in the different tissues varied. No effect of glutamate was observed on liver and kidney phosphate independent glutaminases, while the brain enzyme was strongly inhibited by the addition of product, glutamate (Katunuma,N., et al., 1967, Katunuma, N., et al., 1968).

Effect of maleate derivatives and N-acetyl amino acids

Table 6 shows the activation of the phosphate independent glutaminases by maleate and related compounds.

TABLE 6

Activation of Phosphate Independent Glutaminases
(A) by maleate and analogs

Compounds	Multiples of Activity	Compounds	Multiples of Activity
$CH-COOH$ $CH-COOH$	4-5	$CH-COOH$ $HOOC-CH$	1
$CH-CONH_2$ $CH-COOH$	2-3	$CH-CO$ $CH-CO$ NH	1
$CH-COOC_2H_5$ $CH-COOH$	2-3	$CH-CO-NH$ $CH-CO-NH$	1
CH_2-COOH $CH-COOH$ $NH-COCH_3$	1-1.2	CH_2-COOH $CH-COOH$ NH_2	1
CH_2-CH_2-COOH $CH-COOH$ $NH-COCH_3$	1-1.4	![]-COOH -COOH (pyridine)	1
CH_2-COOH CH_2-COOH	1	![]-COOH -COOH (benzene)	1

(B) by maleate and N-acetyl amino acids

Activator 20 mM	% Activities		
	Brain	Liver (10 mM)	Kidney
None	100	100	100
Maleate	318	390	400
N-acetyl L-aspartate	242	234	120
N-acetyl L-glutamate	260	234	140
N-acetyl DL-methionine	250	412	—
N-acetyl L-phenylalanine	714	234	—
N-acetyl glycine	966	400	—
N-acetyl D-glucosamine	120	167	—
N-benzyl L-glutamate	100	100	—
Acetate	100	100	—

From these results it is deduced that the -CH=CH- structure and at least one free carboxyl group are necessary for significant activation to occur. The activation of phosphate independent glutaminases by maleate is 3-5 fold, but maleate is not a physiological substance. Liver and brain phosphate independent glutaminases are both activated by N-acetyl amino acids (N-acetyl Glu, N-acetyl Asp and N-acetyl Gly) as well as by maleate (Katunuma, N., et al., 1967, Katunuma, N., et al., 1968).

Activation of kidney phosphate independent glutaminase: The enzyme activity rises gradually as the glutamine concentration increases, but reciprocal Lineweaver-Burk plots show non linearity (Fig. 3A). In the presence of the activator maleate, the kinetics became hyperbolic. This suggests that some conformational change in the enzyme molecule takes place at high substrate concentration and the activators facilitate the substrate induced change. Since some activators are known to cause an allosteric polymerization, we have investigated this possibility using ultracentrifugation. The sedimentation constants were determined with and without activator (Fig. 3B). Also the substrate did not influence the molecular weight as determined by sucrose density gradient centrifugation at a low concentration of the enzyme.

These data show that the increase in catalytic activity observed with maleate is not due to polymerization. The Km of the enzyme was 5 mM for glutamine in the presence of 20 mM maleate. The concentration of maleate giving half-maximal activation of the enzyme was 7 mM in the presence of 2.5 mM glutamine (Katunuma, N., et al., 1967).

(A) Kinetics

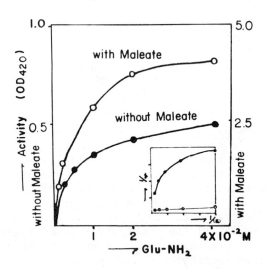

(B) U.C.A.
   (60,000 r.p.m. for
   60 min.at 20°C)

(C) Sucrose density gradient
   pattern (40,000 r.p.m.
   for 300 min. cent.)

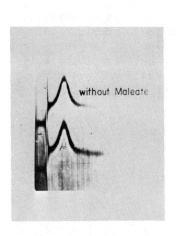

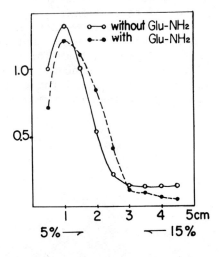

Fig. 3 Activation of phosphate independent glutaminase from kidney

Chemical modification of kidney phosphate independent glutaminase: As shown in Table 7, treatment with diisopropyl phosphofluoridate, photooxidation and 2-hydroxyl-5-nitrobenzyl bromide to the kidney phosphate independent glutaminase did not effect the catalytic activity in the presence and the absence of maleate. These methods modify the hydroxyl groups of serine, histidine and tryptophan residues. Treatment with 2,4,6-trinitrobenzene sulfonic acid, tetranitromethane, N-acetyl-imidazole and glyoxal, which all modify the hydroxyl groups of tyrosine and arginine residues, did not alter the catalytic activity but did inhibit the rate of activation by maleate. Treatment with 1-ethyl-3 (3 dimethylaminopropyl) carbodiimide, which reacts with carboxyl groups, inhibited the catalytic activity in the presence and the absence of activator.

TABLE 7

Chemical Modification on Phosphate Independent Glutaminase from Rat Kidney

Residue	Modifying Reagent	Enzymic Activity	
		with maleate (Activating site)	without maleate (Active site)
Carboxyl group	1-ethyl-3 (3 dimethylamino propyl) carbodiimide	↓	↓
Amino group	Acetic anhydride 2.4.6 TNBS	↓ ↓	→ →
Serine	DFP	→	→
Cysteine	PCMB	→	→
Arginine	Glyoxal	↓	→
Histidine	Photooxidation (Methylene blue)	→	→
Tryptophan	2-hydroxyl-5 nitrobenzyl bromide	→	→
Tyrosine	N-Acetyl imidazole TNM	↓ ↓	→ →

↓ decrease → no change

237

Activation of liver phosphate independent glutaminase:
Substrate saturation plots for phosphate independent glutaminase from liver showed sigmoidal kinetics with L-
glutamine. The presence of 12.5 mM maleate, an activator,
converted the plots to hyperbolic kinetics and resulted in
a decrease in the Km for L-glutamine from 100 mM to 5 mM.
Hillplots gave n values of 3.2 and 1.0 in the absence and
presence of maleate, respectively.

Centrifugation of phosphate independent glutaminase
from liver in a sucrose density gradient revealed an
increase in this sedimentation coefficient in the presence
of 0.1 M L-glutamine (Fig. 4a). Activators such as maleate
and N-acetylglutamate did not alter the sedimentation coefficient (Fig. 4b). These activators reduced the concentration of glutamine from 0.1 M to 0.025 M to induce an
increase in the sedimentation constant (Fig. 4c). The results appear to indicate that the enzyme polymerizes to a
higher molecular weight form in the presence of high concentrations of substrate or in the presence of activator and
lower concentrations of substrate. Polymerization of glutaminases by small ligands has also been observed with the
purified enzymes from pig kidney and brain (Kvamme, E., et
al., 1970, Svenneby, G., 1970).

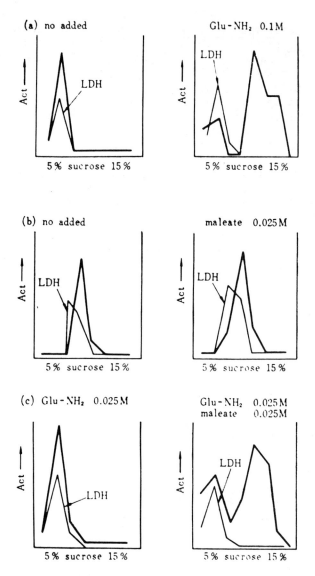

Fig. 4   Ligand induced polymerization of phosphate independent glutaminase from rat liver.
   Sedimentation coefficients were estimated by sucrose density gradient centrifugation in the presence and absence of ligands.

239

Activation mode of brain phosphate independent
glutaminase: The Kinetics and mechanisms of activation for brain phosphate independent glutaminase differ
markedly from those for kidney and liver phosphate independent glutaminase. Substrate saturation plots for brain
phosphate independent glutaminase with L-glutamine showed
Michaelis-Menten kinetics in the presence and absence of
activators, such as maleate, N-acetyl-glutamate, and
N-acetyl-aspartate. The km value of the enzyme was 100 mM
L-glutamine, and in the presence of activator, 20 mM maleate, the Km value was reduced to 40 mM L-glutamine.
Unlike the liver enzyme which polymerizes in the presence
of high substrate concentration or lower substrate concentration plus activator, the brain enzyme does not polyme-
rize under these conditions(Katunuma,N., et al.,1968).

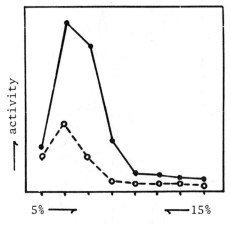

Fig. Sedimentation coefficients of phosphate independent
glutaminase from rat brain by sucrose density gradient
centrifugation in the presence and absence of the substrate.
    ●——● with glutamine plus maleate
    o---o without glutamine

The regulation of glutamine metabolism by phosphate
independent glutaminase in different organs
    As described above, the various properties of these phos-
phate independent glutaminases in kidney, liver and brain
are quite different. These properties are summarized in
Table 8.

TABLE 8
Comparison in Properties of Phosphate Independent
Glutaminases in Liver, Kidney and Brain

	Liver	Kidney	Brain
(A) Protein			
1. solubilization	easy	difficult	difficult
2. S_{20} w		4.1	
3. molecular weight	140,000-200,000	95,000±5,000	140,000-250 000
4. heat stability	unstable	stable	unstable
5. reaction of antibody for kidney PI	−	+	−
(B) Some properties			
1. optimal pH	8.6	particle 7.2 solubilized 8.6	particle 8.6 solublized 8.6
2. inhibition			
PCMB	+	−	+
divalent cation	+	−	+
glutamic acid	−	−	+
3. substrate			
L-Glu.NH₂	+	+	+
D-Glu.NH₂	−	+	+
4. activation			
maleate	+	+	+
N-acetyl amino acid	+	−	+
citrate	+	−	+
5. Km for Glu.NH₂	$1 \times 10^{-1} - 5 \times 10^{-3}$ M*	4.71×10^{-3} M (with maleate)	particle 5×10^{-2} M solublized 2×10^{-3} M (with maleate)
6. polymerization by substrate	+	−	−
(C) Induction	+	−	−

* changeable by substrate (activator) concentration

Metabolic role of kidney glutaminases: The ammonia content of urine represents only about 10-15% of the total
urinary nitrogen, but ammonia excretion from the kidney
plays an important role in the maintenance of acid-base
balance by neutralizing H^+ ions.

The excretion of ammonium ions directly into the urine
has been shown to occur in the kidney where the deamidation

of L-glutamine is catalyzed by two glutaminases. One of the glutaminases is dependent upon phosphate for its activity while the other exhibits activity which is independent of the phosphate concentration. Since the kidney possesses an active tubular transport system for the reabsorption of phosphate, it would not be surprising to find high intracellular concentrations of this ion. It is possible that phosphate ions regulate phosphate dependent glutaminase of kidney in vivo even though the concentration of phosphate required for activity is quite high. Exposure of rats to cold for more than 24 hours resulted in an increase of phosphate dependent glutaminase activity. This increase in activity occurred concomitantly with an increase in urinary ammonia excretion. Also, metabolic acidosis induced by the administration of ammonium salts to rats resulted in an increase of both phosphate dependent and independent glutaminases.

Metabolic role of liver phosphate independent glutaminase: Since inorganic ammonia is a direct substrate of carbamyl phosphate synthetase in the urea cycle, glutamine is not directly used for carbamyl phosphate synthesis. The amide nitrogen of glutamine which is transported to the liver should be hydrolyzed to inorganic ammonia prior to urea synthesis. Liver glutaminase is considered to play an important role in this metabolic pathway. We investigated this problem using suspensions of mitochondria and isolated perfusion of rat livers. A rapid and linear decrease in the concentration of L-glutamine was observed when this amino acid was added to the perfusate of the isolated rat liver. Subsequently, ammonia accumulated in the perfusate and then decreased as urea was formed (Fig. 5). The experiment suggests that the first step of glutamine degradation is deamidation by glutaminase. We studied this phenomenon in more detail using mitochondrial systems.

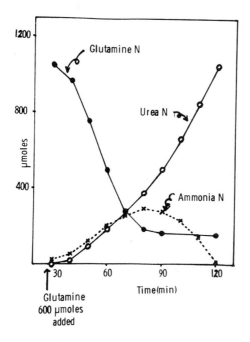

Fig. 5   Urea formation from L-glutamine by using the perfusate of the isolated rat liver

It is well known also that N-acetyl glutamate activates carbamyl phosphate synthetase in the liver urea cycle. Our finding that glutaminase in liver is strongly activated by N-acetyl glutamate is of great significance in view of the present knowledge of ammonia metabolism in the liver. The activation of successive enzymes functioning in the transfer of ammonia to urea by a common activator is considered to be a highly efficient automatic regulatory mechanism. Fig. 6A shows a comparison of the activation of carbamyl phosphate synthesis by N-acetyl glutamate for two nitrogen sources: inorganic ammonia and L-glutamine (using mitochondrial suspensions).

The synthesis of carbamyl phosphate with ammonia as a direct substrate rapidly reaches a maximum velocity with increasing concentrations of N-acetyl glutamate, with glutamine as a substrate higher concentrations of N-acetyl glutamate are required for maximal velocity at 10 mM

243

N-acetyl glutamate, the rate of synthesis of carbamyl phosphate is maximal with inorganic ammonia as a substrate while
the rate is sub-maximal with glutamine as a substrate.
Since the deamidation of L-glutamine is required prior to
the synthesis of carbamyl phosphate, glutaminase appeared
to be the first step in this pathway. High concentrations
of N-acetyl glutamate are required for maximal glutaminase
activation. In order to show glutaminase was responsible
for the lower rate of carbamyl phosphate synthesis observed
with L-glutamine as compared with inorganic ammonia, an experiment using maleate was designed. Maleate is a selective
activator of glutaminase and does not alter the activity of
carbamyl phosphate synthetase. As shown, the synthesis of
carbamyl phosphate in the presence of 10 mM N-acetyl glutamate could be increased by activating glutaminase with maleate (Fig. 6B).

(A)   (B)

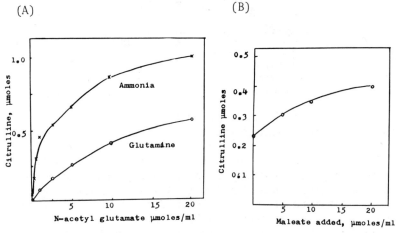

Fig. 6   Activation carbamyl phosphate synthesis by
N-acetyl glutamate and maleate.
    (A) Comparison of carbamyl phosphate synthesis using
        inorganic ammonia or L-glutamine as nitrogen sour-
        ces plotted as a function of N-acetyl glutamate
        concentration. Carbamyl phosphate synthetase acti-
        vity was measured by coupling the reaction with
        ornithine transcarbamylase and measuring colori-
        metrically the formation of citrulline.
    (B) Carbamyl phosphate synthesis plotted as a function
        of maleate concentration in the presence of 10 mM
        N-acetyl glutamate.

If inorganic ammonia was used in place of L-glutamine, maleate would have had no effect. These results demonstrate
that glutaminase is the rate limiting step in carbamyl phosphate synthesis from glutamine.

Since the urea cycle of liver is prominent pathway for
the removal of ammonia liberated from the amide of glutamine,
significant amounts of glutamate are produced in these
reactions. It appears advantageous that liver glutaminase
is not inhibited by glutamate.

In addition, a system of positive feed forward control
participates in the regulation of liver glutaminase. High
concentrations of L-glutamine or lower concentrations of L-
glutamine with an activator induce the polymerization of the
enzyme which is accompanied by an increase of activity.

Metabolic role of brain glutaminase: It is well known
that low concentrations of ammonia are toxic to the brain.
Ammonia in the brain is incorporated into glutamine which
freely passes through the blood-brain barrier and is excreted into the circulating blood. Since the concentration of
L-glutamate in brain tissue is at least 10 mM and brain glutaminase is strongly inhibited by L-glutamate, the enzyme
appears to be inactive. When the concentration of glutamate
in the brain decreases, glutaminase appears to become active
in order to restore the glutamate concentration. In addition brain glutaminase is regulated by N-acetyl aspartic
acid which is also in very high concentration in the brain.

II Ontogeny of glutaminase isozymes

The differentiation of various organ functions in the
mammalian body is an important and interesting problem.
Each organ plays a different metabolic role and each has
specific enzyme patterns. These specific enzymes maintain
the homeostasis of each organ which participates in the
metabolic regulation of whole body. The study of isozyme
patterns and functions in specific organs is a useful tool
in elucidating   regulatory mechanisms. Glutaminase isozymes are formed in many organs, and these enzymes have
characteristic properties. It should be possible to correlate the specific nature of each isozyme with its function
in a given organ. We reported that phosphate independent
glutaminases in adult kidney and adult liver are entirely
different proteins and they have organ specific functions.
In this paper, we discuss the relationship between the

appearance of the organ function and the differentiation of the organ specific isozymes during development.

The evidence for the existence of kidney type glutaminase in fetal liver: As described above, the phosphate independent glutaminases from adult liver and kidney are very different. The phosphate independent glutaminase from adult liver undergoes polymerization in the presence of substrate, but the adult kidney enzyme does not under the same conditions. The phosphate dependent glutaminase from adult liver also undergoes polymerization but requires high concentrations of phosphate (Kvamme, E., et al., 1965).

Glutaminase activity in adult kidney is resistant to heat (50°, 2 min) and PCMB treatment (0.75 mM). The glutaminase activity in adult liver is completely lost by these treatments, but only 70% of activity in fetal liver is lost under these conditions. Two possible hypotheses might be considered to explain this phenomenon. Either the nature of glutaminase from fetal liver is entirely different from that of adult liver, or the unstable activity is like adult liver glutaminase and the stable activity is similar to the adult kidney type enzyme. Experiments were designed to answer this question. If the latter hypothesis were correct then the stable glutaminase activity from fetal liver ought to be readily purified by the same method used for the isolation of adult kidney phosphate independent glutaminase. Indeed, the stable glutaminase activity from fetal liver could be purified by the method used for kidney enzyme as described in Table 2.

Immunochemical analysis was used to assess the similarity between the stable glutaminase from fetal liver and the enzyme from adult kidney. Antiserum to the phosphate independent glutaminase from adult kidney was prepared by standard immunization techniques with rabbits. The antiserum was incubated with purified glutaminase preparations for 30 min at 37°. The incubation mixtures were centrifuged at 105,000 g for 60 min and the glutaminase activity in the supernatant was assayed. No loss of glutaminase activity from adult liver was observed during the reaction with antiserum described above. In contrast, all of the glutaminase activity from the purified adult kidney and heat stable fetal liver preparations disappeared under the antiserum treatment described above.

A similar analysis was performed using the ochterlony method as shown in figure 7.

246

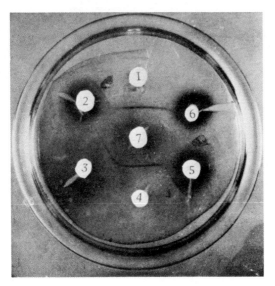

Fig. 7   Immuno diffusion analysis with an ochterlony gel.
The center well, contained antiserum and the surrounding wells contained glutaminases: 1,4, adult kidney phosphate independent glutaminase; 2,5, adult liver phosphate independent glutaminase; 3, fetal liver phosphate independent glutaminase; 6, adult liver phosphate dependent glutaminase.

The antiserum of adult kidney phosphate independent glutaminase was set in the center hole. The holes numberd 1 and 4 contained adult kidney phosphate independent glutaminase and holes numbered 2 and 5 contained phosphate independent glutaminase from adult liver. Hole number 3 contained the heat stable glutaminase from fetal liver and hole number 6 contained phosphate dependent glutaminase from adult liver. No reaction of adult liver phosphate dependent and independent glutaminases was observed with the antiserum. A precipitation band was observed with the enzymes from adult kidney and total liver. The bands from adult kidney and fetal liver clearly fused indicating the proteins are identical immunologically.
Additional evidence for the similarity of phosphate independent glutaminase from adult kidney and the heat stable glutaminase from fetal liver was obtained.   The

two enzymes have identical sedimentation coefficients in sucrose density gradient centrifugation. No change in their sedimentation patterns was observed in the presence of 0.1 M L-glutamine and 0.025 M maleate. Under these conditions the phosphate independent glutaminase from adult liver undergoes polymerization.

The changes of the kidney type glutaminase in fetal liver during development: In the fetal liver, 30% of total glutaminase activity is of the kidney type as shown in figure 8. The amount of the kidney type phosphate independent glutaminase decreased gradually after birth, and disappeared completely after one week.

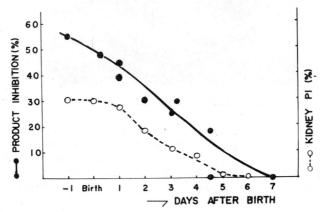

Fig. 8   Amount of kidney type phosphate independent glutaminase in liver during development from fetus to adult.
Total glutaminase activity is shown as 100%, and after PCMB or heat treatment, remaining activity was estimated as the kidney type glutaminase.

Kidney type phosphate independent glutaminase in regenerating liver: Since the regenerating liver grows very rapidly like the fetal, we looked for the presence of kidney type phosphate independent glutaminase in the regenerating liver after partial hepatectomy. No kidney type enzyme was found at any stage of regeneration as shown in figure 9. Both the phosphate dependent and independent glutaminase specific activities in the regenerating liver were decreased 40 to 50% 1 day after partial hepatectomy. The specific activities gradually increased over the next 7 days until the normal levels of enzyme activity were observed.

248

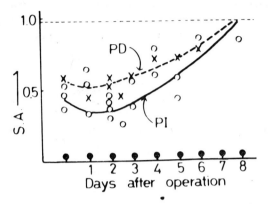

Fig. 9   Glutaminase levels in regenerating liver.
Rats were subjected to partial hepatectomy. Animals
were sacrified from 1 to 8 days following the operation and
the levels of glutaminases determined.   ; liver type phosphate independent glutaminase (PI), ; liver type phosphate
dependent glutaminase (PD), ⵁ ; kidney type phosphate independent glutaminase (PI).

Brain type glutaminase in fetal liver

Since phosphate independent glutaminase from adult
liver is not inhibited by L-glutamate but the enzyme from
adult brain is inhibited, we used this property to look for
the brain type enzyme in fetal liver.   50% of the phosphate
independent glutaminase activity in fetal liver can be inhibited by L-glutamate.   The glutamate inhibited activity is
heat labile and therefore, it can distinguished from the
kidney type activity discussed above.   The glutamate inhibited or brain type activity gradually decreased after birth
and no activity was found after 1 week of gestation.   Figure
10 illustrates the changes in phosphate independent glutaminase isozymes during the development of the liver.   Both the
kidney type and brain type isoenzymes disappear gradually
during the first week of gestation and are replaced by the
adult liver type.

These changes in isoenzyme patterns during development
are not confined to the liver alone.   Studies of the fetal
kidney have shown that 50% of the phosphate independent
glutaminase activity of fetal kidney is heat labile and inhibited by PCMB.   These treatments do not alter the activity

of phosphate independent glutaminase from adult kidney but do inhibit the glutaminase from adult liver. The experiments suggest that fetal kidney contains a phosphate independent glutaminase of the adult liver type.

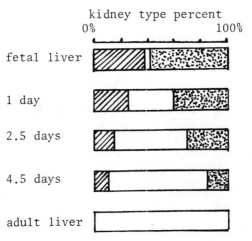

Fig. 10   Changes of glutaminase isozymes during liver development.
;kidney type, ;liver type, ;brain type.

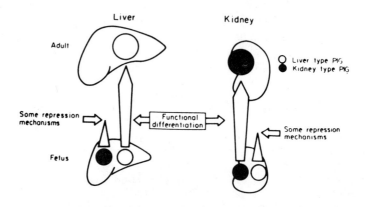

Fig. 11   Differentiation of organ-characteristic glutaminase isozymes.

250

These studies show that fetal liver contains the struc-
tual genes for three types of phosphate independent glutaminases. During development, specific mechanisms of repression seem to emerge and allow for the synthesis of a specific glutaminase isoenzyme, which is suited for the metabolic
function of a given organ (Fig. 11).

Abnormal gene expression of organ specific glutaminases
in hepatomas

Since cancer cells show many features seen in fetal
cells and they grow rapidly like regenerating cells, we
investigated the glutaminase isoenzyme patterns of cancer
cells. The isoenzyme patterns were examined in several
lines of Morris hepatoma cells. Figure 12 illustrates that
some hepatomas contained substantial amounts of kidney and
brain type glutaminases. There was no relationship between
growth rate of the hepatoma cells and the isoenzyme patterns
observed. In addition, there was constant relationship between the amount of kidney type and brain type glutaminase in
these hepatoma cells.

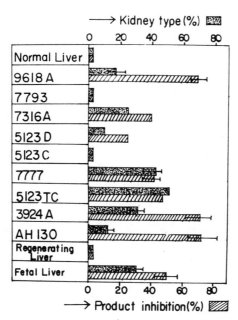

Fig. 12 Existence of kidney type phosphate independent
glutaminase and product inhibition in Morris hepatomas.

These results demonstrate that cancer cells do not show the precise glutaminase isoenzyme patterns of fetal cells, but instead, they have random amounts of the various glutaminase isoenzymes.

III Glutamine ketoacid transaminase isozymes

In 1960, Braunstein, A.B., and Seng, Hsu Ting, reported some properties of rat liver glutamine ketoacid transaminase (GKT) using partially purified preparation   which contained a considerable amount of alanine transaminase activity. We found new isozyme in rat liver of glutamine ketoacid transaminase, one localized in soluble fraction (GKTs) and the other localized in mitochondrial fractions (GKTm). We purified these isozymes to homogeneity and some properties were compared (Yoshida, T., et al., 1968)(Fig. 13).

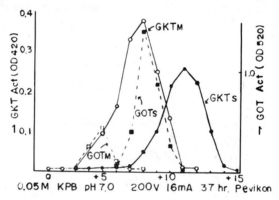

Fig. 13   Isolation of glutamine ketoacid transaminase isozymes by zone electrophoresis

These purified glutamine ketoacid transaminases showed patterns of homogeneous proteins in the analytical ultracentrifuge and were also free from alanine transaminase and glutaminases.

α-Amino group of L-glutamine was the only amino donor for both isozymes and all other amino acids and derivatives such as asparagine, leucineamide, glycineamide, glutamate, aspartate, alanine were unable to act as amino donors. Glyoxylic acid, pyruvic acid, phenyl pyruvic acid, α-ketobutyric acid and oxalacetic acid   acted as amino acceptors in this order. The sedimentation constant (S_{20W}) of both isozymes showed the same value of 5.5.

Many properties of these enzymes are similar, but both isozymes can be separated using DEAE-cellulose column chromatography or zone electrophoresis. Soluble glutamine keto-acid transaminase is eluted in 0.03 M potassium phosphate buffer pH 7.0 from the DEAE-cellulose column, while mito-chondrial glutamine ketoacid transaminase is eluted in the same buffer of 0.1 M and ω-amidase is eluted in 0.005 M of the buffer. The pattern of pevikon zone electrophoresis was shown in Figure 13. Aspartate transaminase isozymes (GOT) were added to the system as a marker. These data indicate the fact that the charge of each isozyme is different.

IV   Phylogenic aspects of glutamine metabolism

It is well known that there are many differences in glutamine metabolism between mammals which are ureoteric animals and birds which are uricoteric animals. In mammals, glutamine is the most important intermediate in detoxication of ammonia liberated from amino acids in various organs, and urea which is directly synthesized from inorganic ammonia in liver is the final excretion form of amino acid nitrogen. On the other hand, glutamine is the direct precursor in the pathway of uric acid synthesis in birds. Both mammals and birds have in common the pathways of purine base synthesis (AMP and GMP) which are components of nucleic acid. Many differences in regulatory mechanisms of glutamine metabo-lism are evident by comparing the control of purine base synthesis for excretion in uricoteric birds with the regula-tion of purine base biosynthesis for nucleic acids and ATP in ureoteric mammals. The feedback control and induction-repression of glutaminases, glutamine synthetase and gluta-mine-PRPP-amido transferase illustrate some of these dif-ferences in regulatory mechanisms.

Figure 14 shows the changes of liver enzyme activities in chickens and rats which were fed on 5 or 50% protein diets. The activities of glutaminases from chicken liver are very low while the activities of glutamine synthetase and glutamine-PRPP-amidotransferase from chicken liver are high. Remarkable increases are observed in activities of glutamine synthetase and glutamine-PRPP-amidotransferase in the liver of chickens fed on high protein diets. These same enzymes in rat liver remained at constant low levels under the same dietary conditions. In contrast, liver glutaminases from rats fed on a high protein diet increased markedly but glutaminases from chicken liver did not increase under the same dietary conditions (Katunuma, N., et al., 1970). These characteristic responses seem to correlate teleologically with the ureoteric and uricoteric nature of the animals.

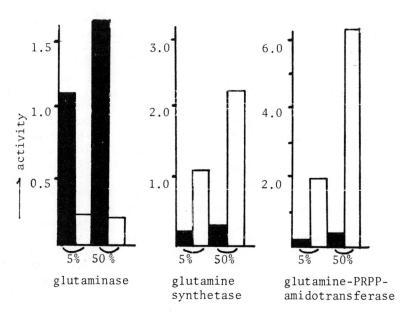

Fig. 14   Effects of high protein diet in chicken and rat liver.
Animals were fed on 5% or 50% casein diet for one week. Closed column (■) indicates the enzyme activity of rat liver and open column (▢) indicates that of chicken liver.

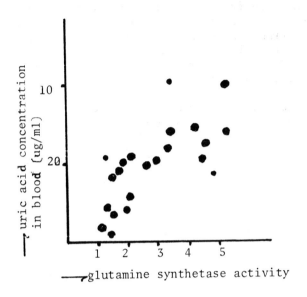

Fig. 15   Relationship between glutamine synthetase activities in chicken liver and uric acid concentration in blood.

In some conditions, the activity of glutamine synthetase itself seems to be one of the rate limiting enzymes in the biosynthetic pathway of uric acid in birds. At fifteen minutes after administration of 300 μmoles/100 g body weight of inorganic ammonia to rats having different strength of glutamine synthetase in liver, the rats were sacrificed. The uric acid level in blood and the activity of liver glutamine synthetase were assayed. The Figure 15 shows the parallel relationship between the amount of uric acid synthesized from ammonia and the activity of liver glutamine synthetase.

Feedback inhibition by products on glutaminase and glutamine-PRPP-amidotransferase: As shown in Figure 16, mammalian liver is not inhibited by the product, glutamate, but glutaminase from avian liver is strongly inhibited by glutamate. On the other hand, the activity of glutamine-PRPP-amidotransferase is inhibited by AMP to different degrees in both chicken and rat liver. The feedback inhibition by AMP in rat liver is much stronger than that in chicken liver for a given concentration of AMP as shown in Figure 17. This is understandable since AMP is one of the main products of the purine pathway in mammals; whereas, the main product of purine synthesis in birds is uric acid.

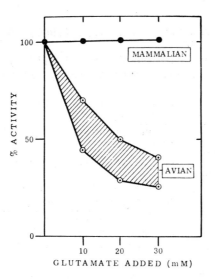

Fig. 16   Effect of glutamate on phosphate independent glutaminase activity in liver.

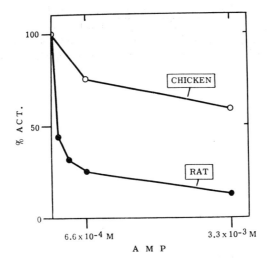

Fig. 17   Effect of AMP on glutamine-PRPP-amidotransferase in rat and chicken liver.

REFERENCES

Braunstein, A.E., and Seng, Hsu Ting. (1960). Biochem. Biophys. Acta. 44, 187.

Katunuma, N., Huzino, A., and Tomino, I. (1967). Advances in Enzyme Regulation. 5, 55. pergamon press, Oxford and New York.

Katunuma, N., Tomino, I.,and Nishino, H. (1966). Biochem. Biophys. Res. Communs. 22(3), 321.

Katsunuma, T., Temma, M., and Katunuma, N. (1968). Biochem. Biophys. Res. Communs. 32(3), 433.

Katunuma, N., Tomino, I., and Sanada, Y. (1968). Biochem. Biophys. Res. Conmuns. 32(3), 426.

Katunuma, N., Kuroda, Y., Sanada, Y., Towatari, T., Tomino,I, and Morris, H.P. (1970). Advances in Enzyme Regulation. 8. 281. pergamon press, Oxford and New York.

Katunuma, N., Matsuda, Y., and Kuroda, Y. (1970). Advances in Enzyme Regulation. 8, 73. pergamon press, Oxford and New York.

Kvamme, E., Tveit, B., and Svenneby, G. (1965). Biochem. Biophys. Res. Communs. 20, 566.

Kvamme E., Tveit, B., and Svenneby, G. (1970). J. Biol. Chem 245, 1871.

Svenneby, G. (1970). J. Neurochem. 17, 1591.

Yoshida, T., Towatari, T., and Katunuma, N. (1968). In " Symposium on Pyridoxal Enzymes "(K. Yamada, N. Katunuma and H. Wada eds.), pp. 63-66. Maruzen Company, Tokyo.

RAT KIDNEY GLUTAMINASE ISOENZYMES

Norman P. Curthoys,[1] Robert W. Sindel
and Oliver H. Lowry

Department of Pharmacology
Washington University School of Medicine
St. Louis, Missouri 63110

INTRODUCTION

In response to metabolic acidosis, the mammalian
kidney exhibits increased synthesis of ammonia. A large
proportion of this ammonia is excreted in the urine and
thereby increases the kidney's capacity to excrete anions
and titratable acid without depleting sodium and potassium
reserves. Use of ^{15}N labeled precursors suggests that
during acidosis 35-51% of the excreted ammonia is derived
from the amide nitrogen of glutamine, 10-26% from the
amine nitrogen of glutamine, 23-48% from arterial
ammonia and less than 10% from the amine nitrogen of
other amino acids (Pitts, 1968). The use of arterial
ammonia is not to be taken to indicate that organs other
than the kidney contribute significantly to urinary ammonia. During acidosis, the kidney actually produces more
ammonia than it excretes and the concentration of ammonia in the venous blood leaving the kidney is greater than
that in its arterial blood supply. The levels of amide and
amine nitrogen which the kidney extracts are sufficient to
account for both the levels of ammonia excreted and the
observed arteriovenous difference (Pitts, et al. , 1963).
Its use merely supports the fact that plasma ammonia
rapidly equilibrates with the pool of ammonia synthesized
within the kidney. Accumulation of ammonia ions in the
urine is thought to be a passive process in that freely
diffusible ammonia is trapped in the more acidic urine

(Balagara and Pitts, 1962).

Mammalian kidneys have two major pathways for the synthesis of ammonia from glutamine (Goldstein, 1967). On one pathway, glutamine is initially deamidated by glutaminase and the resulting glutamate is then oxidatively deaminated by glutamic dehydrogenase. On the second pathway, glutamine transaminase catalyzes the initial reaction, conversion of glutamine to α-ketoglutaramate. The α-ketoglutaramate is then deamidated by a specific ᴡ-deamidase. The relative contribution of the two pathways is uncertain, though many authors ascribe greater significance to the glutaminase pathway (Goldstein, 1966; Weiss and Preuss, 1970).

In addition to the classical phosphate dependent glutaminase (PDG), Katunuma (1966), has reported the occurrence of a phosphate independent glutaminase (PIG). The latter activity neither requires nor is activated by phosphate, but it is strongly activated by maleate. The two isoenzymes have been separated by calcium phosphate gel chromatography and differ in their kinetic properties (Katunuma, et al. , 1967). Only PIG has been extensively purified (Katunuma, et al. , 1968), and therefore, the two isoenzymes have not been sufficiently well characterized to determine if they are structurally related or not.

PDG activity increases 2-4 fold in response to a prolonged feeding of an acidic diet. It has not been established whether this increase represents enzyme activation or an increase in enzyme concentration. Previous experiments which used actinomyocin (Goldstein, 1965; Bignull, et al. , 1968) and ethionine (O'Donovan and Lotspeich, 1968) to inhibit increased activity are contradictory. After 4-7 days of feeding an acidic diet, there is an excellent correlation between increasing ammonia excretion and increasing PDG activity and between the maximal level of activity obtained and acidity of the diet (Rector, et al. , 1955). But, in the initial feeding period, increased ammonia production exceeds the increase in PDG activity (Leonard and Orloff, 1955; Alleyne, 1970). Therefore, increased potential PDG activity, as observed by in vitro assay, may play an important role in the maintenance of the gradual adaptive response to an acidic diet, but it cannot be the only method for triggering increased ammoniagenesis during onset of acidosis.

In addition to its ability to degrade glutamine, rat kidney can also synthesize glutamine from glutamate and ammonia. The biosynthetic reaction is catalyzed by glutamine synthetase and requires ATP hydrolysis. In vivo pulse labeling experiments have shown that in acidosis the kidney extensively degrades glutamine and there is little synthesis of glutamine from glutamate (Damian and Pitts, 1970). During alkalosis the opposite pattern occurs, resulting in a net synthesis of glutamine. This occurs without a change in in vitro activity of glutamine synthetase (Janicki and Goldstein, 1969) and without a decrease in glutaminase activity in alkalosis as compared to normal kidneys. The fact that opposing reactions occur under different metabolic conditions suggests that both renal glutamine synthesis and degradation are under some form of control designed to avoid a futile metabolic cycle which would result only in the loss of ATP.

Kidneys from acidotic rats exhibit reduced levels of glutamate and α-ketogluturate (Goldstein, 1966) and kidney slices from acidotic animals exhibit increased gluconeogenesis from glutamine, glutamate, α-ketoglutarate and oxalacetate, but not from fructose or glycerol (Goodman, et al., 1966). This suggests that during acidosis the carbon skeleton resulting from increased glutamine degradation is readily converted to glucose. Consistent with this idea is the fact that phosphoenolpyruvate carboxykinase activity increases 2-3 fold within 6 hours after administration of an acid load (Alleyne and Scullard, 1969). Increased activity for this rate limiting step in gluconeogenesis could cause the reduction in α-ketoglutarate and

261

glutamate levels. Because physiological levels of gluta-mate inhibit PDG activity greater than 90% when assayed at physiological concentrations of glutamine and phosphate (Goldstein, 1966), it has been proposed that increased gluconeogenesis could stimulate ammoniagenesis by relief of glutamate inhibition. Consistent with this theory is the observation that both gluconeogenesis and ammonia synthesis from glutamine increase in kidney slices from acidotic rats and decrease in slices from alkalotic rats (Goorno, et al. , 1967). Similar experiments with isolated perfused rat kidney have shown that the rate of glucose synthesis is sensitive to small changes in pH of the perfu-sate (Bowman, 1970). At pH 7.23 the rate was approxi-mately four times greater than at pH 7.64.

Although this theory has been provocative and has stimulated considerable research it is by no means proven. The 30% decrease in glutamate concentration observed in acidotic kidneys is sufficient to cause only a 20% increase in glutaminase activity (Goldstein, 1966). This is not sufficient to account for the 6-8 fold increase in ammonia synthesis which occurs during acidosis (Leonard and Orloff, 1955). Also, there is a lack of correlation between glucose and ammonia production early (24 hours) in acid-osis (Alleyne, 1970). This may indicate that the two are not directly linked in their genesis of adaptation. And further, in rat kidney slices, gluconeogenesis from gluta-mine is inhibited by malonate, which does not affect ammonia synthesis, and by phenylpyruvate, which stimu-lates ammoniagenesis (Churchill and Malvin, 1970).

Because of the structural complexity of the kidney, a better definition of these responses might be obtained by assaying PDG and PIG activities and glutamate and glutamine concentrations in the individual tubular struc-tures of the kidney as a function of acidosis. A complete understanding of the regulation of this response will require purification and characterization of the enzymes involved.

METHODS

White male rats (200-240 g) were made acidotic or
alkalotic by providing 0.28 M NH_4Cl or 0.28 M $NaHCO_3$
as their sole source of drinking water. Kidneys were
analyzed after various periods of treatment up to seven
days. Individual tubular structures (weight 5-50 ng) were
dissected freehand from lyophilized 20μ kidney sections
and were weighed on a quartz fiber balance (Lowry and
Passonneau, 1972).

By making use of their different kinetic properties,
specific assays were devised for each of the glutaminase
isoenzymes. The assays were performed in aqueous
droplets under oil (Matschinsky, et al., 1968), and
sufficient sensitivity was obtained by using the glutamate
cycle (Austin et al., 1972) to amplify the concentration of
the product of the glutaminase reaction. The amino
acids were assayed using similar techniques except that
glutamate was measured as the DPNH formed when it was
converted to α-ketoglutarate with the aid of glutamic
dehydrogenase. Glutamine was assayed in the same manner but with the addition of glutaminase. In both cases,
the DPNH was then determined by enzymatic cycling
(Kato and Lowry, 1972).

RESULTS AND DISCUSSION

Distribution of glutaminase isoenzymes. The two
glutaminase activities have essentially complementary
distributions in the various tubular structures of the
normal rat kidney (Fig. 1). PDG activity is high only in
distal straight and distal convoluted tubules and in the
inner stripe region, which consists largely of distal
straight tubules. It is intermediate in value in the inner
zone (papilla), which consists largely of collecting ducts,
and in proximal convoluted tubules. The activity is low
in glomeruli and in proximal straight tubules. Conversely
PIG activity is high only in proximal straight tubules.

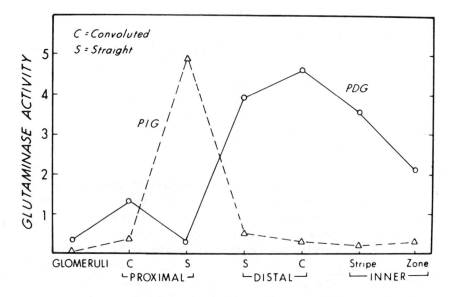

Fig. 1.   Distribution of PDG and PIG activities in various structures of normal rat kidney.   For each structures, 16 samples were assayed for each isoenzyme (S. E. less than 10% mean).   Glutaminase activity is expressed as moles kg^{-1} (dry) hr^{-1} at 20°.   "Inner stripe" indicates a patch cut from inner stripe of outer zone of medulla, and "inner zone" indicates a patch cut from inner zone of medulla (papilla).   Adapted from J. Biol. Chem. manuscript in press.

All the other structures have an activity for this isoenzyme which is less than one tenth that observed in these tubules.

PDG activity increases three fold in whole kidney homogenates from rats made acidotic by giving them NH$_4$Cl for 7 days.   During this period, PDG activity increases exponentially in the proximal convoluted tubules (Fig. 2).   Although no samples were assayed from rats, given NH$_4$Cl for longer than 7 days, previous experiments with homogenates have shown that the increase begins to

264

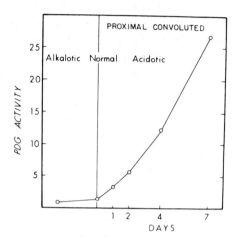

Fig. 2. PDG activity in proximal convoluted tubule. Acidosis and alkalosis (7 days) were induced as described in Methods. Each point represents the mean of 16 samples. Two rats were analyzed for each period of treatment. Activity is expressed as moles kg^{-1} (dry) hr^{-1} at $20°$.

level off after 7-10 days (Rector, et al., 1955). Therefore the 20 fold increase in activity observed is probably close to the maximum. During alkalosis, there is a slight (40%) but significant decrease in PDG activity in the proximal convoluted tubules. None of the remaining structures exhibit a dramatic adaptation in PDG activity in either acidosis or alkalosis. Similarly, PIG activity is not altered in any of the tubular structures.

The ratio of PDG and PIG activity in the acidotic kidney ranged from 0. 05 in proximal straight to 50 in proximal convoluted tubules. Because of this distribution, all of the structural components of the nephron, except glomeruli, have either a high or intermediate level of one or the other glutaminase isoenzyme. This suggests that the ability to synthesize ammonia in significant amounts is distributed throughout the kidney. However, the fact that during acidosis a dramatic increase in PDG activity occurs only in the proximal convoluted tubules suggests that this structure plays the most important role in maintaining increased ammonia synthesis. These conclusions are consistent with previous micropuncture studies which indicate that ammonia is added to the glomerular filtrate along the entire length of the nephron, but that as much as 50% of the ammonia synthesis in both normal and acidotic rat kidney occurs in the proximal convoluted tubule (Glabman, et al. , 1963).

Because the increase in PDG activity is localized within proximal convoluted tubules, what appears to be a 3 fold increase in whole kidney homogenates is in reality a 20 fold increase in this structure. Whether this represents induction of new enzyme or activation of preexisting enzyme, it is challenging to consider what mechanism can account for a 20 fold increase in one group of cells without any change in neighboring cells of the same nephron.

Distribution of glutamate and glutamine. The individual tubular structures of the normal rat kidney also contain different levels of glutamate and glutamine (Fig. 3). Among the various structures, there is a 4. 3 fold range in glutamate concentration and a 2. 6 fold range in glutamine concentration. Glutamate is greatest in the distal convoluted and distal straight tubules where glutamine levels are among the lowest. Conversely, proximal convoluted tubules have about one half the glutamate concentration of whole kidney and proximal straight tubules have even less. In the latter, low glutamate is associated with one of the highest concentrations of glutamine. Because of this distribution, the glutamate to glutamine ratio varies over a 9 fold range among the various tubular structures.

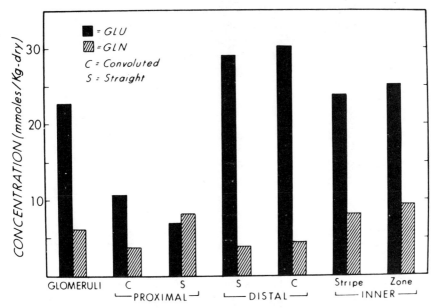

Fig. 3. Distribution of glutamate and glutamine in various structures of normal rat kidney. For each structure 12-16 samples were analyzed (S. E. less than 15% of mean). Adapted from manuscript submitted to Am. J. Physiol.

In all of the individual structures examined, the concentration of glutamine decreases 40 to 60% during the first 2 days of acidosis and then remains at this lower level (Table I). Even the glomeruli showed this decrease. Therefore, the decrease in glutamine levels appears to be a generalized response. This observation combined with the fact that kidney extracts greater amounts of glutamine from the plasma during acidosis (Pitts, 1968), suggests that glutamine is catabolized at a higher rate throughout the various structures of the cortex and outer stripe regions irrespective of the   relative activities of either glutaminase isoenzyme.

Table 1. Effect of acidosis on glutamine concentration in various structures of rat kidney.

Structure[b]	Glutamine-mmoles kg^{-1} (dry)[a]			
	Normal	Acidotic		
		2 days	4 days	7 days
Glom	6. 3	3. 3	2. 7	3. 4
Prox convol	3. 7	1. 5	1. 9	1. 5
Prox st	7. 9	4. 7	5. 1	3. 6
Dist convol	4. 5	1. 7	2. 0	2. 2

[a] For each structure, 8-16 samples were assayed (S. E. less than 15% of mean).
[b] Abbreviations are as used in Fig. 1.

During acidosis, the changes in glutamate concentration differ in each of the structures examined (Table 2). In glomeruli, glutamate decreases 20% within two days

Table 2. Effect of acidosis on glutamate concentration in various structures of rat kidney

Structure[b]	Glutamate-mmoles kg^{-1} (dry)[a]			
	Normal	Acidotic		
		2 days	4 days	7 days
Glom	25. 0	19. 3	21. 2	23. 3
Prox convol	10. 5	9. 2	9. 3	16. 6
Prox st	7. 0	6. 2	8. 4	8. 1
Dist st	29. 0	15. 0	16. 3	22. 0
Dist convol	30. 2	20. 7	19. 4	24. 2

[a] For each structure, 8-16 samples were assayed (S. E. less than 15% of mean)
[b] Abbreviations are as used in Fig. 1.

and then tends to return to normal. In proximal convoluted tubules there is only a slight decrease after 2-4 days. However, between the fourth and seventh day there is an 80% increase in glutamate. In proximal straight tubules there is little or no change during acidosis. The greatest decrease is observed in the distal straight and distal convoluted tubules where glutamate falls 50 and 35%, respectively, after 2 days. But, in both of these structures, glutamate increases significantly between the fourth and seventh day.

The observed decreases in glutamate are similar to those reported previously. Glutamate concentration in whole kidney decreased 30% within 24 hours after an intragastric administration of an acid load (Goldstein, 1966). After 3 days of repeated treatment, the level increased slightly to where it was 20% less than normal. In this study, glutamate concentration reached a minimum in structures which exhibited a decrease, early in acidosis (2-4 days) and increased thereafter. Summation of all the structures assayed could account for approximately a 20% decrease in glutamate after 2 days of acidosis.

During acidosis, phosphoenolpyruvate carboxykinase activity increases only in proximal convoluted tubules. [2] Glutamate concentration shows only a slight decrease in this structure. The largest decrease occurs in distal tubules, which contain low levels of carboxy kinase activity. The fact that the two responses are localized in different tubular structures indicates that the decrease in glutamate during acidosis is not caused by increased gluconeogenesis resulting from increased phosphoenolpyruvate carboxykinase activity.

It has been suggested that decreased levels of glutamate could cause increased ammonia synthesis by relief of inhibition of PDG activity (Goldstein, 1966). The tubular structures which contain a high level of PDG activity do have a high glutamate to glutamine ratio. Also, the largest decrease in glutamate occurs in distal tubules which have the highest PDG activity in normal

269

kidneys. But, according to Goldstein's calculations the 35-50% decrease in glutamate which occurs in these structures could cause only a 30-40% increase in PDG activity. Because of the relatively small mass of these structures, they represent only a small proportion of the total PDG activity. Therefore, it is unlikely that relief of glutamate inhibition could contribute significantly to the 2-4 fold increase in ammonia synthesis which occurs within the first 2 days of acidosis (Leonard and Orloff, 1955).

Purification of phosphate dependent glutaminase. The extensively purified PDG from hog kidney exists in different molecular weight forms (Kvamme, et al., 1970). In a borate-phosphate buffer, the enzyme polymerizes into long double stranded helices of very high molecular weight (Olsen, et al., 1970). Dialysis against a Tris buffer causes dissociation to units of approximately 150,000 molecular weight. It was reasoned that if a similar process occurred with the rat kidney enzyme, it could be readily purified by successive gel filtration of the enzyme in each of its molecular weight forms.

PDG activity is located within the mitochondria, and unlike the hog kidney enzyme, PDG from rat kidney is not readily solubilized by freezing and thawing homogenates. Initial attempts to solubilize the activity using sonication, various detergents and organic solvent extraction of lyophilized homogenates resulted in greater than 90% loss of activity. When solubilization procedures were repeated with borate buffers, 50% of the activity was solubilized with deoxycholate. However, significant activity was also solubilized by aqueous extraction of mitochondria lyophilized in borate buffers. When conditions were optimized with respect to borate, Pi, PPi and pH, 70% of the activity was obtained in a form soluble to centrifugation at 30,000 x g for 1 hour. When this enzyme was dialyzed against a Tris buffer, about 95% of the activity was lost. However, greater than 80% of the activity was recovered by redialysis against borate, Pi,

PPi buffer. To determine if inactivation was accompanied by a change in molecular weight, enzyme which eluted in the void volume of Bio-Gel A-1. 5 m column equilibrated with borate, Pi, PPi buffer was dialyzed against Tris and rechromatographed. The activity was partially included and appeared at the trailing edge of the protein peak which eluted with the void volume.

The specific activity of the Tris form of the enzyme is very protein concentration dependent and the activity is quite labile to storage even at 4°. It was found that low concentrations of dithiothreitol, PPi or ATP stabilize this form of the enzyme. In order to improve resolution of the two forms, a gel of higher inclusion range was employed. When the solubilized extract was chromatographed on Sepharose 4B in borate, Pi, PPi the activity still eluted in the void volume, clearly separated from the majority of the protein. Dialysis of this sample against Tris, PPi, dithiothreitol resulted in only a 3 fold decrease in activity. When rechromatographed in this form, practically all of the absorbance at 280 nm eluted in the void volume whereas the enzymatic activity eluted at a volume 2. 5 times the void volume. After concentration with ammonium sulfate, redialysis against borate, Pi, PPi resulted in a 2. 6 fold reactivation. This procedure (Table 3) resulted in a 1200 fold purification with a 16% yield.

When the Sepharose 4B columns were run in the reverse order, enzyme with about the same final specific activity was obtained. This indicates that reactivation of the purified enzyme is accompanied by polymerization. As opposed to the activity in crude homogenates which has an absolute requirement for an activator such as phosphate, both forms of the purified enzyme show residual activity in the absence of an activator. Also, PPi and ATP, have been found to be potent activators of the purified enzyme. They produce the same maximal activation at concentrations less than one tenth that required of phosphate.

Table 3. Purification of phosphate dependent glutaminase

Step	Protein	Activity	Recovery	Specific Activity
	mg	units[a]	%	units/mg
1. Homogen- ate	10,300	1010	100	0.10
2. Mitochon- dria	2,050	860	85	0.42
3. Soluble extract	1,060	550	55	0.52
4. Ammonium sulfate	710	525	52	0.74
5. Sepharose 4B (borate)	11.7	280	28	24
6. Dialysis vs Tris	7.5	88	--	12
7. Sepharose 4B (Tris)	1.3	64	--	49
8. Dialysis vs borate	1.3	166	16	127

[a] units are expressed as μmoles of glutamate formed per minute

SUMMARY

The two glutaminase isoenzymes have a nearly complementary distribution in the various tubular structures of
the kidney. Only PDG activity increases in response to
acidosis and this increase occurs only in proximal
convoluted tubules. Glutamine concentration decreases
40-60% during acidosis in all the structures examined,
suggesting that increased glutamine catabolism occurs
throughout the kidney cortex and outer stripe regions.
The changes in glutamate concentration do not support
the hypothesis that increased ammoniagenesis during

acidosis is related to increased gluconeogenesis or that relief of glutamate inhibition of PDG activity causes a significant increase in ammonia synthesis.

As with the enzyme from hog kidney, rat renal PDG undergoes an extensive polymerization in the presence of borate. This fact has been used to obtain a 1200 fold purification of PDG by successive gel filtration of the enzyme in each of its molecular weight forms. Characterization of this enzyme, with particular emphasis on its regulation, is the subject for continued research.

Acknowledgments

This research was supported in part by grants from the American Cancer Society (PF-638, P-781) and the National Institutes of Health (NB-05221).

REFERENCES

Alleyne, G. A. O. , and Scullard, G. H. (1969) J. Clin. Invest. , 48, 364.

Alleyne, G. A. O. (1970). J. Clin. Invest. , 49, 943.

Austin, L. , Lowry, O. H. , Brown, J. G. and Carter, J. G. (1972) Biochem. J. , 126, 351.

Balagura, S. and Pitts, R. F. (1962) Am. J. Physiol. , 203, 11.

Bignall, M. C. , Elebute, O. and Lotspeich, W. D. (1968) Am. J. Physiol. , 215, 289.

Bowman, R. H. (1970) J. Biol. Chem. , 245, 1604.

Churchill, P. C. and Malvin, R. L. (1970) Am. J. Physiol. , 218, 353.

Damian, A. C. and Pitts, R. F. (1970) Am. J. Physiol, 218, 1249.

Glabman, S. , Klose, R. M. and Giebisch, G. (1963) Am. J. Physiol. , 205, 127.

Goldstein, L. (1965) Nature, 205, 1330.

Goldstein, L. (1966) Am. J. Physiol. , 210, 661.

Goldstein, L. (1967) Am. J. Physiol. , 213, 983.

Goodman, A. D. , Fuisz, R. E. and Cahill, Jr. , G. F. (1966) J. Clin. Invest. , 45, 612.

Goorno, W. E. , Rector, Jr. , F. C. and Seldin, D. W. (1967) Am. J. Physiol. , 213, 969.

Janicki, R. H. and Goldstein, L. (1969) Am. J. Physiol. , 216, 1107.

Kato, T. and Lowry, O. H. (1972) Submitted to J. Biol. Chem.

Katunuma, N. , Tomino, I. and Nishino, H. (1966) Biochem. Biophys. Res. Commun. , 22, 321.

Katunuma, N. , Huzino, A. and Tomino, I . (1967). Advan. Enz. Reg. , 5, 55.

Katunuma, N. , Katsunuma, T. , Tomino, I. and Matsuda, Y. (1968) Advan. Enz. Reg. , 6, 227.

Kvamme, E. , Tveit, B. and Svenneby, G. (1970) J. Biol. Chem. , 245, 1871.

Leonard, E. and Orloff, J. (1955) Am. J. Physiol. , 182, 131.

Lowry, O. H. and Passoneau, J. V. (1972) "A Flexible System of Enzymatic Analysis," Academic Press, New York.

Matschinsky, F. M. , Passoneau, J. V. and Lowry, O. H. (1968) J. Histochem. Cytochem. , 16, 29.

O'Donovan, D. J. and Lotspeich, W. D. (1968) Enzymologia, 35, 82.

Olsen, B. R. , Svenneby, G. , Kvamme, E. , Tveit, B. and Eskeland, T. (1970) J. Mol. Biol. , 52, 239.

Pitts, R. F. , DeHass, J. C. M. and Klein, J. (1963). Am. J. Physiol. , 204, 187.

Pitts, R. F. (1968) "Physiology of the Kidney and Body Fluids," 2nd edition, Year Book Medical Publisher, Chicago.

Rector, Jr. , F. C. , Seldin, D. W. and Copenhaver, J. H. (1955) J. Clin. Invest. , 34, 20.

Weiss, F. R. and Preuss, H. G. (1970) Am. J. Physiol. , 218, 1697.

Footnotes

1. Present address: Department of Biochemistry, University of Pittsburgh, School of Medicine, Pittsburgh, Pa. 15213.

2. Personal communication from Dr. Helen B. Burch.

GLUTAMINASES AS ANTINEOPLASTIC AGENTS[1]

John S. Holcenberg[2], Joseph Roberts[2] and William C. Dolowy
University of Washington School of Medicine
Seattle, Washington

Abstract

Several enzymes with different ratios of glutaminase to asparaginase activity produce remissions in mouse tumors. Criteria have been established for selection of enzymes suitable as antitumor agents. Tests for selection of susceptible tumors require much further evaluation. Initial studies show that the animal host can tolerate prolonged depletion of plasma glutamine. Combined therapy and insolubilized or encapsulated enzymes may be needed to prevent immune response to the enzymes and to prolong their biologic half-life.

Symbols

A	= Asparagine
Ac GA	= Acinetobacter glutaminase-asparaginase
DON	= 6-Diazo-5-oxo-L-norleucine
DONV	= 5-Diazo-4-oxo-L-norvaline
EC-2	= E. coli asparaginase, enzyme with antitumor activity
G	= Glutamine
IP	= Isoelectric point
Pa GA	= Pseudomonas aeruginosa glutaminase-asparaginase
Pp G	= Pseudomonas putrefaciens glutaminase

Introduction

The parenteral administration of enzymes which degrade amino acids required only for growth of neoplasms offers a potential cancer therapy with marked specificity for the tumor. One such enzyme, L-asparaginase, is already in clinical use and produces remissions in a high percentage of cases of acute lymphocytic leukemia. However, L-asparaginase has little or no therapeutic effectiveness against any other neoplasms in man (Clarkson, et. al., 1970).

In 1964 Greenberg et. al., reported that a
glutaminase-asparaginase preparation with a relatively
high glutamine Km (7 mM) decreased the initial rate of
growth of a number of tumors, including an Ehrlich ascites
carcinoma, but caused no significant increase in the
survival time of tumor-bearing animals. Broome and
Schenkeim (1970) reported that a Pseudomonad glutaminase-
asparaginase preparation caused temporary regression in a
number of mouse lymphomas which were resistant to asparaginase. Bauer et. al., (1971) have also reported that a
glutaminase-asparaginase purified from Pseudomonas
aureofaciens was inhibitory to 9 of 16 tumors in mice and
rats. An asparaginase-resistant mouse adenocarcinoma was
found to be especially sensitive to this enzyme. We have
shown that asparaginase-resistant Ehrlich carcinomas
regressed permanently when tumor-bearing mice were injected
with bacterial glutaminase or glutaminase-asparaginase
preparations (Roberts et. al., 1970, 1971). Thus,
glutaminase treatment offers the potential for cancer
therapy in malignancies other than leukemia. This paper
will review our current ideas on the enzyme characteristics
necessary for antitumor efficacy, the in vivo effects of
these enzymes and potential predictive tests for cellular
sensitivity to glutamine deprivation.

Properties of Glutaminase Enzymes

Microorganisms constitute the most desirable source
for production of the large amounts of enzyme needed for
amino acid depletion therapy. Microorganisms can be
cultured in sufficient quantity at relatively low cost and
little time. Often the yield of enzyme produced may be
greatly increased (10 to 100-fold) by manipulation of
culture conditions or by mutation. The use of bacteria to
produce enzymes suitable for parenteral administration,
however, requires that the final product be exhaustively
purified from various bacterial toxins.

Many different asparaginase preparations have been
tested for antitumor activity (Laboureur, 1969, Adamson and
Fabro, 1968). Evaluation of the literature led to the
following criteria for selection of enzymes suitable for
in vivo testing of antitumor activity:

278

1. Optimal activity and stability at physiological pH.
2. Optimal activity and stability when incubated in vitro with animal and human sera and whole blood.
3. Slow clearance from the circulation when injected into animals.
4. Low Km (10^{-6} to 10^{-5} M range).
5. No inhibition by high concentrations of its products.
6. Derived from a nonpathogenic organism which does not contain excessive endotoxic activity.
7. No significant inhibition by amino acid analogs which might be used in combination treatments.
8. Low immunogenicity.

Asparaginases without antitumor activity fail to meet at least one of the criteria. Enzymes from E. coli and B. coagulans show sharp declines in activity below pH 8. Yeast and Fusarium enzymes are very rapidly cleared from circulation. The B. coagulans and EC-1 enzyme have Km for asparagine over 10^{-3} M (Scheetz et. al., 1971, Broome, 1965, Law and Wriston, 1971).

Riley et. al., (1972) have shown that treatment of mice with a glutaminase-asparaginase enzyme increases the plasma glutamic acid, aspartic acid and ammonia levels 100, 10 and 3-fold, respectively. Thus glutaminases inhibited by these products, like the E. coli glutaminase described by Prusiner (1971), would probably be unsuited for treatment.

The properties of three glutaminase preparations purified in our laboratory are shown in the following tables:

TABLE I

Properties of Glutaminases

Prep.		Km (M)	pH Opt. > 80% act	IP	MW
Ac GA	G	5.8×10^{-6}	5.5 – 9	8.4	138,000
	A	4.8×10^{-6}	$6.5 - 10^{+}$		
Pa GA	G	2.7×10^{-6}	5.5 – 9	6.8	150,000
	A	3.1×10^{-6}	$8.5 - 10^{+}$		
Pp G		3.5×10^{-5}	$7.1 - 9.5^{+}$	—	——

TABLE II

Substrate Specificity

Substrate	Relative Activity (%)		
	Ac GA	Pa GA	Pp G
L-glutamine	100	100	100
L-asparagine	77	55	0
L-asp + L-glut	86	72	100
D-glutamine	35	56	0
D-asparagine	26	52	—
Glu-hydroxamate :			
–hydrolysis	78	90	—
–synthesis	150	71	0
Succinamate	3.1	5	0
DON	0	0.02	0.02
DONV	0.17	0.08	0

280

TABLE III

Inhibition of Glutaminases

	Conc mM	Relative Activity (%)		
		Ac GA	Pa GA	Pp G
DON	0.03	13	77	72
DONV	2	31	100	100
Bromocresol green	1	34	47	—
L-glutamate	5	100	100	100
Ammonia	5	100	100	100
PCMB	0.1	100	100	0
EDTA	0.1	100	100	100

All three enzymes fit our criteria, and possess anti-tumor activity. The Acinetobacter enzyme (Ac GA) had the highest antitumor activity, and, therefore, was crystallized and studied most extensively.

This enzyme has L-glutaminase and L-asparaginase activity in a ratio of 1.2:1. The purification procedure provides an over-all yield of 40-60 percent from crude cell-free extract to homogeneity, and is adaptable to large scale isolation of the enzyme. The highest yields of enzyme were obtained when the organism was grown aerobically in a basal synthetic medium composed of L-glutamic acid, ammonium sulfate, trace minerals and phosphate buffer (Roberts et. al., 1972).

Ac GA caused complete regression of asparaginase-resistant Ehrlich carcinoma in vivo and selectively killed human leukemic leukocytes in tissue culture at about one-hundredth the effective concentration of E. coli asparaginase (Schrek et. al., 1971).

Very similar enzymes have been described in Azotobacter agilis (Ehrenfeld et. al., 1963) and Alcaligenes eutrophus (Allison et. al., 1971).

Two different glutaminase preparations were isolated from Pseudomonas organisms. A glutaminase which has no detectable asparaginase activity has been partially purified from Pseudomonas putrefaciens. In addition, it does not hydrolyze D-glutamine or synthesize glutamyl hydroxamate. This enzyme is inhibited by PCMB and is less stable than the other enzymes in serum.

The second enzyme, a glutaminase-asparaginase from Pseudomonas aeruginosa, has been purified to homogeneity. This enzyme has L-glutaminase and L-asparaginase activity in a ratio of 2:1 at pH 7.5. It produced complete regressions of an asparaginase-resistant Ehrlich carcinoma in 40 percent of treated animals, and more than 300 percent prolongation of mean survival time. Immunodiffusion tests showed that the precipitating antibody present in mice treated with Ac GA does not react with the Pseudomonas enzymes. Similar enzymes have been described by Soda et. al., (1972).

A glutaminase with no asparaginase activity has also been partially purified from a Sarcina species. This enzyme has optimal activity at pH 7.1 - 8.4, a Km of 0.5mM and an isoelectric point below 5. It is completely inhibited by 1 mM EDTA and appears to require either Mn or Mg ions. This enzyme is less effective than the Pseudomonas and Acinetobacter enzymes in treatment of the Ehrlich ascites tumor. In addition, it does not lower plasma glutamine levels as well.

These enzymes have different isoelectric points, but have similar half-lives in plasma of LDH virus treated mice. The half-life is much shorter in mice not carrying tumors or this virus.

Rutter and Wade (1971) have recently shown that profound changes in clearance rates were produced by altering the isoelectric point of Erwinia asparaginase. We are currently investigating the effect of acylation and deamination on clearance. Initial studies show that acylated enzyme has a prolonged plateau plasma concentration followed by a clearance similar to the native enzyme.

Bacterial products in impure enzyme preparations may be important factors in aminal toxicity. Partially purified preparations of Ac GA produced similar toxicity as injections of endotoxin. These effects were not seen with pure samples of this enzyme.

The effects of glutamine and asparagine analogues must also be studied. Asparaginase has been tested in combination with some of these analogues. Unfortunately, some analogues are inhibitors of the asparaginase and glutaminase activity. On the other hand, E. coli asparaginase rapidly hydrolyses the asparagine analogue

DONV (Jackson and Handschumacher, 1970).

Hopefully, some enzymes may elicit little immune response. E. coli asparaginase inhibits various types of immunity. Nevertheless, anaphylactic shock and neutralizing antibodies are frequently seen in patients treated with this enzyme (Capizzi et. al., 1971).

Daily treatment of Swiss mice with Ac GA induces circulating neutralizing and precipitating antibodies. Their appearance is associated with a precipitous shortening of the half-life of the enzyme in circulation. This fall in half-life can be prevented by a single dose of cyclophosphamide on the first day after glutaminase treatment. Thus, combination with immunosuppressants may be needed to prolong the effect of these enzymes.

The molecular size of proteins appear to determine their distribution (Mayerson et. al., 1969). E. coli asparaginase is distributed largely in the vascular space. Movement into the extracellular space is slow and incomplete (Ho et. al., 1971). An enzyme of smaller size might be distributed like albumin in the animal host. This greater distribution may increase the enzyme's ability to deplete the amino acid in the extracellular space.

The molecular weight of Ac GA was found to be less than 100,000 by chromotography on Sephadex and Biogel. Subsequent studies done in collaboration with Dr. David C. Teller, Department of Biochemistry (Holcenberg et. al., 1972) showed that the enzyme consists of four subunits (molecular weight 33,000) by sedimentation equilibrium in 5.5 M guanidine HCl and by electrophoresis in sodium dodecyl sulfate on polyacrylamide gels after cross-linking the protein with dimethyl suberimide.

Moving boundary velocity experiments showed that most of the native enzyme sediments as the tetramer (s_{20w}° = 7.42 ± 0.03S). On the other hand, equivalent boundary calculations always showed a smaller s_{20w}. Analytic sedimentation equilibrium experiments revealed a tetramer-dimer dissociation with a dimer molecular weight of 69,000 ± 3000.

The sedimentation of the active species was determined by zone sedimentation of the enzyme in DONV. The hydrolysis of this substrate was followed with UV optics. Absorbance of protein was cancelled with a new double sector cell in which the enzyme is layered into both sectors simultaneously.

283

$s_{20_w}^o$ for the species which degrade DONV was
7.6 $\pm$ 0.2S. By matching zone sedimentation and active
enzyme experiments, enzyme species smaller than tetramer
were shown to have four percent or less of the activity of
the tetramer. E. coli and Erwinia asparaginases had
approximately the same $s_{20_w}^o$ values for the active species.

The Pseudomonas glutaminase-asparaginase has also
been analysed by this method. The active enzyme has a
sedimentation coefficient of 8.0 $\pm$ 0.3S. This value
agrees with the molecular weight of 150,000 found by gel
filtration. The asparaginases and glutaminases described
by Broome and Schenkeim (1970) and Ramadan et. al., (1964)
are also from Pseudomonas species but have molecular
weights below 90,000. Further studies of the physical
properties of these enzymes are needed, especially the
sedimentation rate of the active species.

Another approach to decrease the immune response and
increase the biologic half-life of these enzymes is to
encapsulate or attach the enzyme to an insoluble matrix.
Initial studies in our laboratory performed in collaboration with Dr. Gottfried Schmer of this institution, and
also studies performed by others (Hasselberger et. al.,
1970; Chang et. al., 1971; and Allison et. al., 1972)
show that these methods are feasible. These approaches
may produce effective glutamine depleting agents from
enzymes which previously were too unstable or rapidly
cleared.

In Vivo Effects of Glutaminases

Plasma glutamine and asparagine are both lowered to
undetectable levels by E. coli L-asparaginase. The glutamine level remains depressed for a shorter time than the
asparagine (Miller et. al., 1969).

We have found that Ac GA injection into mice produces
a depression of glutamine for 3-6 days. Riley et. al.,
(1972) has shown that most plasma amino acids rise
temporarily during enzyme therapy with this enzyme. Per-
sistent elevations of glutamic acid, aspartic acid and
ammonia are seen. Glutamic acid levels rise to about 3 mM
and persist for many days after therapy is stopped. The
effects of these extremely high glutamic acid levels are
unknown.

Very little is known about the effects of prolonged

glutamine depletion on the enzymes involved with glutamine formation and utilization. In tissue culture both human cervical carcinoma cells, strain HeLa, and mouse subcutaneous fibroblast, strain L, increased their levels of glutamyl transferase many fold when glutamine was replaced by glutamic acid in the media (Paul and Fottrell 1963; DeMars 1958). Readdition of glutamine to the L cell cultures produced a very rapid decrease in glutamyl transferase activity. With the human line this activity was just diluted out by further cell growth.

Activity of glutamyl transferase was surveyed in various tissues of the mouse, rat and hamster. The highest levels were found in brain, liver and kidney 0.1-0.7 I.U./ mg protein. Lower levels were found in spleen, skeletal muscle, lung, 0.01-0.03I.U./mg protein. Small intestine had the lowest level, less than 0.01I.U./mg protein. We have studied the effect of three days treatment of Swiss Webster mice with Ac GA 270I.U./kg. These animals had been pretreated with LDH elevating virus to prolong the half-life of the glutaminase-asparaginase. The glutamine depletion therapy produced little change in the specific activity of glutamyl transferase in brain, small intestine, kidney, liver, skeletal muscle, spleen and lung.

A typical experiment with an Ehrlichs ascites tumor in Swiss Webster mice is illustrated in Figure I.

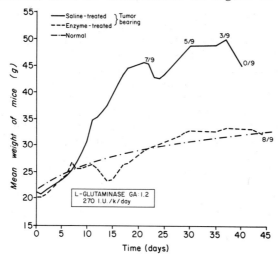

Fig. 1. Treatment of Ehrlichs ascites tumor. From Roberts et. al. (1971).

285

Most of the treated animals have now survived many months without recurrence of the tumor. It is noteworthy that at doses 270-540 I.U./kg/day or 540-1080 I.U./kg three times per week these animals show only transient weight loss as signs of toxicity. On the other hand, female C3H mice showed greater weight loss, decrease in spleen weight to less than 50% of controls, very pale livers, a lowered hematocrit and dilated food-filled stomachs at these doses. Similar high toxicity has been noted in BAF mice (Riley, Personal Communication). Hardy et. al., (1972) have noted toxicity in mice of a Pseudomonas glutaminase-asparaginase at doses over 200 I.U./kg/day. In the dog, five days of treatment with 100 I.U./kg/day produced severe toxicity including depression, anorexia, incoordination, bloody diarrhea, hypoproteinemia, azotemia, hyperphosphatemia and severe weight loss. Autopsy showed cerebral, colonic and renal abnormalities. We have treated two monkeys with the Acinetobacter glutaminase-asparaginase. Glutamine was lowered to nearly undetectable levels for 24 hours with no signs of gross toxicity, diarrhea or EEG abnormalities. Clearly, much more study is needed of the metabolic effects and toxicity of the purified glutaminases.

Tumor regression has been shown in a wide variety of mouse tumors after treatment with glutaminases (Greenberg et. al., 1964; Broome and Schenkeim, 1970; Bauer et. al., 1971; Roberts et. al., 1970, 1971). No published data is available on larger animals, although some of the effects of high levels of asparaginase may be due to glutamine depletion (Sugiura, 1969).

One major difference between the effects of asparaginase and glutaminase on susceptible tumors is that, unlike asparaginase therapy, some tumors that recur fol-lowing glutaminase therapy remain equally sensitive to glutaminase treatment after transfer to fresh hosts. In addition, Broome and Schenkeim (1970) have shown that the tumor cells exposed to the enzyme in tissue culture are apparently protected and able to survive in foci where macrophages have accumulated. It is possible that the macrophages or other cells can act as feeder cells sup-plying enough glutamine for survival of a few tumor cells.

Tumor Testing for Sensitivity to Glutaminase Treatment

The predictive tests which have been evaluated in connection with L-asparaginase therapy have not been useful
for selection of patients for treatment with that enzyme.
Asparagine synthetase is present in tumors resistant to
asparaginase. Unfortunately, this inducible enzyme may be
undetectable in resistant tumors prior to treatment and
appear only after treatment (Haskell et. al., 1970; Prager
and Bachynsky, 1968).
Incorporation of radioactive precursors into protein
and nucleic acids have been studied by several investigators. Although these tests have been useful in many
animal systems, investigators do not agree on their use-
fulness in predicting sensitivity to asparaginase in human
leukemia (Adamson and Fabro, 1968; Oettgen et. al., 1967;
Ho et. al., 1970). Recently, Ohnuma et. al., (1971)
reported that human leukemic cells require cysteine, glutamine, histidine, arginine and tyrosine.
We have done similar studies with animal tumors. The
following tumors were tested for their amino acid requirements for leucine incorporation and for glutamyl transferase and asparagine synthetase activity: Novikoff,
Morris 5132C and Hardy hepatomas; Chang liver cells;
Walker carcinoma; A Snell and C3H spontaneous breast
tumors; 591A, B16 and Apachie mouse melanomas; dog and cat
lymphosarcomas; methylcholanthrene induced mouse sarcoma;
three strains of Ehrlich ascites tumor; Gardner lymphomas
sensitive and resistant to asparaginase; NCTC 2472, RAD,
L1210, FBL3, MBL2, SU 122-TR4, spleen CRF, Lewis lung and
Barrett breast mouse tumors.
Asparagine synthetase (Method of Holcenberg, 1969)
was below 1 nmole/hr/mg protein in only the sensitive
Gardner, mouse spleen CRF, cat and dog lymphosarcoma, two
sensitive RAD strains and Morris 5123C hepatoma. Glutamyl
transferase (Method of Levintow, 1954) was below 10 nmoles/
minute/mg protein in the three Ehrlich ascites strains,
Chang liver, LSTRA, Morris 5123C hepatoma, asparaginase-
sensitive and -resistant RAD, mouse melanomas, methylcholanthrene tumor, Hardy hepatoma, FLB3 and A Snell
breast tumor.
The tumors were tested for leucine incorporation into
protein using a one-hour incubation with radioactive
leucine after one-hour preincubation. RMP1 1640 media

287

without histidine, serine, glutamine, asparagine and arginine was used. Cells were tested for incorporation in
complete media and media deficient in each amino acid.
Glutamine concentrations were varied, glutamate (10-20 mM)
was tested for its ability to replace glutamine, and asparagine was tested for its effect on the glutamine
requirement.

Most cells had low incorporation of leucine into
protein in the absence of glutamine. Nevertheless, some
glutaminase-asparaginase sensitive and resistant cells
require different concentrations of glutamine for maximal
leucine incorporation. For instance, the LSTRA tumor
which is resistant to glutaminase-asparagine treatment has
nearly maximal leucine incorporation at 0.1 mM glutamine
in the presence or absence of asparagine. On the other
hand, two strains of Ehrlichs ascites tumor which are
sensitive to glutaminase treatment have a much greater
glutamine requirement.

EHRLICH   ASCITES CARCINOMA

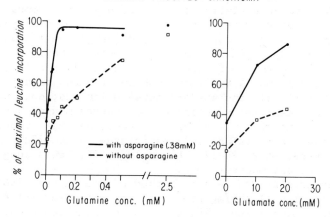

Fig. 2. Effect of L-glutamine and L-glutamate on incorporation of ^{14}C-leucine into tumor cell protein. Washed
Ehrlich cells(2.5 x 10^6-4.0 x 10^6 cells) were incubated
in 2 ml. RMPI 1640 medium. After 1 h, 0.5 µCi of uniformly
labelled L-^{14}C-leucine was added, and incubation continued
for 1 h. Cells were harvested by centrifugation, washed,
heated for 15 min in 5 per cent trichloroacetic acid, collected on glass filter paper, and the radioactivity was
determined. From Roberts et. al. (1970).

288

Figure 2 shows that the Ehrlich's cells have less than 50 percent maximal leucine incorporation at 0.1 mM glutamine when asparagine is absent. The glutamine requirement in the presence of asparagine is similar to that of the LSTRA tumor. The ability of glutamate to replace glutamine is also altered by asparagine depletion. These experiments suggest that greater antitumor effect with some tumors may occur with enzymes that deplete both glutamine and asparagine.

Schrek and Dolowy (1971) have shown a correlation between sensitivity to asparaginase and death of tumor cells in tissue culture. These studies have been extended to glutaminase treatment. Human chronic lymphocytic leukemia cells are much more sensitive in vitro to glutaminase than normal lymphocytes (Schrek et. al., 1971). This method is currently being further refined and extended.

Recently, Chou and Handschumacher (1972) have developed a test which measures the production of asparagine in short term tissue culture. This method may be able to differentiate between asparaginase sensitive and resistant tumor cells even though both have undetectable asparagine synthetase activities. We are currently examining the ability of tumor cells to produce glutamine and glutamate in a similar tissue culture system.

The glutamine requirement for sensitive tumors may not be caused by an inability to synthesize sufficient glutamine, but by a dependence of the cell on glutamine for energy. Glutamate is poorly transported by many tumor cells and glutamine is rapidly converted to glutamate within the cell. Knox et. al., (1969) have shown a correlation between glutaminase in tumor cells and their rate of growth. Recently, Kovacevic and Morris (1972) have shown an alteration in the mitochondrial utilization of glutamine and glutamate in tumor cells.

Glutaminase treatment of animals is clearly an exciting, new type of potential cancer chemotherapy. In addition, it will afford the biochemist and physiologist a chance to examine the effects of complete depletion of circulating levels of an amino acid involved in many tightly controlled metabolic pathways.

Adamson, R. H., and Fabro, S. (1968). Cancer Chemotherapy Reports 52, 617.

Allison, J. P., Mandy, W. J., and Kitto, G. B. (1971). Fed. Eur. Biochem. Soc. Lett. 14, 107.

Allison, J. P., Davidson, L., Gutierrez-Hartman, A., and Kitto, G. B. (1972). Biochem. Biophys. Res. Comm. 47, 66.

Bauer, K., Bierling, R., and Kaufmann, W. (1971) Naturwissenschaften 10, 526.

Broome, J. D. (1965). J. Nat. Cancer Inst. 35, 967.

Broome, J. D. and Schenkeim, I. (1970). Colloque Intern. sur l'Asparaginase CNRS 197, 95.

Capizzi, R. L., Bertino, J. R., Skeel, R. T., Creasey, W.A. Zanes, R., Olayon, C., Peterson, R. G. and Handschumacher, R. E. (1971). Annals Int. Med. 74, 893.

Chang, T. M. S. (1971). Nature 229, 117.

Chou, T. C. and Handschumacher, R. E. (1972). Biochem. Pharm. 21, 39.

Clarkson, B., Krakoff, I., Burchenal, J., Karnofsky, D., Golbey, R., Dowling, M., Oettgen, H. and Lipton, A. (1970). Cancer 25, 279.

DeMars, R. (1958). Biochim. Biophys. Acta 27, 435.

Ehrenfeld, E., Marble, S. J. and Meister, A. (1963). J. Biol. Chem. 238, 3711.

Greenberg, D. M., Blumenthal, G., and Ramadan, M. A. (1964) Cancer Res. 24, 957.

Hardy, W., Iritani, C., Schwartz, M. K., Old, L., and Oettgen, H. (1972). Proc. Amer. Assoc. Cancer Res. 13, 111.

Haskell, C. M., Canellos, G. P., Cooney, D. A., and Hansen, H. H. (1970). J. Lab. Clin. Med. 75, 763.

Hasselberger, F. X., Brown, H. D., Chattopadhyay, S. K., Mather, A. N., Stasiw, R. O., Patel, A. B., and Pennington, S. N. (1970). Cancer Res. 30, 2736.

Ho, D. H. W., Whitecar, J. P., Luce, J. K., and Frei, E. III (1970). Cancer Res. 30, 466.

Ho, D. H. W., Carter, C. J. K., Thetford, B. and Frei, E. III (1971). Cancer Chemotherapy Reports 55, 539.

Holcenberg, J. S. (1969). Biochim. Biophys. Acta 185, 228.

Holcenberg, J. S., Teller, D. C., Roberts, J., and Dolowy, W. C. (1972). J. Biol. Chem. In press.

Jackson, R. C. and Handschumacher, R. E. (1970). Biochemistry 9, 3585.

Knox, W. E., Horowitz, M. L., and Friedell, G. H. (1969). Cancer Res. 29, 669.

Kovacevic, Z., and Morris, H. P. (1972). Cancer Res. 32, 326.

Laboureur, P. (1969). Pathologie-Biologie 17, 885.

Law, A. S., and Wriston, J. C. (1971). Arch. Biochem. Biophys. 147, 744.

Levintow, L. (1954). J. Natl. Cancer Inst. 15, 347.

Mayerson, H. S., Wolfram, C. G., Shirley, H. H., Jr. and Wasserman, K. (1969). Amer. J. Physiol. 198, 155.

Miller, H. K., Salser, J. S., and Balis, M. E. (1969). Cancer Res. 29, 183.

Oettgen, H. F., Old, L. J., Boyse, E. A., Campbell, H. A., Philips, F. S., Clarkson, B. D., Tallal, L., Leeper, R. D., Schwartz, M. K., and Kim, J. H. (1967). Cancer Res. 27, 2619.

Ohnuma, T., Waligunda, J., and Holland, J. F. (1971). Cancer Res. 31, 1640.

Paul, J., and Fottrell, P. F. (1963). Biochim. Biophys. Acta 67, 334.

Prager, M. D., and Bachynsky, N. (1968). Biochem. Biophys. Res. Commun. 31, 43.

Prusiner, S. (1971). Fed. Proc. 30, 1113.

Ramadan, M. E., Asmar, F. E., and Greenberg, D. M. (1964). Arch. Biochem. Biophys. 108, 143.

Riley, V., Spackman, D., Fitzmaurice, M., Roberts, J., Holcenberg, J., and Dolowy, W. (1972). Proc. Amer. Assoc. Cancer Res. 13, 117.

Roberts, J., Holcenberg, J. S., and Dolowy, W. C. (1970). Nature 227, 1136.

Roberts, J., Holcenberg, J. S., and Dolowy, W. C. (1971). Life Sci. Part II Biochem. Gen. Mol. Biol. 10, 251.

Roberts, J., Holcenberg, J. S., and Dolowy, W. C. (1972). J. Biol. Chem. 247, 84.

Roberts, J., Holcenberg, J. S., and Dolowy, W. C. (1972a). Proc. Amer. Assoc. Cancer Res. 13, 25.

Rutter, D. A. and Wade, H. E. (1971). Brit. J. Exp. Path. 52, 610.

Scheetz, R. W., Whelan, H. A., and Wriston, J. C. (1971). Arch. Biochem. Biophys. 142, 184.

Schrek, R., and Dolowy, W. C. (1971). Cancer Res. 31, 523.

Schrek, R., Holcenberg, J. S., Roberts, J., and Dolowy, W. C. (1971). Nature 232, 265.

Soda, K., Ohshima, M., and Yamamoto, T. (1972). Biochem. Biophys. Res. Commun. 46, 1278.

Sugiura, K. (1969). Cancer Chemotherapy Reports 53, 189.

Footnotes

[1] Supported by U.S. PHS Grant CA11881. A preliminary report of some of this work was presented at the meeting of the American Association for Cancer Research, (Roberts et. al., 1972a).

[2] Career Development Awardee, United States Public Health Service.

GLUTAMINASES OF <u>ESCHERICHIA COLI</u>: PROPERTIES, REGULATION AND EVOLUTION

Stanley Prusiner*

Laboratory of Biochemistry, National Heart and Lung Institute, National Institutes of Health, Bethesda, Md. 20014

ABSTRACT

<u>E. coli</u> contain two glutaminases, A and B, which appear to function in cellular catabolism. The level of glutaminase A is controlled by nitrogenous metabolites and cyclic AMP. In contrast, glutaminase B is a microconstitutive enzyme. The activity of glutaminase B is allosterically regulated by adenine nucleotides, divalent cations, and carboxylic acids. Reciprocal controls prevent the coupling of glutaminase B with glutamine synthetase to form a "futile cycle" of amide synthesis and degradation.

Two classes of glutaminases have evolved: 1) catabolic with high rates of catalysis, and 2) anabolic which are subunits of biosynthetic enzymes and which have low rates of catalysis. Of the anabolic glutaminase subunits studied, each appears to be a distinct and different protein with one exception.

---

* Present address: Department of Neurology
University of California, School of Medicine
San Francisco, California 94122

INTRODUCTION

Glutaminases are ubiquitous in biological organisms. These enzymes catalyze the hydrolytic deamidation of L-glutamine resulting in the production of L-glutamate and ammonia. E. coli contain two glutaminase isoenzymes which are easily distinguished by differences in pH optima (Hartman 1968, Prusiner and Stadtman 1971). Glutaminase A has a pH optimum of 5 while glutaminase B is active above pH 7. Glutaminase A was partially purified by Meister and co-workers (Meister, et. al. 1955) and was more extensively purified and studied by Hartman (1973, 1968). Glutaminase B has only recently been studied (Prusiner and Stadtman 1971).

In this communication, the influence of nutritional conditions and of growth phase on the levels of glutaminases A and B is summarized, and the properties and regulation of glutaminase B are described. Lastly, the evolution and function of glutaminases is discussed.

Influence of Growth Condition upon Glutaminases

The influence of nutritional conditions and of culture growth on enzyme levels has been studied extensively in bacteria (Sokatch 1969). As shown in Table I, the level of glutaminase A is markedly changed by the stage of growth and the concentration of ammonia. The level of glutaminase A increases as the organisms enter stationary phase if the culture contains 40 mM NH_4Cl and 11 mM glucose. When the concentration of NH_4Cl is reduced to 4 mM, the level of glutaminase A is diminished and no increase is seen as the bacteria go into stationary phase. The level of glutaminase B is not altered by the stage of growth or the level of ammonia.

TABLE I

Effects of Growth and Ammonia on the Levels of Glutaminases
A and B in E. coli B

Enzyme	NH$_4$Cl*	Logrithmic	Stage of Growth Early Stationary	Late Stationary
	mM	Units/mg	Units/mg	Units/mg
Glutaminase A	40	0.038	0.387	0.468
	4	0.042	0.060	0.067
Glutaminase B	40	0.070	0.065	0.061
	4	0.067	0.079	0.053

One unit of enzyme is the amount that catalyzes conversion of
1.0 µmole of substrate to product per minute under standard assay
conditions (Prusiner and Stadtman 1971).

* Culture medium: 11 mM glucose and minimal salts

The levels of glutaminases A and B in E. coli which were
grown on glutamate or glutamine as nitrogen sources, are shown in
Table II. The level of glutaminase A is increased approximately
5-fold during logarithmic growth by using L-glutamate instead of
NH$_4$Cl as the source of nitrogen (Table I). As illustrated, the
level of glutaminase A did not change appreciably throughout the
growth cycle when L-glutamate was the nitrogen source. In contrast, growth on L-glutamine results in intermediate levels of
glutaminase A during the logarithmic phase with a 10- to 15-fold
increase during the stationary phase of growth. These results are
consistent with the work of Varrichio who examined the levels of
glutaminase A during logarithmic growth (Varrichio 1972). At
present, the mechanism by which nitrogenous metabolites control
the level of glutaminase A is unclear. Attempts to induce
glutaminase A in stationary phase with L-glutamate, D-glutamate,
L-glutamine, D-glutamine and NH$_4^+$ were without success, indic-

ating that the control may be more complicated than simple induction. Glutaminase B levels were unaffected by L-glutamate and L-glutamine in the culture media.

TABLE II

Effects of Glutamate and Glutamine on the Levels
of Glutaminases A and B in E. coli B

Enzyme	Amino Acid *		Logrithmic	Stage of Growth Early Stationary	Late Stationary
		mM	Units/mg	Units/mg	Units/mg
Glutaminase A	L-glutamate	100	0.183	0.301	0.271
		10	0.193	0.176	0.199
	L-glutamine	100	0.122	1.357	1.592
		10	0.072	0.332	0.677
Glutaminase B	L-glutamate	100	0.105	0.080	0.120
		10	0.066	0.062	0.070
	L-glutamine	100	0.068	0.071	0.067
		10	0.090	0.072	0.089

* Culture medium: 11 mM glucose and minimal salts

When E. coli cells were grown on glycerol and 40 mM NH_4Cl instead of glucose and 40 mM NH_4Cl, the rise in glutaminase A levels during early stationary was not observed. Addition of 3', 5' cyclic AMP to the glucose culture also prevented this rise in the level of glutaminase A but had no effect on the glycerol culture. As shown in Table III, cyclic AMP reduced the level of glutaminase A 3 to 5-fold for cells grown on glucose with high (40 mM) or low (4 mM) NH_4Cl. Cyclic AMP had no effect on the level of glutaminase A in glycerol grown cells. These observations appeared to be explained by the demonstration that E. coli cells grown on glycerol have an intracellular cAMP content approximately 100 times higher than those grown on glucose (Makman and Sutherland 1965).

TABLE III

Effects of Glycerol and Cyclic AMP on the Levels
of Glutaminases A and B in E. coli K-12

cAMP	Carbon Source	NH$_4$Cl	Glutaminase A	Glutaminase B
mM		mM	Units/mg	Units/mg
0	60 mM glycerol	40	0.091	0.063
5		40	0.100	0.075
0	30 mM glucose	40	0.466	0.060
5		40	0.103	0.056
0	60 mM glycerol	6	0.032	
5		6	0.018	
0	30 mM glucose	6	0.075	
5		6	0.028	

Cells were grown on a minimal salts medium and harvested in
early stationary phase

The reduction in the level of glutaminase A by cAMP was
abolished by chloramphenicol, an inhibitor of protein synthesis. In
addition, the effect of cAMP was not seen in cells with a deficiency of cAMP receptor protein (Pastan and Perlman 1970). Mixing
experiments eliminated activation or inhibition of the enzyme
directly as a possible explanation for the effect of cAMP. Negative control of enzyme levels in bacteria by cAMP is a new phenomenon and its molecular basis remains to be elucidated. Glutamate
synthase was also found to be under negative control by cAMP
while glutamate dehydrogenase and glutamine synthetase were observed to be under positive control by cAMP (Prusiner et. al. 1972).
Reciprocal effects of cAMP on enzymes have been reported for
glycogen synthetase and glycogen phosphorylase where the enzymes
are phosphorylated by a cAMP dependent protein kinase (Soderling
et. al. 1970). The chloramphenicol experiments, the requirement
for cAMP receptor protein and the slow time course of the cAMP
effect, all suggest that the changes in enzyme levels described

above are not due to covalent chemical modification of the
enzymes. Instead these alterations in enzyme levels appear to involve protein synthesis and/or degradation. The levels of glutaminase B were not altered by cAMP or glycerol.

In all the studies discussed above and other work not presented
here, the level of glutaminase B was not influenced by nutritional
conditions; thus, the enzyme appears to be under constitutive control.

It is noteworthy that the levels of glutaminases A and B were
not affected over the temperature range 30-38° or over the pH
range 6.0 to 7.5. Studies with automatic pH titrators established
that the stationary phase increase in glutaminase A levels occurred
when the pH was maintained constant at 7.0.

The influence of culture conditions on glutaminase activity
has been investigated extensively in Pseudomonas (Mardashev et. al.
1970) and in Achromobacteraceae (Roberts et. al. 1972). Both of
these organisms appear to have a single enzyme which deamidates
both L-glutamine and L-asparagine. The glutaminase from
Pseudomonas is induced by L-glutamine, L-glutamate, L-asparagine
and L-aspartate. The induction was blocked by nitrate or casein
hydrolyzate. D-asparagine, glycine, alanine, valine, histidine,
lysine and tryptophan did not induce the enzyme. The induction
took place during logarithmic growth and was maximal when the
culture media was maintained at pH 7.5. The optimal temperature
for glutaminase induction was 20°-25°. Growth on glucose increased the level of glutaminase two-fold over that found with glycerol.
No studies with cAMP were performed. The glutaminase from
Achromobacteraceae was also induced by L-glutamate during
logarithmic growth. Addition of yeast extract or tryptone decreased the glutamate-induced activity. The optimal temperature for
induction of glutaminase was 20 - 30° with a marked decrease in
the level observed below or above this temperature range.

Purification and Properties of Glutaminase B

As described above, glutaminase B is a constitutive enzyme

298

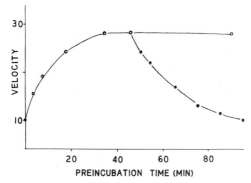

Fig. 1 A. Activation of glutaminase B by preincubation at
23° and inactivation by subsequent preincubation at 4°, plotted as
a function of time. Partially purified glutaminase B from E. coli B
was dialyzed against 500 volumes of 10 mM imidazole chloride
pH 7 at 4° with one change of dialysate. Aliquots of enzyme taken
at times shown, were assayed at 37°. Preincubation at 23° started
at 0 min (o ——— o) and preincubation at 4° begun at 50 min
(● ——— ●) using a portion of the 23° activated enzyme. For
details see (Prusiner and Stadtman 1971).

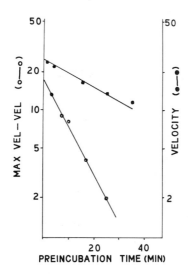

Fig. 1 B. Semi-logarithmic plots for activation and inactivation
of glutaminase B as a function of preincubation time. Preincubation at 23° (o ——— o) and preincubation at 4° (● ——— ●).

whose synthesis is independent of nutritional conditions. Using streptomycin precipitation, ammonium sulfate fractionation, DEAE chromatography, agarose gel filtration and preparative disc electrophoresis, we have purified glutaminase B more than 8000-fold. The purified protein appears approximately 40% pure on polyacrylamide gels and on isoelectric focusing. Glutaminase B in purified and crude preparations has a molecular weight of about 90,000 as determined by gel filtration and polyacrylamide gel electrophoresis. The enzyme is an acidic protein with an isoelectric point of about 5.4.

Preliminary studies with SDS gel electrophoresis suggest that glutaminase B is a multimeric protein composed of 3 or 4 subunits. Assuming the enzyme is composed of 4 subunits and each subunit has one active site, the turnover number of glutaminase B is approximately 22,500 when the enzyme is assayed at 37°. Glutaminase B exhibits maximal activity between pH 7 and 9 with less than 10% of maximal activity at pH 5 which is the pH optimum for glutaminase A.

Glutaminase B is reversibly inactivated by cold. Warming the enzyme at 23° results in reactivation by a first order process with a $t_{\frac{1}{2}}$ of 7.7 min. as shown in Figure 1. The enzyme may be inactivated again by an exposure to 4° and this inactivation is also first order with a $t_{\frac{1}{2}}$ of 28 min. Arrherius plots of the activation process show an energy of activation of 5000 cal/mole which is consistent with a conformational change. No alterations in molecular weight could be detected but as noted above completely pure enzyme was not available so that an inactive species of another molecular weight would not have been observed. Cold inactivation altered both the maximum velocity and the affinity of the enzyme for its substrate. The maximum velocity was reduced 2 to 3 fold by exposure to 4° while the $S_{0.5}$ for L-glutamine was increased from 2.6 mM to 6 mM.

The cold lability of glutaminase B could be prevented by the presence of small ligands such as L-glutamate, L-aspartate and borate. D-glutamate and D-aspartate did not protect against inactivation by cold. The catalytic activity of glutaminase B was stabilized by borate ions at both 4º and 23º. Borate ions have also been found to stabilize mammalian glutaminases (Kvamme et. al. 1970).

Glutaminase B exhibits a high degree of substrate specificity. Only L-glutamine is deamidated at an appreciable rate. The enzyme does not catalyze the hydrolysis of other amides such as D-glutamine, L-asparagine and D-asparagine. In addition, only L-glutamine is a good substrate for the production of an amino acid hydroxamate; D-glutamine and L-glutamate were ineffective substrates in this regard. A high degree of substrate specificity has also been observed with glutaminase A from E. coli (Hartman 1973, 1968) and glutaminase from yeast (Abdumalikov and Nikolaev 1967).

TABLE IV

Substrate Specificity of Glutaminase B Purified
from E. coli B

Substrate	^{14}C Product or NH_4^+ Produced	Amino Acid Hydroxamate
	relative activity	relative activity
1. L-Glutamine	100	100
2. D-Glutamine	2	< 2
3. L-Asparagine	< 0.2	< 2
4. D-Asparagine	< 1	< 2
5. L-Glutamate		< 2
6. L-Isoglutamine		< 2

In contrast, glutaminases from Pseudomonas (Ramadon et. al. 1964) and Achromobacteraceae (Roberts et. al. 1972) catalyze the deamidation of both L-asparagine and L-glutamine at nearly equal rates. It would be of considerable interest to know the differences in active site structure which are required for an enzyme to cata-lyze the deamidation of glutamine or asparagine or both.

Regulation and Function of Glutaminases

Relatively little is known about the function and regulation of glutaminases. The glutaminase isoenzymes of the kidney are the most well characterized of the mammalian glutaminases. These enzymes are localized in the mitochondrion and appear to be allosterically regulated by inroganic phosphate and carboxylic acids (Katunuma 1973, Kvamme et. al. 1970). Renal glutaminases appear to function in the excretion of NH_4^+ and in regulation of acid-base balance (Pitts 1968) while hepatic glutaminases may function in providing ammonia for urea synthesis (Pestana et. al. 1968).

In bacteria, the function and regulation of glutaminase B

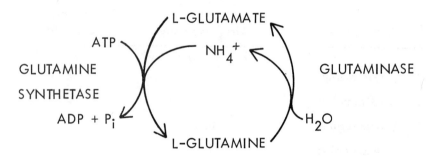

Fig. 2.   "Futile Cycle" of amide synthesis and degradation.

from E. coli has been studied extensively (Prusiner and Stadtman 1971). The regulation of gluatminase B was examined because the enzyme directly opposes the action of glutamine synthetase at neutral pH. In the absence of appropriate controls, the coupling of glutaminase B with glutamine synthesis to form a "futile cycle of amide synthesis and degradation would result in the wasteful consumption of cellular energy stores. Many similar potential "futile cycles" have been recognized in amphibolic pathways of which the gluconeogenic-glycolytic pathway is the best studied (Stadtman 1970, Scrutton and Utter 1968). In general, the anabolic enzyme is regulated by biosynthetic end-products and the catabolic enzyme is regulated by energy metabolites (Sanwal 1970). In the case of glutamine synthetase, 8 biosynthetic end-products from widely divergent pathways feedback to modify the activity of the enzyme (Woolfolk and Stadtman 1964) while glutaminase B is inhibited by nucleoside tri and di phosphates and is activated by nucleoside monophosphates (Prusiner and Stadtman 1971).

As shown in Figure 3, ATP and ADP inhibit glutaminase B in

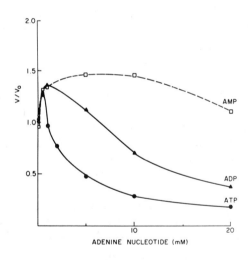

ADENINE NUCLEOTIDE (mM)

Fig. 3. Effect of adenine nucleotides on glutaminase B activity. The enzyme was preincubated with the appropriate concentration of nucleotide at 4° for 30 min. prior to assay. For details see Prusiner and Stadtman (1971). V = velocity in presence of effectors. V_o = velocity in absence of effectors.

concentrations above 1 mM. At concentrations below 1 mM these nucleotides are activators. AMP is also a weak activator. The alterations of glutaminase B activity by nucleotides are seen only when the enzyme is preincubated with the nucleotide prior to assay. Inclusion of the nucleotide in the assay solution has no effect on activity.

Maximal inhibition by ATP was observed when the enzyme was incubated with ATP at 4° for 30 min. The $t_{\frac{1}{2}}$ for inhibition by ATP was 3.5 min. at 4°. During the preincubation period no breakdown of ATP was detected by thin layer chromatography of ^{14}C-ATP. Pre-incubation at higher temperatures or in the presence of activators such as L-glutamate or L-aspartate reduced the extent of inhibition. The inhibition by ATP was reversed upon removal of the ATP using gel filtration or dialysis; however, activators were required for the enzyme to regain full activity. One possible interpretation of the data is that ATP alters the conformation of the enzyme and that this altered conformational state persists even after the removal of ATP unless an activating ligand is present. Alternatively, the binding of ATP with the enzyme may be very tight and activator ligands are necessary for the release of bound ATP.

The inhibition of ATP may be modified by AMP and divalent cations. AMP in concentrations which do not inhibit glutaminase B act synergistically with ATP to increase the inhibition of glutaminase B at levels of ATP between 1 and 5 mM. The synergistic effects of nucleotides and the biphasic inhibition curves strongly suggest that there are at least two nucleotide binding sites on the enzyme.

Divalent cations such as Mg^{++}, Mn^{++}, and Ca^{++} activate glutaminase B when they are preincubated with the enzyme at 4°. The presence of divalent cations in the assay solution has no effect on glutaminase B activity. In Figure 4, the activity of glutaminase B as a function of Mg^{++} concentration is plotted in the absence and presence of ATP. The inhibition of glutaminase B by 5 or 10 mM ATP is readily reversed by Mg^{++}. When 20 mM ATP was used the inhibition was also reversed by Mg^{++} but no change in activity was observed until 10 mM Mg^{++} had been added. The apparent sigmoidal shape of the curve appears to result from the complexing of the Mg^{++} with the ATP which is in excess of that needed to cause maximal inhibition. No data on the binding constants for Mg^{++} or ATP with the enzyme are available. The reversal of ATP inhibition by divalent cations has been observed with several enzymes which are inhibited by ATP such as fumarase and phosphofructokinase (Penner and Cohen 1969). In contrast, enzymes such as glutamine synthetase and hexokinase which utilize ATP as a substrate, require a divalent cation–ATP complex for the reaction to proceed. Also, these enzymes directly bind divalent cations (Stadtman 1970).

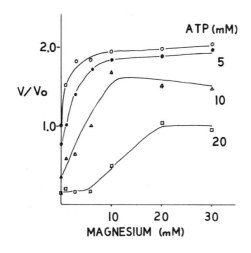

Fig. 4. Magnesium activation of glutaminase B. The enzyme was preincubated with appropriate concentrations of $MgCl_2$ and ATP at 4° for 30 min. prior to assay.

The intracellular concentrations of ATP and other adenine

nucleotides is determined by the balance between biosynthetic reactions which utilize ATP and catabolic reactions which generate
ATP. The regulation of these enzymatic reactions is a complex
function of the levels of divalent cations and adenine nucleotides.
The energy charge hypothesis attempts to examine the influence of
adenine nucleotides on enzymes which determine the level of cellular energy stores (Atkinson 1968). In Figure 5 the energy charge
of glutaminase B and glutamine synthetase are shown where the
total adenylate pool is 10 mM. The energy charge response curve
for glutaminase B is concave downward as seen with catabolic

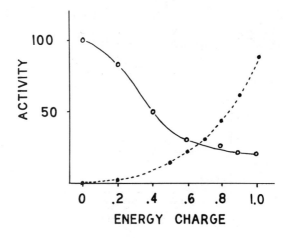

Fig. 5. Energy charge of glutaminase B and glutamine synthetase. Total adenylate pool 10 mM. Glutaminase B (o————— o)
was preincubated with appropriate concentrations of adenine
nucleotides at 4° for 30 min. prior to assay. Glutamine synthetase
(•———---•) was assayed directly in the presence of appropriate
concentrations of adenine nucleotides and 50 mM $MgCl_2$ by Drs.
Amiel Segal and E. R. Stadtman.

enzymes while the curve for glutamine synthetase is concave upward as observed with biosynthetic enzymes. The steep decline in activity of glutaminase B at energy charge levels lower than might be expected appears to be due to the synergistic effects of adenine nucleotides on the enzyme. The energy charge for glutamine synthetase was measured in the presence of 50 mM Mg^{++} by Dr. Amiel Segal while that for glutaminase B was done in the absence of Mg^{++}. Indeed, divalent cations greatly modify the response of both glutaminase B and glutamine synthetase to energy charge and must be considered in any attempt to assess the activity of these enzymes in vivo.

In recent experiments, Schutt and Holzer (1972) have measured the levels of metabolites in E. coli after a pulse of NH$_4$Cl. E. coli cells growing on a glucose-proline media were given 10 mM NH$_4$Cl (Wohlhueter et. al. 1973). Within 20 seconds, the level of glutamine increased 20-fold while 90% of the ATP was consumed. There was a small initial decrease in the level of glutamate, followed by a progressive rise. The activity of glutamine synthetase rapidly decreased probably due to glutamine stimulated adenylylation. Subsequently, the level of glutamine decreased and the level of ATP increased. Presumably the fall in ATP released glutaminase B from an inhibited state which resulted in a reduction of the glutamine level. The deamidation of glutamine in this situation probably serves two purposes: 1) it makes glutamate available for oxidation in the TCA cycle which leads to increased ATP formation, and 2) it reduces biosynthetic processes which utilize the amide nitrogen of glutamine and require ATP.

The reciprocal regulation of glutaminase B and glutamine synthetase is summarized in Figure 6. As shown, glutaminase B is inhibited by ATP and activated by AMP while glutamine synthetase is inhibited by AMP and uses ATP as a substrate (Woolfolk and Stadtman 1964). At present, no data is available on the efficiency of these regulatory processes in the prevention of "futile cycling" that involves amide and degradation. The same reciprocal regulation has been observed with phosphofructokinase and fructose 1,6 diphosphatase which could potentially couple to form a "futile cycle" (Scrutton and Utter 1968). Phosphofructokinase is inhibited by ATP and activated by AMP while fructose 1,6 diphosphatase is

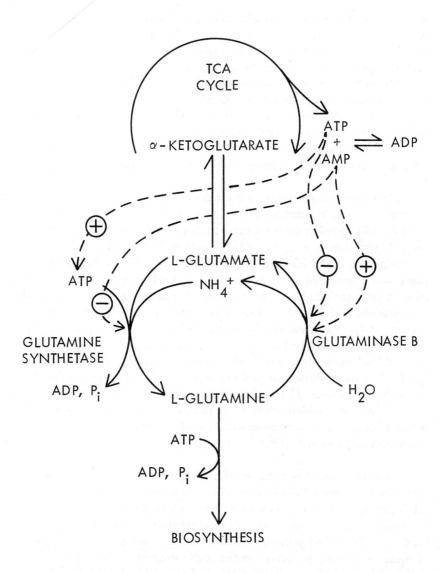

Fig. 6. Reciprocal control of glutamine synthetase and glutaminase B by adenine nucleotides.

inhibited by AMP and activated by ATP.

On the Evolution of Glutaminases

Two distinct classes of glutaminases have evolved in E. coli. One class which contains glutaminases A and B appears to partici- pate in cellular catabolism although the precise role of glutaminase A has not been established. These enzymes specifically catalyze the hydrolysis of the amide bond of L-glutamine at a rapid rate (Hartman 1968). In contrast, a second class of glutaminases partic- ipate in cellular anabolism and these glutaminases are subunits of enzymes or enzyme complexes which catalyze a variety of biosyn- thetic reactions. The rate of catalysis of these reactions is low com- pared with the first class of glutaminases (Levitski and Koshland 1972). Table V summarizes some of the data presently available on the two classes of glutaminases. In addition to the vastly different rates of catalysis, the subunit size is somewhat different both be- tween the two classes and within each class. Glutaminases A and B both have subunits less than 30,000 MW, while the glutaminase subunits of biosynthetic enzymes vary from 40,000 to 60,000 as shown (Table V).

The differences in molecular weight of the glutaminase sub- units of CTP synthetase, carbamyl phosphate synthetase and anthra- nilate synthetase provide firm evidence that a common glutaminase subunit is not produced for all biosynthetic enzymes utilizing the amide nitrogen of glutamine. In addition, the precise and specific stereochemistry of enzyme catalysis (Popjak 1970, Bruice 1970) would also argue against a common glutaminase subunit for the many biosynthetic enzymes which utilize the amide nitrogen of glutamine. The structures of the substrates and the stereochemistry of the specific reactions are so diverse that it is difficult to imagine how a single protein could function as a glutaminase subunit in the biosynthesis of amino acids, nucleotides, amino sugars and enzyma- tic cofactors. In addition, the cooperative interactions between specific subunits which has been demonstrated for CTP synthetase (Levitski 1973) and for carbamyl phosphate synthetase (Trotta et al. 1973) also argue against a common glutaminase subunit for bio- synthetic enzymes utilizing the amide nitrogen of glutamine.

TABLE V: Some Enzymes with Glutaminase Activity in E. coli

Enzyme	Molecular Weight	Subunit Molecular Wt.	Glutaminase Turnover No. min-1	Reference
1. Glutaminase A	10,000	28,000	38,000	(Hartman 1968, 1973)
2. Glutaminase B	90,000	< 35,000	12,000	(Prusiner and Stadt-man in preparation)
3. Anthranilate Synthetase	260,000	62,000	33*	(Ito and Yanofsky 1969, Nagano et al 1970)
4. Carbamyl-P Synthetase	172,000	130,000 42,000	38	(Trotta et al. 1971)
5. CTP Synthetase	210,000	52,000	290	(Levitski and Koshland 1972)

* Value for enzyme from S. typhimurium.

310

Table VI lists some of the various enzymes which utilize the amide nitrogen of glutamine and the location of their genes on the E. coli chromosome (Taylor 1970). The map positions are quite different, as expected, since the amide nitrogen of glutamine is used in a wide variety of biosynthetic reactions.

Also noted in Table VI are the levels of glutaminases A and B in mutants deficient in the various enzymes listed. In all of the mutants examined, both glutaminases A and B were present at the same level as found in the parent strains. Only in the case of anthranilate synthetase has the gene location of the glutaminase subunit been determined (Ito and Yanofsky 1966). This gene has been designated Try D and, as shown, the levels of glutaminases A and B were unaffected in mutants lacking the gene. Since the individual subunits of the other biosynthetic enzymes have not been mapped, we cannot prove genetically that glutaminases A or B are not subunits of one or more of these biosynthetic enzymes. However, the lack of any alterations of glutaminase A and B levels is consistent with the data presented above on structural and catalytic differences between the two classes of glutaminases.

It is noteworthy that the glutaminase subunit of anthranilate synthetase in S. typhimurium and probably E. coli is a distinct portion of a single polypeptide chain which also contains phospho ribosyl transferase activity (Grieshaber and Bauerle 1972). In contrast, the glutaminase subunit of anthranilate synthetase in B. subtilis contains no phosphoribosyl transferase activity. This glutaminase subunit in B. subtilis has been assigned to an extra-operonic gene which also codes for the glutaminase subunit of PABA synthetase (Kane et. al. 1972). Further studies are needed to establish whether or not the glutaminase subunits of both anthranilate synthetase and PABA synthetase do have identical amino acid sequences. This extra-operonic gene has been shown not to code for the glutaminase activity of CTP synthetase in B. subtilis (Kane et.al. 1972). There are no data available on glutaminases which deamidate glutamine at high rates for catabolic processes in B. subtilis.

Other examples of different proteins with common subunits have been reported for the pyruvate and α-ketoglutarate

311

TABLE VI: Glutamine-dependent Biosynthetic Enzymes in E. coli

Deficient Enzyme	Gene Symbol	Map Position* (Min)	Glutaminase Level in Mutants‡ A	B
1. Anthranilate Synthetase	Try A-E	25	+	+
	Try D	25	+	+
	Try E	25	+	+
2. Phosphoribulosyl Formimino Amidotransferase	His H	39	+	+
3. Carbamyl-P Synthetase	Pyr A	0	+	+
4. CTP Synthetase	Pyr G	48	+	+
5. FGAR Synthetase	Pur G	48		+
6. PRPP Amidotransferase	Pur F	44	+	+
7. GMP Synthetase	Gua A	48	+	+
8. Glutamate Synthase	Glu S			
9. Fructose 6-P Amidotransferase	Glm S	74		+

+ = No significant difference from wild type controls

* See Taylor (1970)

‡ Mutants of E. coli (gene symbol/strain, source) 1. Try A-E/AE del 1, Try D/D9778, Try E/E 5972, C. Yanofsky; 2. His H/H750, P. Hartman; 3. Pyr A/P4 X Jef 8, A. Pierard; 4. Pyr G/LD-300, J. Neuhard; 5. Pur G/4497, P. DeHaan; 6. Pur F/4526, B. Bachmann; 7. Gua A/4509, 4522, B. Bachmann; 8. Glu S/AB 1450, E. Adelberg

312

dehydrogenase complexes where each appears to have the same flavoprotein as characterized by several physical methods (Pettit and Reed 1967) and by a single genetic locus (Guest 1972). Mutants of E. coli which lack this flavoprotein are deficient in pyruvate and α-ketoglutarate dehydrogenases but both activities may be restored by addition of the purified flavoprotein to crude extracts (Guest 1972). Recent studies on the amino acid sequences of the C_1 subunit of bovine lutenizing hormone and of the α subunit of bovine TSH have shown the two subunits to be identical except for differences in sugar moieties (Pierce et.al. 1971). Undoubtedly other examples of shared subunits by different proteins will arise, but the present information suggests that enzymes with shared subunits will have relatively similar substrates or catalyze similar types of reactions.

In conclusion, it is of considerable interest to ask why glutamine has evolved as the storage moiety for utilizable nitrogen in living organisms. Why not asparagine or urea? We know of no instance where asparagine or urea are able to donate nitrogen for a biosynthetic reaction except where asparagine is an amide nitrogen donor in the synthesis of glutaminyl tRNA (Wilcox and Nirenberg 1968). Since the amide bonds of glutamine and asparagine appear to have very similar free energies (Levintow and Meister 1954), there is no apparent thermodynamic explanation for the specific use of L-glutamine. We can only assume that glutamine provides not only some advantageous mode for ammonia storage but also an advantage in synthetase and transferase reactions which utilize the amide nitrogen. Preliminary studies suggest that the cell could probably survive without glutamine except for protein synthesis and use NH_4^+ directly for the various biosynthetic reactions instead of glutamine (Gibson et.al. 1967).

It is a pleasure to thank Dr. Earl Stadtman for his guidance and encouragement throughout these studies. Author also wishes to thank Drs. C. Yanofsky, P. Hartman, A. Pierard, J. Neuhard, P. DeHaan, B. Bachmann and M. Berberich for mutants of E. coli.

REFERENCES

Abdumalikov, A. K., and Nikolaev, A. (1967) Biokhimiya, 32, 859.
Atkinson, D. (1968) Biochemistry, 7, 4030.
Bruice, T. C. (1970) In "The Enzymes" (P. D. Boyer, ed.), Vol. II, 3rd Edition, pp. 217-279, Academic Press, New York.
Dietrich, J., and Henning, U. (1970) Eur. J. Biochem., 14, 258.
Gibson, F., Pittard, J., and Reich, E. (1967) Biochim. Biophys. Acta, 136, 573.
Grieshaber, M., and Bauerle, R. (1972) Nature New Biol., 236, 232.
Guest, J. R. (1972) Personal Communication.
Hartman, S. (1968) J. Biol. Chem., 243, 853.
Hartman, S. (1973) In "The Enzymes of Glutamine Metabolism" (S. Prusiner and E. R. Stadtman, eds.), Academic Press, New York.
Ito, J., and Yanofsky, C. (1966) J. Biol. Chem., 241, 4112.
Kane, J. F., Holmes, W., and Jensen, R. (1972) J. Biol. Chem., 247, 1587.
Katunuma, A. (1972) In "The Enzymes of Glutamine Metabolism" (S. Prusiner and E. R. Stadtman, eds.), Academic Press, New York.
Kvamme, E., Tveit, B., and Svenneby, G. (1970) J. Biol. Chem., 245, 1871.
Levintow, L., and Meister, A. (1954) J. Biol. Chem., 209, 265.
Levitski, A. (1973) In "The Enzymes of Glutamine Metabolism" (S. Prusiner and E. R. Stadtman, eds.), Academic Press, New York.
Levitski, A., and Koshland, Jr., D. E. (1972) Biochemistry, 11, 241.
Makman, R. S., and Sutherland, E. W. (1965) J. Biol. Chem., 240, 1309.
Mardashev, S. R., Eremenko, V. V., and Nikolaev, A. (1970) Mikrobiologiya, 39, 11.
Meister, A., Levintow, L., Greenfield, R. E., and Absendschein, P. A. (1955) J. Biol. Chem., 215, 441.
Pastan, I., and Perlman, R. (1970) Science, 169, 339.
Penner, P. E., and Cohen, L. (1969) J. Biol. Chem., 244, 1070.

Pestana, A., Marco, R., and Sols, A. (1968) Febs Letters, 1, 317.
Pettit, F., and Reed, L. J. (1967) Proc. Nat. Acad. Sci., 58, 1126.
Pierce, J., Liao, T., Carlsen, R., and Reimo, T. (1971) J. Biol. Chem., 246, 866.
Pitts, R. (1968) "Physiology of the Kidney and Body Fluids", pp.195 -211 Year Book Medical Publishers, Chicago.
Popjak, F. (1970) In "The Enzymes" (P. D. Boyer, ed.), Vol. II, 3rd Edition, pp. 115-215, Academic Press, New York.
Prusiner, S., Miller, R., and Valentine, R. (1972) Proc. Nat. Acad. Sci. 69, 2922.
Prusiner, S., and Stadtman, E. R. (1971) Biochem. Biophys. Res. Comm., 45, 1474.
Ramadan, M., Asmar, F., and Greenberg, D. (1964) Arch. Biochem. Biophys., 108, 143.
Roberts, J., Holcenberg, J. S., and Dolowy, W. (1972) J. Biol. Chem., 247, 84.
Sanwal, B. D. (1970) Bact. Rev., 34, 20.
Schutt, H., and Holzer, H. (1972) Eur. J. Biochem., 244, 1070.
Scrutton, M. C., and Utter, M. F. (1968) Ann.Rev. Biochem., 37, 249.
Soderling, T. R., Hickenbottom, J. P., Reimann, E. M., Hunkeler, F. L., Walsh, D. A., and Krebs, E. G. (1970) J. Biol. Chem., 245, 6317.
Sokatch, J. R. (1969) "Bacterial Physiology and Metabolism," pp.1-368, Academic Press, New York.
Stadtman, E. R. (1970) In "The Enzymes" (P. D. Boyer, ed.), Vol. II, 3rd Edition, pp. 398-460, Academic Press, New York.
Taylor, A. (1970) Bact. Rev., 34, 155.
Trotta, P., Burt, M., Haschemeyer, R., and Meister, A. (1971) Proc. Nat. Acad. Sci., 68, 2599.
Trotta, P., Pinkus, L., Wellner, V., Estis, L., Haschemeyer, R., and Meister, A. (1973) In "The Enzymes of Glutamine Metabolism" (S. Prusiner and E. R. Stadtman, eds.) Academic Press, New York.
Varrichio, F. (1972) Arch. Mikrobiol., 81, 234.
Wilcox, M., and Nirenberg, M. (1968) Proc. Nat. Acad. Sci., 61, 229.

Woolfolk, C. A., and Stadtman, E. R. (1964) Biochem. Biophys. Res. Comm., 17, 313.

Wohlhueter, R. M., Schutt, H., and Holzer, H. (1973) In "The Enzymes of Glutamine Metabolism" (S. Prusiner and E. R. Stadtman, eds.), Academic Press, New York.

SECTION II
BIOSYNTHETIC ENZYMES UTILIZING THE AMIDE NITROGEN OF GLUTAMINE

RELATIONSHIPS BETWEEN GLUTAMINE AMIDOTRANSFERASES AND GLUTAMINASES

Standish C. Hartman

Department of Chemistry
Boston University
Boston, Massachusetts

The amide NH_2 group of glutamine serves as the immediate source of nitrogen atoms in a wide range of biochemical substances, including purines and pyrimidines; the amino acids, tryptophan, histidine, asparagine, glutamic acid, and arginine; amino sugars; and the coenzymes, folic acid (via p-aminobenzoic acid) and nicotinamide nucleotides. Through its incorporation into the α-amino position of glutamic acid it gains entry into most of the other amino acids and innumerable other metabolites. About a dozen enzymes are now known which catalyze such transfer of the NH_2 group and to which the trivial name "amidotransferase" has been applied. Although variations in the stoichiometry of these reactions exist, most of them can be represented by the equation

$$
\begin{array}{llll}
\begin{matrix} CONH_2 \\ | \\ CH_2 \\ | \\ CH_2 \\ | \\ CHNH_3^+ \\ | \\ COO^- \end{matrix}
& + \quad
\begin{matrix} X \\ \| \\ C\text{-}OH \\ | \\ Y \end{matrix}
\; + \; H_2O \;\rightarrow\;
\begin{matrix} COOH \\ | \\ CH_2 \\ | \\ CH_2 \\ | \\ CHNH_3^+ \\ | \\ COO^- \end{matrix}
& + \quad
\begin{matrix} X \\ \| \\ C\text{-}NH_2 \\ | \\ Y \end{matrix}
\end{array}
$$

Displacement of the hydroxyl group in the acceptor is usually facilitated by interaction with a nucleoside triphosphate or by prior formation of a phosphoryl derivative.

319

STANDISH C. HARTMAN

It is highly likely that these several amination reactions initially occurred in evolution with ammonia rather than glutamine as the direct reactant. Indeed, the kinship of ammonia and glutamine persists as evidenced by the fact that almost every one of the amidotransferases can utilize ammonia as an alternate substrate (glucosamine synthetase is an exception (1)). The pH-rate profiles of several of the reactions suggest that the free amine rather than ammonium ion is the chemically reactive species, a result consistent with the nucleophilic nature of the processes. But ammonia must be a very inefficient substrate since only about one percent of the total ammonium is present in basic form in the neutral pH range. Evidently Nature required a better aminating agent, especially to operate in times of nitrogen limitation for microorganisms and in compartments of animal tissues where ammonium ion concentration is always low.

The glutamine-utilizing amidotransferases, acting in concert with glutamine synthetase, present a thermodynamically favorable route for the incorporation of inorganic ammonia into nitrogenous intermediates in which glutamic acid acts cyclically as the ammonia carrier. This indirect process may be viewed as effecting deprotonation of ammonium ion at the expense of ATP hydrolysis, according to reactions 1-3:

1. Glutamine synthetase $(\Delta G_1^{0\prime} = -3.9 \text{ kcal/mole})$

NH_4^+ + Glutamate + ATP $\rightarrow$ Glutamine + ADP + P_i

2. Glutaminase action of amidotransferase
$(\Delta G_2^{0\prime} = -0.3 \text{ kcal/mole})$

Glutamine + H_2O $\rightarrow$ Glutamate + $[NH_3]$ + H^+

3. Sum of 1 + 2 $(\Delta G_3^{0\prime} = -4.2 \text{ kcal/mole})$

NH_4^+ + ATP + H_2O $\rightarrow$ $[NH_3]$ + H^+ + ADP + P_i

where the $\Delta G^{0'}$ values are for pH 7 (2). [NH_3] represents
an enzyme-bound form of ammonia, released by action of the
amidotransferase at the site of attack upon the activated
acceptor. $\Delta G_2^{0'}$ comes from the free energies of hydrolysis
of glutamine and of ionization of ammonium at pH 7, with
no special interactions between the enzyme and the NH_3
assumed. Whether or not this formulation is mechanisti-
cally valid, it accounts thermodynamically for production
of reactive ammonia of high chemical potential.

It is theoretically and experimentally proper to consider one function of the amidotransferases to be that of
a glutamine hydrolase. The nitrogen atom in amide linkage
is inherently a poor nucleophile owing to hybridization of
its free electron pair with the carbonyl group, so that
cleavage of the C-N bond in all probability precedes
attack by nitrogen on the acceptor molecule. Significantly
several amidotransferases do exhibit glutaminase activity
in that they hydrolyze glutamine and related glutamyl
derivatives, especially under conditions in which the
transfer of the leaving group to the normal acceptor is
prevented (3-6). The related acyl transfer to hydroyl-
amine has also been observed (6,7). These events have
been formulated as proceeding through acyl enzyme intermediates analogous to those occurring with the proteolytic enzymes, chymotrypsin and papain.

The chemical state of the nitrogen atom during its
transfer is of interest although little experimentation
has been directed to this question. It has been suggested
that a pyridoxal coenzyme (8), or perhaps an electrophilic
center such as occurs in histidine deaminase (9), might
act as intermediate carrier of the amino group, but the
evidence so far does not support this notion. It should
be noted that any combination of this type would reduce
the nucleophilic reactivity of the group being transfered.
A transition metal ammine complex is another possibility,
one which would suggest a role for the iron in PRPP amidotransferase (7,10). However, related metal ions have not
been detected in other enzymes of this group. The
difficulty of expelling an amide ion (NH_2^-) from amides

makes it unlikely that this entity is the true intermediate. In view of the facile reaction of free ammonia with the activated acceptor in several of the cases in point, I am inclined to view that the transient species is NH_3, protected in a non-aqueous pocket on the enzyme from protonation by solvent and solute molecules. A similar proposal has been made by Khedouri, et al. (11).

One of the most exciting findings now emerging from several laboratories working in this field is that certain of the amidotransferases contain dissociable subunits, the function of which is to bind and react with glutamine essentially as implied by Reaction 2. After dissociation the remaining enzyme components often are able to catalyze amination with ammonia, but not with glutamine, as substrate. Not only do these results have significance for understanding the construction and function of multi-substrate enzymes, but they have led various workers to the postulate that, in evolution, acquisition of reactivity with glutamine may have arisen by association of diverse ammonia-dependent enzymes with a primordial protein having the ability to mobilize glutamine, perhaps a glutaminase. The clearest case for this view comes from the work of Kane, et al., who showed that the same glutamine binding subunit, of M = 16,000, interacts with two different enzymes of B. subtilis involved in the amination of chorismic acid to o- and p-aminobenzoate, respectively (12).

There are some problems with accepting this hypothesis as being valid for the whole class of amidotransferases, however. First, the possibility must be admitted that considerable structural modification of the hypothetical subunit could have occurred, since components ranging in molecular weight from 9000 (p-aminobenzoate synthetase of E. coli (13)) to 42,000 (carbamyl phosphate synthetase (5)) have been observed. In other instances the glutamine binding regions appear to be covalently linked to larger catalytic entities (e.g., FGAR amidotransferase (14), the phosphoribosyl transferase component of anthranilate synthetase in Salmonella (15), and CTP synthetase (16)). Also, the amination reactions leading to guanosine-5'-phosphate (17) and asparagine (18) require ammonia in bacteria while the similar processes in animal cells are catalyzed by glutamine amidotransferases (18, 19). It would seem that the ability to use glutamine was acquired at a different

stage in evolution in these cases than for the bacterial
amidotransferases.

In another approach to investigating homology, the use
of the glutamine analogues, azaserine, diazo-oxonorleucine
(DON), and 2-amino-4-oxo-5-chloropentanoic acid allows
selective labeling of groups at the glutamine binding
sites. So far, these agents have been reported to alkylate
cysteine sulfhydryl groups in four amidotransferases: FGAR
amidotransferase (20), CTP synthetase (16), anthranilate
synthetase (3), and carbamyl phosphate synthetase (21).
Extension of these studies to other systems and to larger
polypeptide stretches around the key cysteine moieties are
anticipated with great interest as they should provide a
critical test of the "common origin" hypothesis.

A composite scheme for the action of a hypothetical
amidotransferase is shown in Fig. 1 which synthesizes the
main features outlined in this discussion. The sulfhydryl
group reactive with glutamine analogues is appropriately
placed for nucleophilic displacement of the NH_2 group, as
shown, to form a thioacyl intermediate which is subsequently hydrolyzed. Ammonia arising by displacement from glutamine or by diffusion from the medium gains access to the
activated acceptor (for which the phosphorylating agents
have been omitted). Blocking the glutamine site with an
inhibitor moiety or dissociating the glutamine binding
component entirely leaves the remaining complex still able
to react with ammonia.

It should be clearly noted that, while this discussion
has focused on the numerous common features exhibited by
enzymes of glutamine metabolism, it would be a mistake to
assume at this stage that they all function by totally
homologous mechanisms. In fact, divergences in behavior
from the basic pattern have been detected which will be
dealt with by the contributors to these symposia. It
remains valid to expect, however, that reinforcement of our
understanding of these systems, as well as of enzymic
mechanisms generally, will derive from continued comparisons of the fundamental similarities they display.

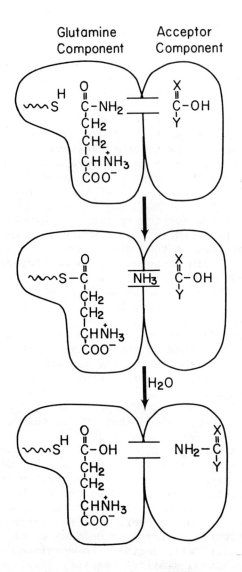

Fig. 1. Proposed scheme for function of a generalized amidotransferase.

Glutaminase A from E. coli

General Properties. Our laboratory has been examining
a glutaminase from E. coli with a view to determining
whether this somewhat simpler system constitutes a model
for the related functions of the biosynthetic amidotransferases (22). Its substrates include amides, esters, and
thioesters of L-glutamic acid with relatively non-bulky
substituents, compounds which are also hydrolyzed by FGAR
amidotransferase (4). The γ-glutamyl moiety is transfered
to hydroxylamine as well as to water with a constant
partitioning ratio for all substrates. This result implies that the reaction proceeds through an intermediate
common to all substrates which may reasonably be identified
as a γ-glutamyl enzyme. As with most other glutamine
metabolizing enzymes, this one reacts covalently and irreversibly with the substrate analogue, DON.

The native enzyme has a molecular weight of about
110,000 according to gel filtration. A single protein band
is observed after polyacrylamide gel electrophoresis in the
presence of sodium dodecyl sulfate in a position corresponding to M = 28,000. The molecule therefore consists of
four subunits, apparently identical by this criterion and
also as judged by the stoichiometry of ^{14}C-DON binding
(one equivalent per 30,000 daltons).

Glutaminase A thus displays at least superficial
resemblances to the glutamine component of amidotransferases, including an acylation-deacylation mechanism of
action, inactivation by DON, and presence of glutamine subunits of moderate molecular weight. Unfortunately it has
not yet been possible to identify the group to which DON
binds, so that comparisons in this important respect
cannot be made.

Cooperativity. The enzyme is most effective in the
acidic pH range and exhibits normal hyperbolic kinetics
between pH 3.5 and 5.5. A transition occurs above about
pH 5.4 to a state unable to bind substrates, as judged by
the fact that the K_M values increase rapidly with pH while
the V_{max} values do not change (see Fig. 2). A plot of the
change in K_M vs. pH in the form of a Hill function has a
slope of about 4, a result indicating a cooperative phase
transition mediated by protonic dissociation (22). An

325

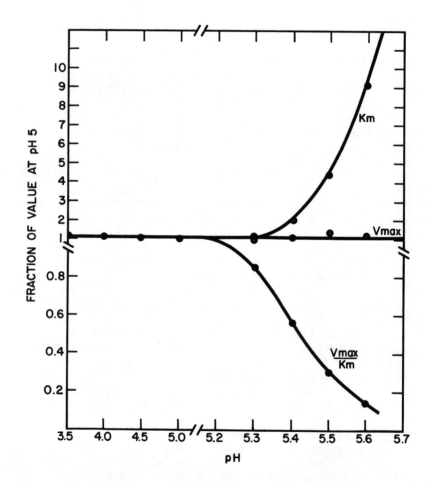

Fig. 2. Effect of pH on kinetic parameters. Essentially
similar normalized values were obtained with
either γ-methyl glutamate or glutamyl γ-methylamide
as substrate.

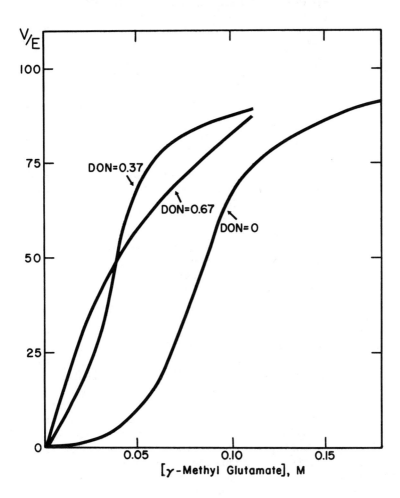

Fig. 3. Effect of DON on cooperativity of glutaminase.
Velocities of hydrolysis of γ-methyl glutamate
were determined in a pH-stat at pH 5.80 and normalized to the same amount of active enzyme.
Values given for each curve are the fraction of
the total sites occupied covalently by DON.

implication of these observations is that the enzyme saturation curve should be sigmoid at a sufficiently high pH
since both protons and substrate stabilize the same form
of the enzyme. This in fact is observed at pH 5.8 as shown
in Fig. 3. It is of interest that when the pseudo-
substrate, DON, occupies a fraction of the binding sites
the homotropic cooperativity is lost, presumably owing to
conservation of the active conformation of the remaining
unoccupied sites. Thus under appropriate conditions of pH
and substrate concentration, irreversible binding of the
"inhibitor" leads to an apparent activation of the enzyme.

Speculations on Function in vivo. Glutaminase A is an
enzyme without an established biological role, although
it seems reasonable to assume that its allosteric properties in vitro are the basis for a regulatory function in
the cell. The enzyme is essentially absent in rapidly
growing cells but is synthesized de novo during a two-hour
period following depletion of energy sources (glucose or
glycerol) and cessation of growth. During this time the
pH of the culture medium drops from about 6.5 to 5.3. If
the pH is held at 6.5 during the stationary phase, about
50% of the maximum amount of glutaminase is still formed.
This suggests that the trigger for glutaminase synthesis
is related to depletion of the energy source and not to the
pH change (23). From the behavior observed in vitro it
would appear that the signal for turning on the glutaminase activity is a local drop of pH below about 5.8 to 5.6.
What is the possible reason for activating glutaminase?
The consequence presumably is to destroy the glutamine in
the cell, the substrate for all amidotransferases (the
alternate substrate, ammonia, must be present in insig-
nificant concentrations in the existing pH range). In one
stroke all of these biosynthetic activities would be
turned off, including vital reactions in the formation of
amino acids, purines, pyrimidines, and amino sugars. Cell
growth would cease. The advantage of this situation to
cell survival is that the limited metabolic energy sources
can now be used for maintenance rather than for continued
unwanted biosynthesis. While this argument is largely
speculative and many details need to be filled in, it does
serve to tie the enormously divergent metabolic activities
of glutamine to the dramatic changes which must occur in
the transition between logarithmic and stationary phases of
bacterial growth.

328

References

1. Winterburn, P.J. and Phelps, C.F., Biochem. J., 121 701 (1971)

2. Jencks, W.P., in Sober, H.A., "Handbook of Biochemistry and Molecular Biology," Chemical Rubber Co., Cleveland, 1968, p. J-144

3. Nagano, H., Zalkin, H., and Henderson, E.J., J. Biol. Chem., 245, 3810 (1970)

4. Li, H.C. and Buchanan, J.M., J. Biol. Chem., 246, 4713 (1971)

5. Trotta, P.P., Burt, M.E., Haschemeyer, R.H., and Meister, A., Proc. Nat. Acad. Sci. U.S., 68, 2599 (1971)

6. Tamir, H. and Srinivasan, P.R., J. Biol. Chem., 247, 1153 (1972)

7. Hartman, S.C., J. Biol. Chem., 238, 3024 (1963)

8. Baker, R.B., Biochem. Pharmacol., 2, 161 (1959)

9. Givot, I.L., Smith, T.A., and Abeles, R.H., J. Biol. Chem., 244, 6341 (1969)

10. Rowe, P.B. and Wyngaarden, J.B., J. Biol. Chem., 243, 6373 (1968)

11. Khedouri, E., Anderson, P.M., and Meister,A., Biochemistry, 5, 3552 (1966)

12. Kane, J.F., Holmes, W.M., and Jensen, R.A., J. Biol. Chem., 247, 1587 (1972)

13. Huang, M. and Gibson, F., J. Bacteriol., 102, 767 (1970)

14. Frère, J.M., Schroeder, D.D., and Buchanan, J.M., J. Biol. Chem., 246, 4727 (1971)

15. Hwang, L.H. and Zalkin, H., J. Biol. Chem., 246, 2338 (1971).

16. Long, C.W., Levitzki, A. and Koshland, D.E., Jr., J. Biol. Chem., 245, 80 (1970).

17. Moyed, H.S. and Magasanik, B., J. Biol. Chem., 226, 351 (1957).

18. Meister, A., "Harvey Lectures," Series 63, Academic Press, New York, 1969, p. 139.

19. Lagerkvist, U., J. Biol. Chem., 233, 143 (1958)

20. Dawid, I.B., French, T.C., and Buchanan, J.M., J. Biol. Chem., 238, 2178 (1972)

21. Pinkus, L.M., Wellner, V.P. and Meister, A., Federation Proc., 31, 474Abs. (1972).

22. Hartman, S.C., J. Biol. Chem., 243, 853, 870 (1968).

23. Smith. A.L. and Hartman, S.C., unpublished results.

INVOLVEMENT OF GLUTAMIC ACID AND GLUTAMINE
IN PROTEIN SYNTHESIS AND MATURATION

Alan Peterkofsky

Laboratory of Biochemical Genetics
National Heart and Lung Institute
Bethesda, Maryland 20014

The process of incorporation of amino acids into proteins is characterized by the specific assembly of amino
acids into polypeptides (protein synthesis). Many proteins
are modified after the assembly phase, as in the conversion
of zymogens to enzymes or the formation of polymers from
monomer proteins (maturation). The involvement of glutamic
acid and glutamine in these two processes will be treated
separately.

PROTEIN SYNTHESIS

Generally, each amino acid commonly found in proteins
is characterized by the presence of a unique aminoacyl-tRNA
synthetase and unique species of tRNA. Muench and Saffille
(1968) have presented evidence for 56 isoacceptors in $\underline{E}$.
$\underline{coli}$ tRNA. The characteristic reaction is:

amino acid + ATP + tRNA $\rightleftarrows$ aminoacyl-tRNA + AMP + PPi

A. Separate enzymes and tRNA species for glutamic acid
and glutamine. At one time, the suggestion was made (Zubay,
1962) that the occurrence of glutamic acid in proteins was
the result of the deamidation of protein-bound glutamine.
Studies in cell-free extracts were interpreted to support
a pathway in which there was no aminoacyl-tRNA synthetase
for glutamic acid. Instead, glutamine could be enzymatically attached to tRNA and subsequently incorporated into
protein. A mechanism was envisaged in which certain
selected glutamine residues in protein were deamidated to
form glutamic acid.

Subsequent experiments made this hypothesis untenable - Coles and Meister (1962) demonstrated that extracts of yeast or E. coli could aminoacylate radioactive glutamic acid to tRNA (see Table 1). This incorporation was diluted out by cold glutamic acid but not cold glutamine.

TABLE 1

Incorporation of glutamate into ribonucleic acid

Reaction mixture	Incorporation (cpm)
[^{14}C]-glutamate (0.2 µmole)	371
[^{14}C]-glutamate (0.2 µmole) + [^{12}C]-glutamate (10 µmoles)	20
[^{14}C]-glutamate (0.2 µmole) + [^{12}C]-glutamine (10 µmoles)	241

Reproduced from Coles and Meister (1962).

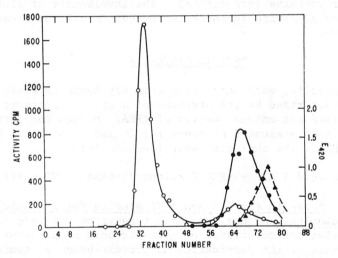

Fig. 1. Separation of glutamyl and glutaminyl-tRNA synthetases of E. coli. Elution pattern from DEAE-cellulose under a gradient of increasing phosphate concentration. The open and solid circles represent activity obtained with glutamine and glutamate, respectively. Reproduced by permission of the copyright owner from Lazzarini and Mehler, 1964.

The existence of a unique activity responsible for the attachment of glutamic acid to tRNA was established by Lazzarini and Mehler (1964). Chromatography on DEAE-cellulose led to a clear separation of the aminoacyl-tRNA synthetases for glutamic acid and glutamine (see Figure 1).

The solution of the genetic code produced data clearly demonstrating that glutamic acid and glutamine utilized distinct and separate triplet codons and isoacceptor tRNA's. Glutamic acid utilizes the codons GAA and GAG, while CAA and CAG serve to recognize the tRNA's which bind glutamine (Caskey et al., 1968). Reverse phase-column chromatography has resolved two isoacceptors each for glutamic acid and glutamine in E. coli (Muench and Saffille, 1968). It is thus clearly established that in E. coli a separate mechanism exists for the attachment of glutamic acid and glutamine to tRNA. A unique exception to this situation will be discussed below.

B. Properties of the aminoacyl-tRNA synthetases for glutamic acid and glutamine. The glutamyl-tRNA synthetase of E. coli has been purified (Lazzarini and Mehler, 1964; Ravel et al., 1965; Lapointe et al., 1972). The recent studies of Lapointe et al. (1972) indicate that the enzyme from E. coli K12 is unusual. After purification of the enzyme to homogeneity, it could be resolved by isoelectric focusing into a completely active core enzyme and an inactive protein factor. While the core enzyme had the same specificity as the holoenzyme for amino acids and tRNA's, its activity was modified by the protein factor. In the presence of the factor, the Km of the core enzyme for glutamic acid was reduced 15-fold and that for ATP 6-fold, while that for tRNA^{Glu} was increased 6-fold. The significance of the capacity of this aminoacyl-tRNA synthetase to be regulated in this unique way is not yet understood.

A homogeneous preparation of E. coli glutaminyl-tRNA synthetase has been prepared by Folk (1971). In contrast to the situation for glutamyl-tRNA synthetase, this enzyme appears to consist of a single polypeptide chain. Sucrose gradient centrifugation was used to demonstrate a 1:1 complex between glutaminyl-tRNA synthetase and its cognate tRNA.

C. <u>The general mechanism of the aminoacyl-tRNA synthetase reaction</u>. The aminoacyl-tRNA synthetase reaction has
generally been formulated as represented by Fig. 2.

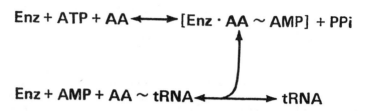

Fig. 2. The mechanism of the aminoacyl-tRNA synthetase
reaction.

Substantial evidence has accumulated to support this
scheme. Most of the enzymes studied catalyze an amino acid-
dependent exchange of pyrophosphate into ATP (for a complete
review, see Loftfield, 1972). Experiments have also been
presented that some enzymes use chemically synthesized amino-
acyl-adenylates to form ATP (e.g., DeMoss <u>et al</u>., 1956).
In addition, the tryptophanyl-tRNA synthetase has been shown
to form stoichiometric amounts of enzyme-bound tryptophanyl-
adenylate when incubated with ATP and tryptophan (Kingdon
<u>et al</u>., 1958). On the basis of this type of evidence and a
variety of kinetic studies according to the procedures of
Cleland (1963a,b,c), the mechanism is generally formulated
as "ping-pong", wherein the product PPi is released from
the enzyme before tRNA binds to the enzyme.

D. <u>The tRNA requirement for the PPi ↔ ATP exchange re-
action</u>. It was observed some years ago (Ravel <u>et al</u>., 1965)
that the aminoacyl-tRNA synthetase for glutamine will not
carry out the glutamine-dependent PPi ↔ ATP exchange reaction in the absence of tRNA (see Figure 3). A similar
tRNA requirement was subsequently shown for the arginyl-tRNA
synthetase of <u>E</u>. <u>coli</u> (Mehler and Mitra, 1967). The studies
of Ravel <u>et al</u>. (1965) indicated that the glutamyl-tRNA synthetase was also unusual in that the tRNA requirement for
the PPi ↔ ATP exchange reaction could be abolished at very
high levels of glutamic acid. These observations have made
it difficult to accommodate the aminoacyl-tRNA synthetases
for arginine, glutamic acid and glutamine into the general
reaction scheme outlined in Figure 2.

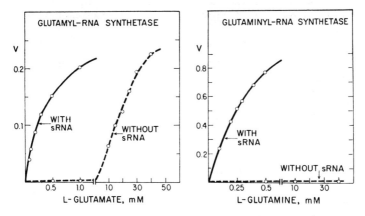

Fig. 3. The effect of sRNA on the glutamyl and gluta-
minyl-dependent ATP ↔ PPi exchange. Reproduced by permission of the copyright owner from Ravel et al. (1965).

While insufficient studies have been done on the aminoacyl-tRNA synthetases for glutamic acid and glutamine, the
picture has been substantially clarified with the arginyl-
tRNA synthetase of E. coli. Recent studies, primarily involving kinetic studies of various exchanges at steady state
have generated data which support a random sequential mechanism for the reaction as shown in Fig. 4 (Papas and Peterkofsky, 1972):

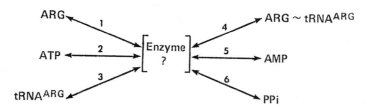

Fig. 4. A random substrate binding mechanism for E.
coli arginyl-tRNA synthetase.

In this model, all substrates are bound randomly to form
a central complex which breaks down to randomly release the
products arginyl-tRNA, AMP and PPi. At present, it is not
possible to decide if the tRNA requirement for the PPi ↔ ATP
exchange is due to (a) the promotion by tRNA of the release

335

of tightly bound PPi (Rochovansky and Ratner, 1967) or (b)
the concerted nature of the reaction (Wedler and Boyer,
1972).

E. Glutamine is not directly attached to tRNA in gram-
positive organisms. Wilcox and Nirenberg (1968) discovered
an unusual mechanism in extracts of Bacillus whereby glutamine was synthesized from glutamic acid while attached to
tRNA as outlined in the following scheme:

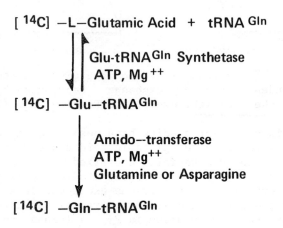

No activity could be found which resulted in the attachment of free glutamine to tRNA. However, tRNA acylated
with $[^{14}C]$-glutamic acid was resolvable by reverse-phase
chromatography into two species (see Fig. 5).

While the first peak of $[^{14}C]$-Glu-tRNA responded in a
ribosome-binding assay to the glutamic acid codons GAA and
GAG, the second peak responded to the glutamine codons CAA
and CAG. Further, when the $[^{14}C]$-amino acid attached to the
two fractions in the presence of $[^{14}C]$-glutamic acid and unlabeled glutamine was released by incubation at pH 8.0 and
characterized by paper chromatography, it was found that
fraction 1 was predominantly glutamic acid, while fraction 2
was mainly glutamine. When fraction 2 was acylated with
$[^{14}C]$-glutamic acid in the absence of other amino acids, it
was shown to contain solely glutamic acid (see Table 2).

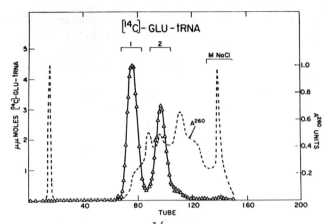

Fig. 5. Fractionation of [^{14}C]-glu-tRNA (B. megaterium) by reverse phase-column chromatography. △———△, [^{14}C]-Glu-tRNA precipitated by TCA at 3°C; - - - -, absorbancy at 260 mμ. Reprinted with permission from Wilcox and Nirenberg, 1968.

TABLE 2

[^{14}C] PRODUCTS AFTER DEACYLATING
[^{14}C]-GLU-tRNA FRACTIONS

tRNA Fraction	Amide Donor	[^{14}C]-Products (%)	
		[^{14}C]-GLU	[^{14}C]-GLN
1	+	92	1
2	–	88	4
2	+	13	79

Reproduced with permission from Wilcox and Nirenberg (1968).

Fractionation of the extracts led to the separation of glutamyl-tRNA synthetase from the amido-transferase which catalyzed the conversion of [^{14}C]-Glu-tRNA to [^{14}C]-Gln-tRNA. A study of the specificity for amide donor indicated that the enzyme could utilize either glutamine or asparagine as a source of the amide nitrogen of glutamine. Thus, as shown in the scheme above, the synthesis of glutamine destined for protein is accomplished via a glu-tRNA intermediate.

An elegant study by Wilcox (1969) provided evidence on the nature of ATP utilization during the amido-transferase reaction. It was shown that ATP was split during the course of the reaction to yield ADP and Pi. In the absence of amide donor, radioactivity from $[^{32}P]-\gamma$-ATP was bound to Glu-tRNA^{Gln}. The interpretation of this and other data was that there was an intermediate involving a phosphorylation of the γ-carboxyl group of Glu-tRNA^{Gln}. The amido-transferase reaction was pictured as follows:

(1) $[^{14}C]$-Glu-tRNA^{Gln} + enzyme + ATP $\xrightarrow{Mn^{++}}$

$\quad [P-[^{14}C]$-Glu-tRNA^{Gln}:enzyme] + ADP

(2) $[P-[^{14}C]$-Glu-tRNA^{Gln}:enzyme] + L-glutamine $\xrightarrow{Mn^{++}}$

$\quad [^{14}C]$-Gln-tRNA^{Gln} + enzyme + Pi + glutamic acid

MATURATION OF PROTEINS

A. Pyroglutamic acid formation. Many proteins (e.g., collagen and immunoglobulins, Dayhoff, 1969) are characterized by the presence of a blocked N-terminal amino acid, pyroglutamic acid (pGlu). The mechanism by which pGlu arises has been the subject of considerable discussion and is not yet completely clarified. Blomback (1967) has discussed the evidence that pGlu is artifactually derived from the spontaneous cyclization of proteins containing N-terminal glutamine. Bernfield and Nestor (1968) showed that glutaminyl-tRNA could be cyclized to pGlu-tRNA by an enzyme from papaya latex, glutamine cyclotransferase. These authors suggested that N-terminal pGlu was derived from pGlu-tRNA^{Gln}.

The above two hypotheses were rendered unlikely by the recent experiments of Twardzik and Peterkofsky (1972). Cell suspensions derived from a mouse plasmacytoma that secretes an immunoglobulin containing N-terminal pyroglutamic acid were allowed to synthesize protein in the presence of labeled glutamic acid or glutamine. Chromatographic examination of an enzymatic digest of protein labeled with glutamic acid showed the presence of only labeled glutamic acid and pyroglutamic acid; when cells were labeled with glutamine, the protein contained labeled glutamine, glutamic acid and pyroglutamic acid (see Fig. 6).

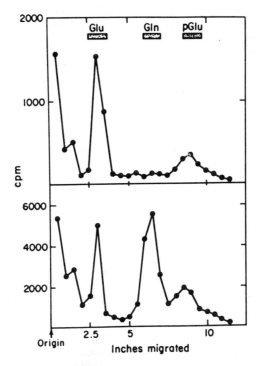

Fig. 6. Ascending paper chromatography in 88% phenol –
12% H_2O of radioactive protein hydrolysate from cells labeled
with [^{3}H]-glutamic acid (upper panel) or [^{14}C]-glutamine
(lower panel). The relative mobilities of glutamic acid
(Glu), glutamine (Gln) and pGlu in this solvent are indicated in the upper panel. Reproduced with permission from
Twardzik and Peterkofsky, 1972.

These data indicate that N-terminal pGlu is derived from
glutamic acid and not by some mechanism involving a direct
cyclization of free or bound glutamine. Fig. 7 outlines two
possible schemes for the biosynthesis of pGlu from Glu. The
first mechanism is completely analogous to the formation of
formylmethionyl-tRNA in bacteria; in this case pGlu is
formed from Glu while attached to tRNA. The second mechanism involves the direct incorporation of glutamic acid destined for pGlu formation into protein, followed by an unde-
fined post-translational modification. Which of these two
mechanisms actually occurs in pGlu formation is not yet

339

delineated.

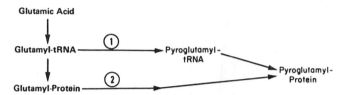

Fig. 7. Mechanisms for pGlu formation from glutamic acid. Reprinted with permission from Twardzik and Peterkofsky, 1972.

B. <u>Transglutaminase and protein crosslinking</u>. One of the mechanisms found in nature for covalently linking protein subunits together consists of the formation of γ-glutamyl-amide bonds. The production of insoluble fibrin clots is due to such a process, the formation of ε(γ-glutamyl) lysine crosslinks (Pisano <u>et al</u>., 1968). The insolubility of hair proteins probably results from such a crosslinking reaction to form ε(γ-glutamyl)lysine bonds (Chung and Folk, 1972a).

The crosslinking reactions are catalyzed by the enzymes known as transglutaminases (Chung and Folk, 1972b). The reaction mechanism of the enzyme is shown in Fig. 8.

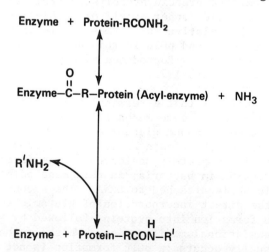

Fig. 8. The mechanism of the transglutaminase reaction.

Folk (1969) has demonstrated that the enzyme operates by a ping-pong mechanism. Transglutaminase reacts with a peptide-bound glutamine designated protein-($RCONH_2$) in Fig. 8 to form an acyl-enzyme and release ammonia. The acyl-enzyme can then react with a variety of substrate amines ($R'-NH_2$) to form γ-glutamyl amides. When plasma transglutaminase (often called Factor XIII) acts on fibrin, the protein-bound glutamine of one fibrin monomer is covalently attached to the $\varepsilon-NH_2$ group of a lysine residue of another monomer to form a $\varepsilon(\gamma$-glutamyl)lysine bond. A similar crosslinking mechanism probably occurs in the formation of hair protein.

Acknowledgements: I thank Dr. John Folk for a helpful discussion and access to unpublished material on transglutaminase. My appreciation also goes to Mrs. Helmi Carpenter for her preparation of this manuscript.

REFERENCES

Bernfield, M. R., and Nestor, L. (1968). _Biochem. Biophys. Research Commun._ 33, 843.

Blomback, B. (1967. _In_ "Methods in Enzymology" (C. H. W. Hirs, ed.), Vol. XI, pp. 398. Academic Press, New York.

Caskey, C. T., Beaudet, A., and Nirenberg, M. (1968). _J. Mol. Biol._ 37, 99.

Chung, S. I., and Folk, J. E. (1972a). _Proc. Nat. Acad. Sci. USA_ 69, 303.

Chung, S. I., and Folk, J. E. (1972b). _J. Biol. Chem._ 247, 2798.

Cleland, W. W. (1963a). _Biochim. Biophys. Acta_ 67, 104.

Cleland, W. W. (1963b). _Biochim. Biophys. Acta_ 67, 173.

Cleland, W. W. (1963c). _Biochim. Biophys. Acta_ 67, 188.

Coles, N., and Meister, A. (1962). _Proc. Nat. Acad. Sci. USA_ 48, 1602.

Dayhoff, M. O. (1969). "Atlas of Protein Sequence and Structure", Vol. 4. National Biomedical Research Foundation, Silver Spring, Md.

DeMoss, J. A., Genuth, S. M., and Novelli, G. D. (1956). _Proc. Nat. Acad. Sci. USA_ 42, 325.

Folk, J. E. (1969). _J. Biol. Chem._ 244, 3707.

Folk, W. R. (1971). _Biochem._ 10, 1728.

Kingdon, H. S., Webster, L. T., and Davie, E. W. (1958). _Proc. Nat. Acad. Sci. USA_ 44, 757.

Lapointe, J., Seno, T., and Söll, D. (1972) Fed. Proc. 31, 449.

Lazzarini, R. A., and Mehler, A. H. (1964). Biochem. 3, 1445.

Loftfield, R. B. (1972) In "Progress in Nucleic Acid Research",(J. N. Davidson and Waldo E. Cohn, eds.), Vol. 12, p. 87. Academic Press, New York.

Mehler, A. H., and Mitra, S. K. (1967). J. Biol. Chem. 242, 5495.

Muench, K. H., and Saffille, P. A. (1968). Biochem. 7, 2799.

Papas, T. S., and Peterkofsky, A. (1972). Biochem., in press.

Pisano, J. J., Finlayson, J. S., and Peyton, M. P. (1968). Science 160, 892.

Ravel, J. M., Wang, S., Heinemeyer, C., and Shive, W. (1965). J. Biol. Chem. 240, 432.

Rochovansky, O. and Ratner, S. (1967). J. Biol. Chem. 242, 3839.

Twardzik, D. R., and Peterkofsky, A. (1972). Proc. Nat. Acad. Sci. USA 69, 274.

Wedler, F. C., and Boyer, P. D. (1972). J. Biol. Chem. 247, 984.

Wilcox, M., and Nirenberg, M. (1968). Proc. Nat. Acad. Sci. USA 61, 229.

Wilcox, M. (1969). Cold Spring Harbor Symposium 34, 521.

Zubay, G. (1962). Proc. Nat. Acad. Sci. USA 48, 894.

THE INFLUENCE OF SUBSTRATES AND MODIFIERS ON L-GLUTAMINE D-FRUCTOSE 6-PHOSPHATE AMIDOTRANSFERASE

Peter J. Winterburn and Charles F. Phelps

Department of Biochemistry, University College,
Cardiff, U.K. and Department of Biochemistry,
University of Bristol, Bristol, U.K.

Abstract

The history of L-glutamine D-fructose 6-phosphate amidotransferase is reviewed and some problems relating to the preparation of the enzyme are discussed. The actions of substrates, modifiers and inhibitors are described and used to develop plausible models for the catalytic mechanism and the interaction between the ligand binding sites.

Introduction

Hexosamine containing polymers are an important feature of the extracellular environment where they occur either in soluble form, glycoproteins and glycosamino-glycans, or as structural components of membranes, glyco-proteins, glycolipids and peptidoglycans. These amino sugar containing polymers are responsible in part for conferring antigenic properties to membrane surfaces and more recently have been implicated in recognition phenomena involving soluble glycoproteins (Winterburn and Phelps, 1972). The hexosamines present in the polymers are usually N-acetylated which suppresses the ionisation of the amino group by the formation of an amide function. Hydrogen bonding between this amide and other functional groups contributes to the preservation of the steric inter-relationships necessary for the function of the polymer (Atkins et. al., 1972).

The first proposal that glutamine was the source of

343

the amino group of polymeric hexosamines was made by
Lowther and Rogers (1956) which implicated an enzyme
previously described by Leloir and Cardini (1953) which
aminated hexose 6-phosphate to form glucosamine-6-P.
Subsequently the identity of the hexose phosphate was
established by Ghosh et. al. (1960) as fructose-6-P and the
reaction catalysed by the enzyme L-glutamine D-fructose 6-
phosphate amidotransferase (EC 2.6.1.16) as:-

L-glutamine + D-Fru-6-P $\longrightarrow$ D-GlcN-6-P + L-glutamate

An alternative route of glucosamine-6-P formation by the
direct amination of fructose-6-P by ammonia catalysed by
glucosamine 6-phosphate deaminase (EC 5.3.1.10) was
proposed by Comb and Roseman (1959). However, this reaction is believed to be primarily a catabolic one concerned
with the hydrolysis of glucosamine-6-P.

Since hexosamines are of widespread distribution and
this amidotransferase is uniquely responsible for the
formation of hexosamines whether glucosamine, galactosamine
or mannosamine – the latter usually occurring as sialic
acid – it is not surprising that the amidotransferase has
been found in a diverse selection of organisms. After the
initial detection of the enzyme in Neurospora crassa
(Leloir and Cardini, 1953), Blumenthal et. al. (1955)
reported the presence of the activity in a wide variety of
fungi and bacteria. Some partial purifications from microorganisms have been reported (Ghosh et. al., 1960; Endo et.
al., 1970). The rat liver amidotransferase has been the
most extensively studied (Pogell and Gryder, 1957; Ghosh
et. al., 1960; Kornfeld et. al., 1964; Kornfeld, 1967;
Winterburn and Phelps, 1970; Kikuchi et. al., 1971a; Miyagi
and Tsuiki, 1971; Winterburn and Phelps, 1971a,b,c)
although other mammalian sources have been investigated
e.g. bovine tracheal linings (Ellis and Sommar, 1971),
thyroid glands (Trujillo et. al., 1971) and retina (Mazlen
et. al., 1969) and various tumours (Kikuchi et. al., 1971a;
Kaufman et. al., 1971).

Kornfeld et. al. (1964) presented evidence of a feedback control operating on the hexosamine pathway in rat
liver. The end product of the first segment of the pathway
is UDP-GlcNAc and it was shown in vivo and in vitro that

344

this metabolite was a potent negative feedback modifier of the amidotransferase (Bates et. al., 1966). This control over hexosamine production operates in all eukaryotic tissues so far investigated (Kornfeld, 1967; Ohta et. al., 1968; Mazlen et. al., 1969; Endo et. al., 1970; Ellis and Sommar, 1971; Kikuchi et. al., 1971a; Trujillo et. al., 1971) but not in prokaryotic organisms (Kornfeld, 1967). It was later shown by Winterburn and Phelps (1970) that the amidotransferase was susceptible to the actions of other modifiers and that UDP-GlcNAc probably was not the primary regulator of the activity.

The main theme of past work on the amidotransferase has been the economic importance to the cell of controlling an enzyme which is responsible for hexosamine synthesis and is a drain on the hexose monophosphate pool. Quite apart from these considerations the reaction is chemically of great interest. The sugar undergoes a change in ring size from furanose to pyranose accompanied by a ketose to aldose transition while the amination by glutamine, unlike the reactions catalysed by other glutamine amidotransferases, does not require the concommitant hydrolysis of a pyrophosphate linkage.

The object of the present paper is to develop a model for the catalytic action of this enzyme and to discuss the effects of non-substrate ligands.

Experimental Procedures

Preparation of the Amidotransferase
All workers have observed that the amidotransferase is unstable and hence difficult to purify. This has led to a situation where the kinetic studies have been performed mainly with crude fractions contaminated with activities which degraded substrates, products and modifiers. Meaningful interpretation of such work in terms of the operation of the enzyme is difficult. It is hoped that this report on the purification will assist workers to prepare an enzyme which we believe approximates to the native protein.

After death there was a rapid decline in the recoverable activity from a tissue, especially metabolically

active ones. This contrasted with the noted in vivo
stability of the amidotransferase (Bates and Handschumacher,
1969). The activity recovered from calf liver was
negligible when the delay between slaughter and homogenisation was 2 hr although rapid treatment (less than 30 min)
revealed an activity twice that of rat liver (S.B. Thomas
and P.J. Winterburn, unpublished data). The degradation of
glucose-6-P and UDP-GlcNAc, stabilisers of the enzyme
(Winterburn and Phelps, 1971c) was probably the cause of
this inactivation. Gentle homogenisation with a loose-
fitting pestle gave the best yields and freezing was
avoided at these early stages, recommendations made
previously by Bates and Handschumacher (1969) and Gryder
and Pogell (1960) respectively, to prevent the rupture of
the lysosomal membranes.

The majority of workers have used initially the
ammonium sulphate fractionation of Gryder and Pogell (1960)
followed by removal of salt by gel-filtration. This
procedure was subject to variable and often large losses
unless the buffers contained either glucose-6-P or
fructose-6-P (Winterburn, 1969). These ammonium sulphate
fractions possessed among other contaminating activities
phosphoglucose isomerase which depleted the fructose-6-P
and generated glucose-6-P, a modifier of the feedback
inhibitor (Winterburn and Phelps, 1971b; Miyagi and Tsuiki,
1971). Miyagi and Tsuiki (1971) demonstrated that much of
the confusion in interpreting the action of UDP-GlcNAc has
been a result of phosphoglucose isomerase contamination.
Several workers have used DEAE ion-exchangers to remove the
isomerase, a procedure originally used by Ghosh et. al.
(1960). However the duration of contact between the amidotransferase and the exchanger had to be minimal because
catalytic inactivation or desensitisation to UDP-GlcNAc
resulted from prolonged binding (Kornfeld, 1967; Winterburn
and Phelps, 1971c; Ellis and Sommar, 1972). An improvement
has been the use of DEAE-Sephadex in place of the cellulose
derivative by Miyagi and Tsuiki (1971). This permitted a
continuous elution gradient scheme lasting 20 hr with
retention of UDP-GlcNAc sensitivity. A variety of thiol
protecting agents have been employed by various workers.
Miyagi and Tsuiki (1971) who used high concentrations of
dithiothreitol suggested that the properties of the amidotransferase were modified by changes in the level of thiol
protection. The use of the correct thiol agent and

concentration will remain unresolved until the redox state of cytosolic thiols is known.

The two purification procedures used for the present studies on the rat liver amidotransferase have been described in detail previously (Winterburn and Phelps, 1971a). These schemes were rapid, taking only a few hours, and yielded a preparation free of contaminating activities which retained sensitivity to UDP-GlcNAc. The calf liver (S.B. Thomas and P.J. Winterburn, unpublished data) and neonatal rat skin enzymes (Winterburn, 1973) have since been purified by the same methods. One of the two procedures yielded a preparation free of stabilising effectors while the other was stabilised with 1mM fructose-6-P. Attempts to remove the fructose-6-P after preparation by methods based on separation by molecular size were unsuccessful, the recoveries being low and variable. The amidotransferase prepared by the two methods exhibited only minor differences in properties.

Assay of the Amidotransferase
    The amidotransferase has always been assayed by the colorimetric determination of glucosamine-6-P either directly or by the Morgan-Elson reaction after acetylation. For impure enzyme fractions the latter procedure was preferable to avoid inaccuracies caused by enzymic acety-lation of glucosamine-6-P by the active specific N-acetyl-transferase. The incubation conditions and assays were described by Winterburn and Phelps (1971a).

    The sensitivity of the enzyme to UDP-GlcNAc was measured by determining the activity in the presence and absence of 0.1mM UDP-GlcNAc in the standard incubation. The ratio of these respective velocities, termed the control ratio, was taken as indicative of the sensitivity of the amidotransferase to its feedback inhibitor. This ratio varied slightly between preparations, thus all studies were performed on preparations exhibiting a control ratio of less than 0.55. Similarly an index of the sensitivity of the enzyme to other modifiers was expressed in terms of the modifier ratio. The velocity was measured in the standard assay in the presence of the modifier and 0.1mM UDP-GlcNAc. The ratio of this velocity to that with the modifier omitted was termed the modifier ratio. The concentrations of AMP, glucose-6-P and UTP employed in the incubations were 2.0, 1.5 and 2.5mM respectively.

Other procedures and chemicals have been detailed previously (Winterburn and Phelps, 1971a,b,c).

Results and Discussion

Effect of substrates
The kinetic parameters were investigated by varying the concentration of one substrate at fixed concentrations of the second substrate. The measured velocities were treated graphically by the v against v/A linear transform of the two substrate rate equation and the apparent V and Km values used to determine the true V, Km and Ki by secondary plots as described by Winterburn and Phelps (1971a). Computer programs were used to calculate the kinetic constants from data which was graphically linear (Haarhoff, 1969).

Using an enzyme fraction prepared without fructose-6-P stabilisation at the optimum pH of 7.5 the Km was 0.24mM for fructose-6-P and the value of the Hill coefficient determined at ten fixed glutamine concentrations was 0.95 $\pm$ 0.08 signifying no homotropic interactions between fructose-6-P binding sites. Fructose-6-P protected preparations consistently gave slightly lower Km values in the range 0.08 - 0.15mM. For comparison other workers have reported the Km for fructose-6-P of the rat liver amidotransferase as 0.22mM (Kornfeld, 1967) and 0.31mM (Miyagi and Tsuiki, 1971). The apparent Km for fructose-6-P approximates to the physiological range of this metabolite in rat liver, 0.04 - 0.14mM (Start and Newsholme, 1968). Thus, the amidotransferase will be sensitive to fluctuations in the level of this metabolite caused by alterations in the flux of material into and out of the hexose monophosphate pool. The fructose-6-P concentrations quoted represent total fructose-6-P. The effective free concentrations may be considerably lower than these values if one considers that fructose-6-P will be sequestered by the various enzymes which bind this metabolite.

The Km for glutamine was 0.69mM for the unstabilised preparation and only slightly higher at 0.75mM for the stabilised one. This compared closely with values reported by other workers (Kornfeld, 1967; Bates and Handschumacher,

1969; Miyagi and Tsuiki, 1971). The Hill coefficient was unitary for both preparations. Ellis and Sommar (1971) reported that the amidotransferase from bovine tracheal lining exhibited substrate inhibition by glutamine at concentrations greater than 1.6mM. In the absence of the feedback inhibitor the rat liver enzyme did not show this effect at the highest concentrations tested (6mM). Since the concentration of glutamine in bovine trachea was 0.62 μmole/gm wet weight, Ellis and Sommar (1971) proposed that glutamine played a regulatory role in hexosamine synthesis. The possible control of rat liver amidotransferase by changes in glutamine concentration is harder to assess. In vivo the amidotransferase probably exists primarily as a complex with UDP-GlcNAc (Winterburn and Phelps, 1971b) under which conditions glutamine is an inhibitor as well as a substrate. Thus at the intracellular glutamine levels corresponding to 4.7 μmole/gm wet weight (Brosnan et.al., 1970) the balance between the two responses could result in glutamine being a regulator.

Investigation into substrate binding

Information on the addition order of substrates was obtained using the substrates as ligands to alter the inactivation rate of the enzyme under controlled conditions (Winterburn and Phelps, 1971c). The inactivation of the amidotransferase in the absence of protecting agents at pH 7.5 was a biphasic process at the three temperatures tested; at 10°C the first phase was complete within 5 hr of enzyme preparation. All stabilisation experiments were conducted at 10°C on the second phase of the inactivation which was an apparent first-order reaction. The first order rate constants were used to estimate the ligand dissociation constants using a model based on that described by London et. al. (1958).

6mM glutamine decreased the rate constant by 8% corresponding to a dissociation constant of approximately 75mM. The slight stabilising influence of glutamine on amidotransferase preparations had been noted by other workers (Endo et. et., 1970; Miyagi and Tsuiki, 1971). The dissociation constant calculated for this weak binding is approximately 30 times greater than the apparent glutamine Km at this temperature (estimated by extrapolation of the temperature dependence of Km between 25 and 38°C, Winterburn, 1969).

349

Stabilisation by fructose-6-P had been reported by Endo et. al. (1970) but the preparation of Miyagi and Tsuiki (1971) did not respond to this substrate. In the system used by Winterburn and Phelps (1971c) fructose-6-P exhibited a complex effect on the denaturation process with an initial increase in activity before the resumption of the inactivation phase. The rate constant for the inactivation was dependent on the fructose-6-P concentration, low levels (less than 0.05mM) increasing the rate of inactivation but higher levels stabilised the enzyme preparation. The complexity of the strong binding of fructose-6-P precluded the estimation of the dissociation constants by the simple model employed. However, the behaviour was interpreted as an interaction of this substrate at two or more sites of differing affinities, each binding event influencing the denaturation rate constant.

The computer analysis of the kinetic data at pH 7.5 and 37° in addition to calculating apparent Km values fitted the velocities to the ordered Bi Bi or ping pong Bi Bi rate equations (Cleland, 1963a). This treatment revealed that the value of Ki was approximately zero and the data had the closer fit to the ping pong Bi Bi equation (Winterburn and Phelps, 1971a). Graphically the lines obtained at the several concentrations of the second substrate converged to a point coincident with the abscissa of the v against v/s plot i.e. V/Km was constant. The analysis of data derived at other pH values and temperatures showed that as the conditions were varied the value of Ki became positive. On the basis of the kinetic evidence it was concluded that there was an ordered addition of substrates prior to the release of the first product and that the approach of the Ki value to zero under certain conditions probably reflected changes in the rate constants which made the addition of the first substrate effectively irreversible. The participation of a substituted enzyme intermediate was considered unlikely. Coupled with the studies on substrate binding it is proposed that fructose-6-P binds prior to glutamine, a suggestion originally made by Gryder and Pogell (1960).

Ghosh et. al. (1960) reported that 6-diazo-5-oxo-L-norleucine (DON) was a potent glutamine analogue of this amidotransferase. This competitive inhibition for the glutamine site has been confirmed for the bovine tracheal

lining (Ellis and Sommar, 1972) and retinal enzymes (Mazlen
et. al. 1970). Bates and Handschumacher (1969) working
with the rat liver enzyme considered that the inhibition by
this analogue was an irreversible alkylation of an active
site residue rather than the formation of a reversible
enzyme-inhibitor complex. They proposed that glutamine
binding was mainly through the α-amino-carboxylate
functions. The studies of Winterburn and Phelps (1971c)
using a variety of potential analogues confirmed the role
of this part of the molecule. However, the failure of the
aliphatic amino acids to inhibit leads us to conclude that
enzyme-glutamine complex formation requires the simultaneous binding of both ends of the molecule.

Model for the reaction mechanism

This amidotransferase differs from other amidotransferases in not using ammonia as an alternative substrate
and not requiring the cleavage of a pyrophosphate bond,
the large negative ΔG of this essentially irreversible
reaction (Ghosh et. al., 1960; Winterburn and Phelps,
1971a) being provided by the hydrolysis of the amide
linkage. However, the change from the furanose ring of the
ketose phosphate to the pyranose of the aldose, glucosamine
-6-P, bears similarities to the Lobry de Bruyn - Alberda
van Eckenstein reaction. The enzymic catalysis of this
rearrangement, which has been described for phosphoglucose
isomerase (Dyson and Noltmann, 1968) and phosphomannose
isomerase (Gracy and Noltmann, 1968) coupled with an
earlier proposal of Davidson (1966), is developed here as
an hypothesis of the catalytic mechanism of the amidotransferase.

The enzyme-fructose-6-P complex is produced by the
formation of a hydrogen bond between a protonated amino
group and the furanose ring oxygen (Fig. 1a). From an
investigation into the effect of pH on the kinetic
parameters, Winterburn and Phelps (1971a) speculated that
fructose-6-P binding involved a group with an apparent pK
of 7.8 which was in the acidic form at pH 7.5. This group
could be an α-amino group, the pK value being low compared
with the lysine amino group which serves this function in
phosphoglucose isomerase. Phosphomannose isomerase
possesses an active site α-amino group with a pK of 7.8 but,
because the reactive sugar phosphate is the straight chain
form, this group is assigned to phosphate binding. A

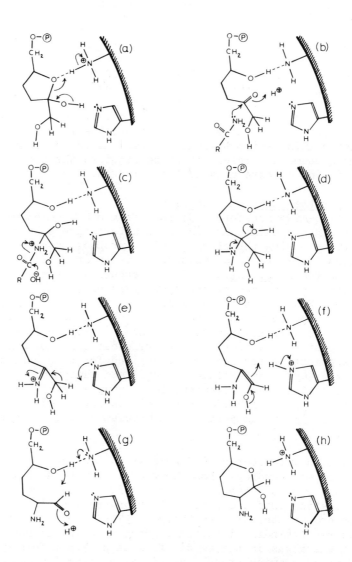

Fig. 1a–h. Proposed mechanism for the amidotransferase-catalysed reaction.
$$R = -CH_2-CH_2-CH(NH_2)-COOH$$

conformational change permits glutamine to bind through its
α-amino and carboxyl groups and amide function. The
migration of the hydrogen bond from hydrogen-oxygen to
nitrogen-hydrogen effects ring opening (Fig. 1b) and
liberates the carbonyl group. Chemically a nucleophilic
attack by an amide nitrogen on a carbonyl is rare but may
be facilitated by several factors: (a) steric considerations, for example see Cohen and Witkop (1955), which are
important features of enzymic catalysis, (b) an increase in
the polarisation of the carbonyl functions by hydrogen
bonding, and (c) a simultaneous attack at the amide
carbonyl function. Mizobuchi et. al., (1968) proposed that
a thiol in the active site of phosphoribosyl-formylglycine
amidine synthetase together with an electron withdrawing
agent assisted the labilisation of the amide and amination
of the carbonyl. The presence of such a group in the
fructose-6-P aminating system is uncertain although the
alkylated residue proposed by Bates and Handschumacher
(1969) may serve this function. The amination (Figs. 1c
and 1d) leads to imine formation (Fig. 1e). The production
of the aldose is by an acid-base catalysis using a histidine residue as described for the two isomerases (Fig. 1f
and 1g). The histidines of phosphoglucose isomerase and
phosphomannose isomerase have pK values of 6.6 compared
with a group in the amidotransferase which ionises with an
apparent pK of 6.95 (Winterburn and Phelps, 1971a).
Unfortunately other information on the identity of this
group is not available therefore assignment as a histidine
residue must be speculative. Pyranose ring formation is by
a reversal of the furanose ring opening mechanism (Fig. 1h).

Inhibition by UDP-GlcNAc

The feedback inhibition of the amidotransferase by
UDP-GlcNAc was first reported by Kornfeld et. al. (1964).
Kornfeld (1967) observed that even high UDP-GlcNAc
concentrations did not completely inhibit the enzyme, this
has since been confirmed by other workers for the amidotransferase from rat liver (Winterburn and Phelps, 1971b)
and other sources (Endo et. al., 1970; Ellis and Sommar,
1972).

The nature of the inhibition was examined in greater
detail using a three dimensional matrix of substrates and
inhibitors (Winterburn and Phelps, 1971b). The velocity
was measured at each of five concentrations of each

substrate and at each of five concentrations of UDP-GlcNAc.
The type of inhibition was determined graphically and
expressed in the nomenclature of Cleland (1963b). This
analysis revealed that UDP-GlcNAc was a linear uncompetitive inhibitor towards glutamine with an estimated Ki of
70 μM. In addition, substrate inhibition by the glutamine
was induced, a finding originally reported by Bates and
Handschumacher (1969) and confirmed by Mazlen et. al.,
(1970) for the bovine retinal enzyme. A representative

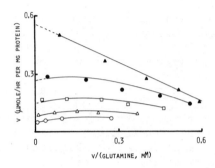

Fig. 2.   Inhibition of the amidotransferase by UDP-GlcNAc.
Glutamine was the variable substrate and fructose-6-P was
0.36mM (for further details see Experimental Procedures).
The UDP-GlcNAc concentrations were: ▲ , none; ● , 25 μM;
□ , 50 μM; Δ , 100 μM; ○ , 150 μM.

graph depicting the substrate inhibition superimposed on
the uncompetitive inhibition is shown in Fig. 2. With
respect to the other substrate, fructose-6-P, the nature of
the inhibition was dependent on the glutamine concentration.
At high levels of glutamine UDP-GlcNAc was a non-competitive inhibitor (Fig. 3a) - the lines on the v against v/s
plot converging to a point to the left of the ordinate. As
the glutamine concentration was decreased so the inhibition
tended to an apparent competition between fructose-6-P and
UDP-GlcNAc for a common site (Fig. 3b) - the graph showing
convergence to a point on the ordinate. Such a change in
behaviour was probably because at high glutamine concentrations the effect of the feedback inhibitor towards
fructose-6-P was measured under conditions of glutamine
substrate inhibition while the data obtained at low
glutamine was under "ideal" two substrate conditions.

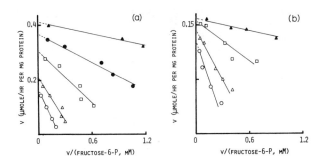

Fig. 3. Inhibition of the amidotransferase by UDP-GlcNAc.
Fructose-6-P was the variable substrate and the UDP-GlcNAc
concentrations were: ▲ , none; ● , 25 μM; ☐ , 50 μM; Δ , 100 μM;
○ , 150 μM. The glutamine concentrations were: (a) 1.33mM,
and (b) 0.25mM.

The basically competitive nature of the UDP-GlcNAc
inhibition for fructose-6-P signified that these two
ligands bound to the same enzyme form, which the fructose-
6-P binding studies had shown to be the free enzyme. Using
the ligand binding procedure as described above, the UDP-
GlcNAc binding to the free enzyme was investigated. In
agreement with other reports on its stabilising influence
(Bates and Handschumacher, 1969; Endo et. al., 1970), 0.1mM
UDP-GlcNAc decreased the first order rate constant from
which was calculated a dissociation constant of 35 μM (Fig.
4). Although the free enzyme was only weakly stabilised by
glutamine, in the presence of UDP-GlcNAc glutamine enhanced
the protection provided by the feedback inhibitor (Fig. 5).
This gave an apparent dissociation constant for glutamine
of 3.5mM which compared with 75mM for the binding of this
ligand in the absence of UDP-GlcNAc. These findings confirmed the kinetic evidence that UDP-GlcNAc induces a
glutamine – binding site on the enzyme.

Since the various parts of a protein are stabilised by
different combinations of forces, careful denaturation by
physical or chemical agents can lead to selective unfolding
of the polypeptide chain. Thus, an effector binding site on
an allosteric protein can be partially or fully denatured

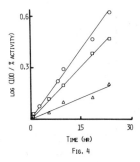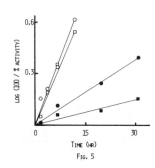

Fig. 4.   Influence of UTP and UDP-GlcNAc on the denaturation rate at 10°C in 50mM Tris - 10mM cacodylate - 5mM EDTA -
5mM GSH - 100mM KCl, pH 7.5.   Samples were removed at
intervals and assayed for remaining activity (see Experimental Procedures). O, control; Δ, 0.1mM UDP-GlcNAc, □,
2mM UTP + 0.1mM UDP-GlcNAc.
Fig. 5.   Influence of glutamine on the denaturation rate in
the presence and absence of UDP-GlcNAc.   The conditions
were as detailed for Fig. 4. O, control; □, 6mM glutamine;
●, 0.1mM UDP-GlcNAc; ■, 0.1mM UDP-GlcNAc + 6mM glutamine.

while catalytic activity is retained.   This process,
desensitisation, was applied to the amidotransferase
(Winterburn and Phelps, 1971c) using thermal denaturation
and thiol inactivation.   Under both of these conditions the
sensitivity to UDP-GlcNAc declined faster than the loss in
catalytic activity.

Kornfeld et. al. (1964) noted that other UDP-sugars
were inhibitory although the potency of these actions was
considerably lower than that of UDP-GlcNAc.   Kinetically
UDP-Xyl, UDP-Glc and UDP-GlcA have been shown to inhibit in
the same manner as UDP-GlcNAc (Winterburn and Phelps,
1971b) and competition experiments were consistent with
there being a common binding site for all the UDP-sugars.

Several pieces of evidence indicate that although UDP-
GlcNAc behaves as a competitive inhibitor towards one of
the substrates the binding site is distinct from the
catalytic one, as originally proposed by Kornfeld (1967).
Firstly, both forms of desensitisation cause an alteration

in the control ratio which is an index of the relative activities of the sites and, secondly, it is possible to
prepare catalytically active amidotransferase which is
devoid of regulation by UDP-GlcNAc (Kornfeld, 1967;
Winterburn and Phelps, 1971c; Ellis and Sommar, 1972). The
binding site has a high specificity for the nucleotide
moiety although GDP-sugars are slightly inhibitory
(Kornfeld et. al., 1964). All UDP-sugars except the 2-
acetamido derivatives exhibit approximately equal poten-
cies but are considerably less inhibitory than UDP-GlcNAc.
UDP-GalNAc shows an inhibitory capacity intermediate
between that of UDP-GlcNAc and the other UDP-sugars
(Kornfeld, 1967). These findings imply that the UDP-GlcNAc
site is specific for the nucleotide and acetamido functions
but is non-specific for the remainder of the sugar. The
formation of an enzyme – UDP-GlcNAc complex probably
induces a conformational change which increases the
affinity of the enzyme for glutamine. This glutamine site
may be a second site separate to the catalytic glutamine
site or may represent a slight rearrangement of the active
site which permits glutamine to bind prior to fructose-6-P.
Mazlen et. al. (1970) preferred the former mechanism but
Bates and Handschumacher (1969) suggested the latter
arrangement whereby UDP-GlcNAc lowered the Km for glutamine
at the active site. In terms of the reaction mechanism the
inhibitory effect of lowering V could be the movement of
the amide-labilising nucleophilic group. The binding data
does not discriminate between the two possibilities. However, the kinetic evidence discussed by Winterburn and
Phelps (1971c) favours UDP-GlcNAc inducing the alternative
pathway of ternary complex formation via an enzyme-glutamine complex i.e. glutamine binds only at the active site.

Activation by UTP
    Although on its own UTP had little effect, the pres-
ence of UTP in a system inhibited by UDP-GlcNAc modified
the action of the inhibitor by raising V and removing the
substrate inhibition by glutamine (Winterburn and Phelps,
1971b). Both of these activating effects were compatible
with a decreased affinity of the enzyme for UDP-GlcNAc.
The ligand binding studies (Fig. 4) indicated that in the
presence of 0.1mM UDP-GlcNAc, 2mM UTP increased the
apparent dissociation constant for UDP-GlcNAc from 35 µM to
290 µM. Thus these studies were in agreement with the
kinetic behaviour but the mode of interaction of this

357

ligand with the enzyme was unknown. Thermal desensitisation studies demonstrated that the amidotransferase lost
sensitivity to UTP at the same rate as UDP-GlcNAc
(Winterburn and Phelps, 1971c). Similarly, although the
effects were more complex, p-chloromercuribenzoate
inhibited the actions of UTP and UDP-GlcNAc to the same
degree.

Inhibition of the amidotransferase by chemicals
possessing anti-inflammatory activity had been noted by
Bollet and Shuster (1960) and Schönhöfer (1966). The
complex inhibitory action of a range of these compounds on
the amidotransferase was studied by Winterburn (1969). Two
of the anti-inflammatory drugs, phenylbutazone and 4-iso-
butylphenylpropionic acid (ibuprofen), were tested as
potential desensitising agents by measuring the control and
modifier ratios. The results, shown in Table 1, signified
that the response of the enzyme towards UDP-GlcNAc and the
modification of this response by UTP were unaffected by
these inhibitors.

TABLE 1

Effects of phenylbutazone and ibuprofen on the
control and modifier ratios

	Control	With 2.5mM phenylbutazone	With 5mM ibuprofen
UDP-GlcNAc control ratio	0.46	0.45	0.45
UTP modifier ratio	1.35	1.32	1.45
G6P modifier ratio	0.45	0.65	0.57
AMP modifier ratio	0.57	0.73	0.76

The control and modifier ratios were determined as described in Experimental Procedures with the stated additions of
the anti-inflammatory agents.

It is concluded that the evidence favours UTP competing
for the UDP-GlcNAc site and thus it activates by being an
inhibitor of an inhibitor.

Inhibition by glucose-6-P and AMP
As noted for the action of UTP, neither glucose-6-P
nor AMP exhibited significant kinetic effects in the

absence of UDP-GlcNAc. Winterburn and Phelps (1970)
reported that glucose-6-P and AMP increased the inhibition
induced by the feedback inhibitor. A more detailed analysis of the actions revealed that they modified the UDP-
GlcNAc inhibition towards both substrates (Winterburn and
Phelps, 1971b). With glutamine as the variable substrate,
the glucose-6-P + UDP-GlcNAc or AMP + UDP-GlcNAc combinations created essentially the same effects as UDP-GlcNAc on
its own except that the affinity of the enzyme for glutamine at the substrate inhibition site was increased (Fig.
6). UDP-GlcNAc had behaved as a competitive inhibitor
towards fructose-6-P with a tendency to non-competitive
inhibition at high glutamine concentration. The addition

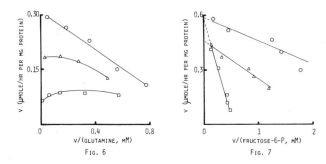

Fig. 6. Effect of glucose-6-P on the inhibition by UDP-
GlcNAc. Glutamine was the variable substrate and fructose-
6-P was 0.17mM. To the incubation mixtures were added: O,
0.5mM glucose-6-P; Δ, 25 µM UDP-GlcNAc; □, 25 µM UDP-GlcNAc
+ 0.5mM glucose-6-P.
Fig. 7. As for Fig. 6 except that fructose-6-P was the
variable substrate and glutamine was 6mM.

of glucose-6-P or AMP eliminated the non-competitive
element so that the glucose-6-P + UDP-GlcNAc and AMP + UDP-
GlcNAc combinations were competitive for fructose-6-P (Fig.
7). This led to confusing effects since at high fructose-
6-P the AMP or glucose-6-P could have either slight activation or no effect while lower concentrations of fructose-
6-P revealed the inhibition. The apparent Ki values were
calculated for these combinations and compared with the Ki
for UDP-GlcNAc. From the data of Fig. 2c of Winterburn and
Phelps (1971b) the Ki with respect to fructose-6-P for the
abscissa intercept - the "slope" using Cleland (1963b)

nomenclature – was calculated as 15 µM. The inhibition
created by 25 µM UDP-GlcNAc + 0.5mM glucose-6-P yielded a
Ki of 3 µM while 25 µM UDP-GlcNAc + 1.0mM AMP gave a Ki of
8 µM. Thus, glucose-6-P is more potent than AMP and these
metabolites increase the apparent affinity of the enzyme
for its feedback inhibitor.

The change in the apparent affinity for UDP-GlcNAc was
observed in the ligand binding studies also (Winterburn and
Phelps, 1971c). Although in the absence of the feedback
inhibitor no binding by AMP was observed, 1mM AMP lowered
the denaturation rate constant corresponding to a decrease
in the apparent dissociation constant from 35 µM to 9.5 µM.
1mM glucose-6-P exhibited a similar effect on UDP-GlcNAc
binding but because glucose-6-P protected the enzyme as
well, with an apparent dissociation constant of 0.44mM, the
interpretation of this ligand binding data required a more
complex model than that used.

The effect of thermal and chemical denaturation on the
interaction between the AMP or glucose-6-P sites and the
UDP-GlcNAc site indicated that these sites were distinct
(Winterburn and Phelps, 1971c). These findings were
supported by the action of the two anti-inflammatory
agents. The inhibitors caused an increase in the glucose-
6-P and AMP modifier ratios (a ratio of 1 was complete
desensitisation) signifying a decrease in the effectiveness
of these ligands without a similar influence on the UDP-
GlcNAc inhibition (Table 1).

The evidence presented is compatible with the operation of distinct sites for binding glucose-6-P and AMP which
are separate from the UDP-GlcNAc and catalytic sites.
Binding of either modifier increases the affinity of the
amidotransferase for UDP-GlcNAc and the similarity of the
actions of glucose-6-P and AMP suggests that their effects
may operate by the same mechanism. Because these potent-
iators are only altering the affinity of the inhibitor, the
nature of the inhibition is unaltered, merely enhanced
Glucose-6-P is without apparent effect on the inhibition by
two other UDP-sugars, UDP-Glc and UDP-GlcA (Winterburn and
Phelps, 1971b) a finding which further supports a role for
the acetamido function in inducing the change in conformation at the active site.

Model for effector binding

Collation of the conclusions drawn in the preceding
sections leads us to propose a model for the binding order
of the modifiers to the amidotransferase. This model which
takes no account of substrate bindings is shown diagrammat-
ically in Scheme 1 - permitted transitions are shown as
solid arrows and those not yet investigated as broken
arrows.

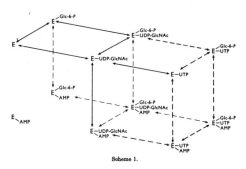

Scheme 1.

Thus the free enzyme, E, can bind either glucose-6-P or
UDP-GlcNAc and having bound one of these can then bind the
other to form a ternary complex, E-glucose-6-P-UDP-GlcNAc.
UTP may bind to E-UDP-GlcNAc but in so doing displaces the
UDP-GlcNAc. The free enzyme is apparently unable to bind
AMP so the only enzyme form so far found to which this
metabolite can bind is E-UDP-GlcNAc.

The physiological consequences of this complex series
of interactions in terms of hexosamine biosynthesis have
been discussed by Winterburn and Phelps (1971b). However,
it is worth mentioning that all of the effects reported in
this paper for UDP-GlcNAc, UTP, AMP and glucose-6-P occur
within the quoted cellular concentration ranges of these
metabolites.

REFERENCES

Atkins, E.D.T., Phelps, C.F., and Sheehan, J.K. (1972).
    Biochem. J., 128, 1255.
Bates, C.J. and Handschumacher, R.E. (1969). Advan. Enzyme
    Regulation, 7, 183.
Blumenthal, H.J., Horowitz, S.T., Hemerline, A., and
    Roseman, S. (1955). Bacteriol. Proc., 137

Bollet, A.J., and Shuster, A. (1960). J. Clin. Invest., 39, 1114.
Cleland, W.W. (1963a). Biochim. Biophys. Acta, 67, 104.
Cleland, W.W. (1963b). Biochim. Biophys. Acta, 67, 173.
Cohen, L.A., and Witkop, B. (1955). J. Am. Chem. Soc., 77, 6595.
Comb, D.E., and Roseman, S. (1958). J. Biol. Chem., 232, 807.
Davidson, E.A. (1966). In "The Amino Sugars" (E.A. Balazs and R.W. Jeanloz, eds.), Vol. IIB, pp. 1-44. Academic Press, New York.
Dyson, J.E.D., and Noltmann, E.A. (1968). J. Biol. Chem., 243, 1401.
Ellis, D.B. and Sommar, K.M. (1971). Biochim. Biophys. Acta, 230, 531.
Ellis, D.B., and Sommar, K.M. (1972). Biochim. Biophys. Acta, 267, 105.
Endo, A., Kakiki, K., and Misato, T. (1970). J. Bacteriol., 103, 588.
Ghosh, S., Blumenthal, H.J., Davidson, E., and Roseman, S. (1960). J. Biol. Chem., 235, 1265.
Gracy, R.W., and Noltmann, E.A. (1968). J. Biol. Chem., 243, 5410.
Gryder, R.M., and Pogell, B.M. (1960). J. Biol. Chem., 235, 558.
Haarhoff, K.N. (1969). J. Theoret. Biol., 22, 117.
Kaufman, M., Yip, M.C.M., and Knox, W.E. (1971). Enzyme, 12, 537.
Kikuchi, H., Kobayashi, Y., and Tsuiki, S. (1971). Biochim. Biophys. Acta, 237, 412.
Kornfeld, R. (1967). J. Biol. Chem., 242, 3135.
Kornfeld, R., Kornfeld, R., Neufeld, E.F., and O'Brien, P.J. (1964). Proc. Nat. Acad. Sci. U.S.A., 52, 371.
Leloir, L.F., and Cardini, C.E. (1953). Biochim. Biophys. Acta, 12, 15.
London, M., McHugh, R., and Hudson, P.B. (1958). Arch. Biochem. Biophys., 73, 72.
Lowther, D.A., and Rogers, H.J. (1956). Biochem. J., 62, 304.
Mazlen, R.G., Muellenberg, C.G., and O'Brien, P.J. (1969). Biochim. Biophys. Acta, 171, 352.
Mazlen, R.G., Muellenberg, C.G., and O'Brien, P.J. (1970). Exp. Eye Res., 9, 1.
Meister, A. (1965). In "Biochemistry of the Amino Acids", Vol. 1, 2nd Ed., Academic Press, New York.

Miyagi, T., and Tsuiki, S. (1971). Biochim. Biophys. Acta, 250, 51.

Mizobuchi, K., Kenyon, G.L., and Buchanan, J.M. (1968). J. Biol. Chem., 243, 4863.

Ohta, N., Pardee, A.B., McAuslan, B.R., and Burger, M.M. (1968). Biochim. Biophys. Acta, 158, 98.

Pogell, B.M., and Gryder, R.M. (1957). J. Biol. Chem., 228,

Schönhöfer, P. (1966). Med. Pharmacol. Exp., 15, 491.

Start, C., and Newsholme, E.A. (1968). Biochem. J., 107, 411.

Trujillo, J.L., Horng, W.J., and Gan, J.C. (1971). Biochim. Biophys. Acta, 252, 443.

Winterburn, P.J. (1969). Ph.D. Thesis (Bristol).

Winterburn, P.J. (1973). in preparation.

Winterburn, P.J., and Phelps, C.F. (1970). Nature, 228, 1311.

Winterburn, P.J., and Phelps, C.F. (1971). Biochem. J., 121, 701.

Winterburn, P.J., and Phelps, C.F. (1971). Biochem. J., 121, 711.

Winterburn, P.J., and Phelps, C.F. (1971). Biochem. J., 121, 721.

Winterburn, P.J., and Phelps, C.F. (1972). Nature, 236, 147.

GLUTAMINE PHOSPHORIBOSYLPYROPHOSPHATE AMIDOTRANSFERASE

James B. Wyngaarden

Department of Medicine, Duke University Medical Center
Durham, North Carolina

The first committed reaction of purine biosynthesis involves the interaction of phosphoribosylpyrophosphate (PP-ribose-P), glutamine, and water to form phosphoribosylamine. The reaction is catalyzed by the enzyme glutamine PP-ribose-P amidotransferase,* (Goldthwait, 1956; Hartman and Buchanan, 1957), and is the site of end-product inhibition by purine ribonucleotides (Wyngaarden and Ashton, 1959; Caskey et al, 1964). This article will summarize present knowledge of this enzyme and of the regulation of its activity.

I.   Kinetics and Mechanism of Action

A.   Reaction and Assays
     The enzyme catalyzes the following reaction:

$$\text{Glutamine + PP-ribose-P + } H_2O \xrightarrow{Mg^{2+}} \text{Glutamate +}$$

$$\text{5-phosphoribosylamine + } PP_i$$

Assays have been developed which depend upon the disappearance of PP-ribose-P (Goldthwait, 1956; Henderson and Khoo, 1965) or the appearance of glutamate, pyrophosphate, or phosphoribosylamine (PRA). Glutamate may be determined by spectrophotometric assay with glutamic dehydrogenase employing NAD (Hartman and Buchanan, 1957) or 3-acetyl-pyridine NAD (Wyngaarden and Ashton, 1959) as hydrogen

---

*The term glutamine PP-ribose-P amidotransferase has been employed in place of ribosylamine-5-phosphate:pyrophosphate phosphoribosyltransferase (glutamate amidating), EC 2.4.2.14, the name recommended by the Enzyme Commision of the International Union of Biochemistry.

acceptor, or by detection of radiochemical yield when glutamine-^{14}C is employed (Hartman, 1963; Hill and Bennett, 1969; Rowe et al, 1970). Hydroxylamine will replace water in the reaction. Determination of γ-glutamylhydroxamic acid has also been employed as an assay (Hartman, 1963a). Phosphoribosylamine may be determined by its further conversion to glycinamideribonucleotide (GAR) in the presence of glycine, ATP, and GAR synthetase, followed by the generation of a diazotizable amine in a subsequent indicator reaction in which GAR is transformylated from inosinic acid with the generation of 5´-phosphoribosyl-5-amino-4-imidazolecarboxamide (Hartman et al, 1956; Nierlich and Magasanik, 1965).

B. Substrates and Inhibitors

PP-ribose-P cannot be replaced by α-D-ribofuranosyl 1,5-diphosphate, ribose 5-phosphate (Goldthwait, 1956) or ribose 5-phosphate plus ATP (Wyngaarden and Ashton, 1959). PP-ribose-P is bound as the magnesium complex. Km or $[S]_{0.5}$ values of PP-ribose-P range from 0.05 to 0.5 mM, and of Mg^{2+} from 0.2 to 0.5 mM, in preparations from different species (Wyngaarden, 1972). Ammonium sulfate (Hill and Bennett, 1969) and orthophosphate (Holmes et al, 1972) inhibit competitively with PP-ribose-P.

α-Ketoglutaramic acid will not react with PP-ribose-P (Goldthwait, 1956), but certain alkylamides can replace glutamine in the reaction (Hartman, 1963a). These include γ-glutamylmethylamide, γ-glutamylhydroxymethylamide, γ-glutamyl-n̲-butylamide, and γ-glutamylcyclohexylamide. The reaction rate in the presence of these analogs is one-tenth to one-fifth of that observed with an equivalent concentration of glutamine. Km or $[S]_{0.5}$ values for glutamine range from 0.5 to 5.0 mM with enzyme from different species (Wyngaarden, 1972).

Highly purified chicken liver enzyme also catalyzes reactions between PP-ribose-P and certain alcohols (methanol, ethanol, 1-propanol, t̲-butyl alcohol), amines (ammonium chloride, methylammonium chloride, hydroxyamine), and hydrazine or phenylhydrazine, as detected with the pyrophosphate assay (Hartman, 1963a). The 5-phosphoribosyl derivative formed in the presence of ammonium ions is 5-phosphoribosylamine, since it can be converted to GAR (Goldthwait, 1956). The K_m for ammonium at pH 8.0 is 0.4 M. It is suggested that different reaction sites exist for glutamine and for alcohols and amines. Levels of DON

(10^{-4} M) that completely inhibit the reaction with glutamine have no inhibitory effect upon the reaction with
methanol, ammonia, or phenylhydrazine (Hartman, 1963b).

O-Diazoacetyl-L-serine (azaserine) and 6-diazo-5-oxo-
L-norleucine (DON) inhibit competitively with glutamine by
binding reversibly at a site specific for that substrate
and once bound react covalently with a functional group at
a site by virtue of their reactive diazo substituents
(Hartman, 1963b; Mizobuchi and Buchanan, 1968). From a
study of the kinetics of competition of the two inhibitors
with glutamine in the PP-ribose-P amidotransferase reaction, K_i values of 4.2×10^{-3} have been obtained for
azaserine and of $1.9 \pm 0.4 \times 10^{-5}$M for DON (Hartman, 1963b).

Progressive irreversible inhibition is observed when
the enzyme is incubated with the diazo analogs of glutamine. The initial rate of inactivation is accelerated more
than 10-fold in the presence of PP-ribose-P and Mg^{2+}, but
neither alone has any effect upon the rate of inactivation.
The concentration of PP-ribose-P required for half-maximal
effect in promoting the inactivation reaction is 2.2 ± 0.5
$\times 10^{-5}$M.

A constant maximal amount of ^{14}C-DON is eventually
bound by the enzyme, up to concentrations of DON of at
least 10^{-3}M. The ultimate amounts of ^{14}C taken up are
identical in the presence or the absence of PP-ribose-P
and Mg^{2+}. Purified chicken liver enzyme binds 1 mole of
DON per $198,000 \pm 6000$ gm of enzyme (Hartman, 1963b).

Glutamine does not inhibit the attachment of DON-^{14}C
to the enzyme in the absence of PP-ribose-P and Mg^{2+}. Its
effect in the presence of these substrates could not be
tested because the catalytic rate of the enzyme was about
10^4 times greater than the rate of its inactivation by DON.
Although these observations could indicate that the site
of covalent attachment of DON is distinct from the site to
which glutamine and DON bind reversibly, an alternative
explanation is that glutamine is bound to the enzyme only
when PP-ribose-P and Mg^{2+} are present. The latter interpretation accords with conclusions based on kinetic studies.
Although some ability of the enzyme to bind glutamine analogs must exist even in the absence of PP-ribose-P and Mg^{2+},
the reversible binding of DON is much enhanced by prior
attachment of PP-ribose-P and Mg^{2+}. The K_i of covalent
attachment is $2.4 \pm 1.0 \times 10^{-5}$M in the presence of PP-
ribose-P and Mg^{2+}, and at least 100-fold larger in the
absence of these substrates. Hartman (1963b) concludes

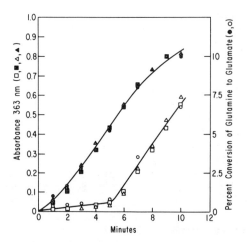

Fig. 1. Kinetic curves demonstrating the lag phase of the PP-ribose-P amido-transferase reaction with enzyme as prepared (open symbols) and with enzyme preincubated for 15 minutes with PP-ribose-P ($2.5 \times 10^{-4} M$) and Mg^{2+} ($3 \times 10^{-3} M$) at 4°C (filled symbols). Data obtained with the coupled assay and continuous recording are represented by squares. Data obtained by measurement of total glutamate in consecutive aliquots by an enzymatic spectrophotometric method are shown by triangles, and those obtained by a radiochemical method by circles. All assays were performed at 37°C. This experiment was run to establish that the lag phase, whether long or short, was not an artifact of the kinetically coupled (standard) assay. From Rowe *et al.* by permission of the American Chemical Society.

that both the reversible and irreversible processes of inhibition occur from the same binding site, and that this site is identical with the one specific for glutamine.

C.   Sequential Binding of Substrates

Several additional lines of evidence support the conclusion that binding of PP-ribose-P and Mg^{2+} precedes binding of glutamine. K_m or $[S]_{0.5}$ values for PP-ribose-P are independent of the concentration of glutamine, but K_m or $S_{0.5}$ values for glutamine fall progressively with increasing concentrations of PP-ribose-P (and Mg^{2+}), in studies with amidotransferase from chicken liver (Hartman, 1963a), adenocarcinoma 755 cells (Hill and Bennett, 1969) and Schizosaccharomyces pombe (Nagy, 1970).

Enzyme preparations from pigeon liver show a highly variable lag phase before maximal velocity is achieved (Rowe et al, 1970; Wyngaarden and Ashton, 1959). It can be shortened considerably by incubation with PP-ribose-P and Mg^{2+}, but not by incubation with either substance alone, nor with glutamine, nor with glutamine plus Mg^{2+}. This lag is not an artifactual property of the coupled kinetic assay system which involves continuous measurement of glutamate production with glutamic dehydrogenase and 3-acetyl pyridine NAD. Identical lag phases are observed when alternative assays of glutamate are employed, in which aliquots are removed consecutively and analyzed for total glutamate (Fig. 1) (Rowe et al, 1970).

D. Reaction Mechanism

Remy et al (1955) established that the pyrophosphate bond in PP-ribose-P is an α-linkage. In all natural purine nucleotides the glycosidic bond is a β-linkage. Since in the reaction between phosphoribosylamine, glycine, and ATP in the formation of GAR, the C—N bond of phosphoribosylamine is not involved, it is concluded that phosphoribosylamine is already in the β-configuration.

A glutamyl derivative of phosphoribosylamine should have an amideglycosidic linkage with stability comparable to that of GAR. A search for such a compound has been unsuccessful (Goldthwait et al, 1955). The liberation of PP_i and glutamic acid in equivalent amounts during the reaction is evidence that such an intermediate is not accumulating in significant quantities. Finally, in a displacement reaction forming such a glutamine derivative, inorganic pyrophosphate should exchange easily with PP-ribose-P. No exchange of ^{32}-P labeled inorganic pyrophosphate with PP-ribose-P was found when both substrates were incubated with enzyme with (Goldthwait, 1956) or without (Hartman and Buchanan, 1957) glutamine.

The question of an enzyme-bound phosphorylated intermediate has been debated (Buchanan et al, 1959). The absence of exchange between inorganic pyrophosphate-^{32}P and the pyrophosphate group of PP-ribose-P has been interpreted by Hartman and Buchanan (1957) as ruling out the existence of a reversible reaction of the type:

$$PP\text{-ribose-}P + E \rightleftharpoons 5\text{-phosphoribosyl-}E + PP_i$$

They propose a single displacement reaction involving a nucleophilic attack of the amide nitrogen of glutamine upon the 1-carbon of PP-ribose-P, activated, or rendered electrophilic, by the strongly electron-attracting pyrophosphate group. This hypothesis requires that the entire

369

reaction take place at one enzyme site without the formation of a definite covalent intermediate such as the
glutamyl derivative of phosphoribosylamine.

II. Structure

A. Molecular Weight
 From determinations of sedimentation and diffusion
coefficients, Hartman (1963b) calculated a molecular weight
of 210,000 ± 21,000 for chicken liver enzyme. In sucrose
density gradient studies of pigeon liver enzyme, Caskey et
al (1964) found s values of 8.8-9.3 S (mean 9.0S) corresponding to a molecular weight of 208,000.
 In the preparative procedure of Rowe and Wyngaarden
(1968) fractions of pigeon liver extracts are eventually
applied to either a DEAE-cellulose or DEAE-Sephadex A-25
column, and subjected to linear chloride gradient elution.
The enzyme appears in a single sharp symmetrical peak. For
maximal recovery, inclusion of 60 mM β-mercaptoethanol is
required. The peak of the elution curve may contain amidotransferase of 2500-4000 U/mg, representing 800-1000-fold
purification from the original supernatant. Fractions are
then concentrated with a Diaflo ultrafiltration cell.

B. Subunits
 When a concentrated sample of enzyme is applied to an
upward flow Bio-Gel P-300 column and eluted with Tris-Cl-
Mg buffer in the presence of 10 mM mercaptoethanol, three
peaks appear, each of which as amidotransferase activity
(Fig. 2). The molecular weights of the enzyme contained
in these three peaks are approximately 210,000, 108,000,
and 52,000. If mercaptoethanol is omitted, the smallest
molecular weight peak does not appear, and both earlier
peaks are correspondingly larger (Fig. 2B). Gel electrophoresis shows that the enzyme component having a MW of
210,000 is composed of four electrophoretically identical
subunits. Sedimentation velocity studies of the material
in the center of peaks B and C (Fig. 2) yield sedimentation
coefficients, $s_{20,w}$, of 6.02 and 3.95 S, corresponding to
molecular weights of approximately 100,000 and 50,000.
 The amidotransferase of 200,000 MW readily dissociates
into molecules of 100,000 MW on dilution, but dissociation
of the 100,000 MW enzyme occurs only in the presence of
relatively high concentrations of thiols, and does not
occur in 8 M urea alone. No further reduction in size of

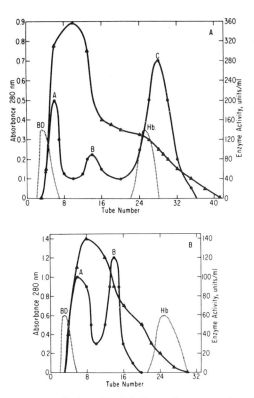

Fig. 2 Reverse (upward) flow Bio-Gel P-300 chromatography of PP-ribose-P amidotransferase on a column, 50 × 2.5 cm. ▲, absorbance at 280 nm. ●, enzyme activity in units/ml; BD, blue dextran marker (void volume, 29.0 ml); Hb, hemoglobin marker. The flow rate was 8 ml per hour; 2-ml fractions were collected. (A) Chromatography of thiol-prepared enzyme with peak A (MW, approximately 210,000), peak B (MW, approximately 108,000), and peak C (MW, approximately 52,000). (B) Chromatography of thiol-free enzyme preparation with peak A (MW, approximately 210,000) and peak B (MW, approximately 108,000). From Rowe and Wyngaarden   by permission of the American Society of Biological Chemists.

the 50,000 MW species occurs in thiols plus urea. The three species of amidotransferase appear to represent tetramer, dimer, and monomer forms of the enzyme. Each is active enzymatically. We have not been able to demonstrate reassociation of subunits in sedimentation velocity (Rowe and Wyngaarden, 1968) or sucrose density gradient studies (Caskey et al, 1964) in the presence of various combinations of substrates, even in the absence of thiols. These studies suggest that monomer and dimer may function as enzyme without reaggregation during the assay.

371

C. Nonheme Iron

Purified amidotransferase is yellow brown and contains iron, but no other metal in significant amounts (Hartman, 1963a). Our analyses (Rowe and Wyngaarden, 1968) indicate 6 gm atoms per mole of dimer prepared without β-mercapto-ethanol, and 4.4—4.8 gm atoms per dimer prepared in the presence of 60 mM β-mercaptoethanol. Up to one-third of the iron may be removed during routine preparation of PP-ribose-P amidotransferase in 60 mM mercaptoethanol, by treatment with 8.0 M urea, or by denaturation with 4 M ethanol.

Iron chelators such as 1,10-o-phenanthroline, batho-phenanthroline, Tiron, and α,α'-dipyridyl slowly remove 1 of 3 iron atoms (per monomer) and removal is accelerated in 8.0 M urea. Removal of the remaining 2 atoms of iron requires the presence of reducing agents such as β-mercap-toethanol (4M) or dithionate, and results in the formation of a white protein precipitate. Incubation with PP-ribose-P and Mg^{2+}, AMP and Mg^{2+}, or GMP and Mg^{2+} completely pro-tects the enzyme against removal of iron and inhibition by 1,10-o-phenanthroline. No such protection occurs with glutamine, or if Mg^{2+} is omitted in the experiments cited above. Protection by the ribonucleotide is observed whether or not the enzyme is sensitive to inhibition by these agents (Rowe and Wyngaarden, 1968; Rowe et al, 1970).

Highly purified dimer prepared in the absence of thiol shows the expected absorption peak at 279 nm plus a smaller peak at 415 nm preceded by a shoulder at 360 nm (Rowe and Wyngaarden, 1968). Extinction coefficients of 8.18×10^4 and 1.02×10^4 M^{-1} cm^{-1} were obtained at 279 and 415 nm, respectively, based on a molecular weight of 100,000. Addition of 0.3 M β-mercaptoethanol results in a marked decrease in the absorption at 415 nm within 45 minutes. This treatment is known to remove iron and con-firms that the 415 nm peak is in fact due to protein bound iron.

These studies suggest at least two different roles for iron. One atom per monomer may be involved in the catalytic function. However, the observation that inhibi-tion by phenanthroline may be prevented by PP-ribose-P and Mg^{2+}, or by AMP, or GMP (plus Mg^{2+}) even with enzyme which is not sensitive to nucleotide inhibition, indicates that this iron is not directly involved as a binding group of either the substrate or inhibitor site. It is more likely that this iron is involved in establishing the protein

372

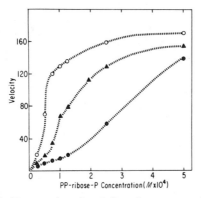

FIG. 3 Michaelis-Menten plot of activity of enzyme preincubated with PP-ribose-P and Mg^{2+}, illustrating the sigmoid kinetics with respect to PP-ribose-P. ○, No added AMP; ▲, $1.0 \times 10^{-3} M$ AMP; and ●, $2 \times 10^{-3} M$ AMP. From Rowe et al. by permission of the American Chemical Society.

conformation essential for enzyme function. The other two iron atoms per monomer appear to be concerned with tertiary structure, and their removal results in irreversible denaturation. The requirement of reducing conditions for their removal suggests the presence of —Fe—S— bonds. However, there is no acid-labile sulfur in the enzyme (Rowe and Wyngaarden, 1968).

III. Regulation of Activity

A. Effects of Substrates
When the reaction is initiated by addition of enzyme (pigeon liver) which has not been preincubated with substrate, the substrate-velocity plots form rectangular hyperbolas; no evidence of sigmoidicity is observed, even in the presence of nucleotide inhibitors (Caskey et al, 1964; Wyngaarden and Ashton, 1959). However, in studies with enzyme which has been preincubated with PP-ribose-P and Mg^{2+} for 15 minutes, the substrate-velocity plot is sigmoidal, and more obviously so in the presence of nucleotide inhibitors (Rowe et al, 1970) (Fig. 3). Enzymes prepared from adenocarcinoma 755 cells (Hill and Bennett, 1969) or S. pombe (Nagy, 1970) show sigmoidal plots under standard assay conditions. The enzyme from human placenta shows hyperbolic kinetics in the absence of nucleotides, but sigmoidal plots in their presence (Holmes et al, 1972).
With Hill plots, a value of n = 1 was repeatedly

373

obtained by Caskey et al (1964) with pigeon liver enzyme
studied in the standard assay. With enzyme preincubated
with PP-ribose-P and Mg^{2+}, values for n ranged from 1.7 to
3.2 (Rowe et al, 1970; Wyngaarden, 1972). A value for n
of 1.9 was reported by D. L. Hill and Bennett (1969) for
the enzyme for adenocarcinoma cells, and of 1.5 by Shiio
and Ishii (1969) with enzyme from B. subtilis studied in
presence of AMP. Values of n increasing from 1.7 to 2.5
with increases of glutamine were reported by Nagy (1970)
for the enzyme from S. pombe. Values of n as high as 3
were found with a mutant form of the enzyme from S. pombe
(aza-I). With enzyme from human placenta, values of n
increase from 1 to 2.7 in the presence of 3 mM AMP (Holmes
et al, 1972).

Velocity-substrate plots form typical hyperbolic
curves for glutamine with all enzymes studied, irrespective of the concentration of PP-ribose-P (Wyngaarden,
1972).

With pigeon liver enzyme the Hill coefficient is 1
with preincubated as well as with standard enzyme, and is
unchanged in the presence of 2 mM AMP (Wyngaarden, 1972).
However, with S. pombe enzyme, the value of 1 increases to
1.8 and 2.2 in the presence of 0.2 or 0.4 mM IMP, indicating cooperativity of glutamine binding in this preparation (Nagy, 1970).

B. Effects of Purine Ribonucleotide Inhibitors

The pigeon liver enzyme is competitively inhibited by
purine 5'- ribonucleotides (Fig. 4) but not by (2',3')
ribonucleotides, 5'- deoxyribonucleotides, ribonucleosides
or free bases, nor by pyrimidine compounds. Effective
inhibitors include AMP, ADP, GMP, GDP, IMP, 4-amino-5-
imidazolecarboxamide ribonucleotide, and ATP, but not GTP,
IDP, or ITP (Wyngaarden and Ashton, 1959). In addition,
the enzyme is inhibited by 5'-phosphoribosyl derivatives
of 6-thiopurine, 6-thioguanine, 8-azaguanine, and allo-
purinol (McCollister et al, 1964), and by the corresponding
derivative of cordycepin (Rottman and Guarino, 1964).
Half-maximal inhibition with AMP or GMP occurred at concentrations ranging from 0.09 to 2.5 mM with various
preparations; indeed some were totally desensitized to
nucleotides during purification (see below).

Limited studies of the three enzyme fractions obtained
from Bio-Gel P-300, or from sucrose density gradient separations, disclosed inhibition of all fractions by AMP and

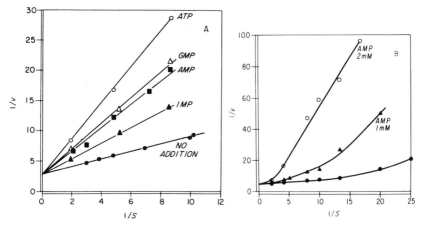

FIG. 4 Competitive inhibition of PP-ribose-P amidotransferase by purine ribonucleotides. (A) Enzyme prepared according to Wyngaarden and Ashton, fraction V , not preincubated with substrate. Nucleotide concentration 0.2 mM. (B) Enzyme prepared according to Rowe and Wyngaarden, fraction 5 , preincubated 15 minutes with PP-ribose-P and Mg^{2+}. $v = \triangle$ $OD_{\lambda 383 nm}/10$ minutes. $S = mM$. Figure A from Wyngaarden and Ashton by permission of American Society of Biological Chemists.

GMP (Rowe et al, 1970). Approximate $[I]_{0.5}$ values of 1 mM were observed for each nucleotide for all three fractions. These studies suggest that PP-ribose-P amidotransferase is both active as enzyme and sensitive to nucleotides in tetramer, dimer and monomer states.

Studies with PP-ribose-P amidotransferase from other sources have disclosed some interesting differences. The enzyme from A. aerogenes is inhibited by AMP, ADP, GMP, IMP, and GTP but not by ATP (Nierlich and Magasanik, 1965). The enzyme from B. subtilis is weakly inhibited by GMP and GTP (Momose et al, 1965; Shiio and Ishii, 1969), more strongly by AMP and ADP (Shiio and Ishii, 1969) and 5'-phosphoribosyl cordycepin (Rottman and Guarino). The amidotransferase from S. pombe is sensitive to GMP and IMP but less well inhibited by AMP (Nagy, 1970). PP-ribose-P amidotransferase from rat liver is inhibited by GMP, AMP, and ATP, (Caskey et al, 1964). The PP-ribose-P amidotransferase of mouse spleen, induced by Friend leukemia virus, appears to be more sensitive to guanyl ribonucleotides than to adenyl compounds (Reem, 1968; Reem and Friend, 1967). The enzyme from adenocarcinoma 755 cells is sensitive to some extent to every nucleotide tested. Inhibition by ribonucleoside and deoxyribonucleoside triphosphates can be overcome by additional magnesium (Hill and Bennett, 1969). This is not the case with inhibition of the pigeon liver enzyme by ATP (Wyngaarden and Ashton, 1959). Enzyme from human lymphoblasts is inhibited by AMP and GMP (Wood and Seegmiller, 1971). The enzyme from human placenta is equally sensitive to AMP and GMP but is also weakly inhibited by a number of pyrimidine nucleotides (Holmes et al, 1972). With all enzymes studied, suitable concentrations of effective ribonucleotides produce 100% inhibition (Wyngaarden, 1972).

In studies of cooperativity between inhibitor binding sites (Johnson et al, 1942; Taketa and Pogell, 1965) with pigeon liver enzyme (Caskey et al, 1964) in the tetramer form, Hill plots yield values of n' ranging from 1.3 to 3.2 for AMP, ADP, ATP, GMP, GDP, $\overline{IMP}$, and 5-phosphoribosyl-6-mercaptopurine at $[I]_{0.5}$ concentrations. The amidotransferase from B. subtilis (Shiio and Ishii, 1969) has $\underline{n}$ values of 3.1 for AMP and 2.5 for ADP. With enzyme from S. pombe (Nagy, 1970), n' values for IMP and GMP were 2.5 at saturating values of PP-ribose-P. With enzyme from adenocarcinoma 755 cells (Hill and Bennett, 1969), $\underline{n'}$ values were independent of PP-ribose-P concentrations.

In studies of end-product inhibition of PP-ribose-P amidotransferase from pigeon liver, several preparations lost all sensitivity to ATP while undergoing as much as an 8-fold gain in specific activity. In other experiments enzyme preparations lost and regained sensitivities to AMP and GMP, but the variations with respect to the two nucleotides occurred independently (Caskey et al, 1964). The lack of parallelism of responses to AMP and GMP suggested the possibility of more than one type of inhibitor binding site. Studies were therefore conducted with pairs of ribonucleotide inhibitors in which the effects of each ribonucleotide alone were compared with inhibitions caused by two ribonucleotides acting together. In the cases of AMP plus GMP, AMP plus IMP, AMP plus GDP, and AMP plus 6-thiopurine ribonucleotide, the effects of the two inhibitors acting together were significantly greater than the predicted additive effects (Fig. 5) (Caskey et al, 1964). In Hill plots, values of $\underline{n'}$ as high as 4.8 were obtained at total concentrations of two nonhomologous inhibitors (AMP plus 6-mercaptopurine ribonucleotide) which reduced the velocity by 50%. In the cases of AMP plus ADP, GMP plus IMP, GMP plus GDP, and AMP plus ATP, the inhibitory effects were additive. These results indicate that the

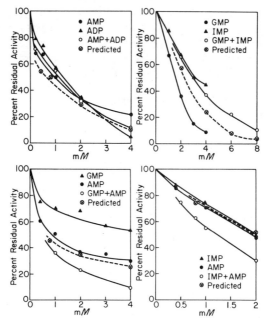

Fig. 5. Synergistic inhibition of PP-ribose-P amidotransferase by pairs of non-homologous purine ribonucleotides. In all four frames, the observed activity values in the presence of two nucleotides are plotted in terms of *total* nucleotide concentrations, and the two inhibitors are present in equimolar concentrations. The *predicted* curve is determined by the product of the fractional inhibitions caused by each inhibitor alone. For example, in the upper left-hand panel, there was 57% residual activity in the presence of 1 mM ADP, and 50% residual activity in the presence of 1 mM AMP. The predicted activity in the presence of 1 mM ADP and 1 mM AMP is (0.57 × 0.50) 100 or 28.5%. Thus the predicted residual activity curve represents the minimal activity (maximal inhibition) anticipated if each inhibitor acts independently of the other. Observed inhibitions greater than those found with equimolar *total* concentration of either inhibitor alone, and greater than the individual inhibitions (predicted residual activity curves) represent synergistic effects of two inhibitors. As shown here, AMP + GMP (lower left-hand panel) and AMP + IMP (lower right-hand panel) act synergistically. The velocities in the absence of inhibitor were 0.023–0.030 optical density unit/minute. From Caskey *et al.* by permission of the American Society of Biological Chemists.

pigeon liver enzyme contains separate binding sites for 6-amino and 6-hydroxypurine ribonucleotide compounds, and that these sites are interacting when occupied by non-homologous inhibitors. The term "synergistic inhibition" appropriately describes this type of allosteric regulation.

Very similar control features have also been observed by Nierlich and Magasanik (1965) with the PP-ribose-P amidotransferase of A. aerogenes, by Jha (1972) with enzyme from Neurospora crassa, and by Holmes et al (1972) with the human placental enzyme, all of which show synergistic inhibition by 6-hydroxy- and 6-aminopurine compounds. The synergistic inhibition is competitive and can be completely overcome by high concentrations of PP-ribose-P. Synergistic inhibition can also be discerned in studies of the amidotransferase of S. pombe (Nagy, 1970) which show inhibitory effects of equimolar mixtures of AMP plus IMP, and of AMP plus GMP, that are greater than the calculated additive effects. However, preparations from adenocarcinoma 755 cells do not show synergistic effects of combinations of inhibitory ribonucleotides (Hill and Bennett, 1969).

The purified amidotransferase from an 8-azaguanine resistant mutant of S. pombe (Aza-I) shows a 10-fold reduction in sensitivity to IMP and GMP, but normal inhibition by AMP (Nagy, 1970). A mutant strain of ascites cells which is resistant to 6-methylmercaptopurine also shows reduced sensitivity to 6-hydroxypurine compounds and normal inhibition by 6-aminopurines (Henderson et al, 1967).

The synergistic nature of the inhibitions by 6-amino- and 6-hydroxypurine ribonucleotides on the first step in their biosynthesis should permit the more effective curtailment of purine biosynthesis when both types of inhibitors are present simultaneously, but allow for a more moderate control when only one kind of purine is in excess.

Sensitivity to nucleotide inhibitors may change in the course of enzyme purification. With pigeon liver enzyme (Caskey et al, 1964) desensitization occurred most often during dialysis of an ammonium sulfate fraction against distilled water. In this procedure the enzyme precipitates, and on solution in buffer, the enzyme although catalytically active, is frequently nucleotide insensitive. Nevertheless, deliberate desensitization of nucleotide-sensitive enzyme could not be achieved by measures which have been successfully employed with other allosteric enzymes, such as treatment with heat, metabolic inhibitors, or urea. The highly purified enzyme from chicken liver prepared by Hartman (1963a) is insensitive to nucleotides. The amidotransferase from S. pombe retains its activity and its sensitivity toward inhibitors for about a week when stored at 0°, but loses sensitivity toward inhibitors 2-3 days at $-20^{\circ}C$ (Nagy, 1970). The enzymes from A.

378

aerogenes Nierlich and Magasanik, 1965), B. subtilis (Rottman and Guarino, 1964), adenocarcinoma 755 cells (Hill and
Bennett, 1969) and human placenta (Holmes et al 1972)
retain nucleotide sensitivity during partial purification.

As pointed out above, the nucleotide-insensitive PP-
ribose-P amidotransferase is protected against inhibition
by 1,10-o-phenanthroline by preincubation with AMP plus
Mg^{2+}, or GMP plus Mg^{2+}, but not by nucleotide or Mg^{2+} alone,
nor by glutamine (Rowe et al,1970). These results suggest
that the nucleotide-insensitive enzyme is still capable of
binding AMP or GMP. In this respect the enzyme resembles
the first enzyme of histidine biosynthesis, phosphoribosyl-
ATP-pyrophosphorylase, in which Martin (1963) has shown by
equilibrium dialysis that insensitive enzyme binds the
"inhibitor" histidine.

IV. Enzyme Derepression and Repression

Nierlich and Magasanik (1963) have demonstrated
repression and derepression of six enzymes of purine biosynthesis in A. aerogenes, including three of the pathway
leading to IMP. Changes of activity of PP-ribose-P amidotransferase are coordinate with those of FGAR-amidotransferase but noncoordinate with changes of activity of the
other enzymes, including that of GAR synthetase, the second
enzyme of the pathway.

Pur D mutants of Salmonella typhimurium (Westby and
Gots, 1969) which are deficient in activity of GAR synthetase, contain wild-type levels of PP-ribose-P amidotransferase when grown on xanthine (5 mg/ml). By contrast Pur
F mutants, which lack PP-ribose-P amidotransferase, produce
higher, derepressed levels of GAR synthetase (88-180 nmoles
total GAR per milligram of protein) compared with wild-type
levels (35 units/mg) under similar growth conditions.

Reem and Friend (1967) find that mouse spleen normally
contains no detectable activity of PP-ribose-P amidotransferase. Following infection with Friend leukemia virus,
activity of PP-ribose-P amidotransferase appears by day 4,
increases rapidly to a maximum by days 6-9, and thereafter
declines over 2-4 weeks. The enzyme activity is subject to
ribonucleotide feedback inhibition both in vivo and in
vitro. Whether PP-ribose-P amidotransferase is coded by
derepressed host genome or by viral genome is not known.

379

V. Summary

Kinetic studies do not unequivocally provide a determination of the minimal number of binding sites, since rate
kinetic effects alone can give ligand concentrations of
powers higher than the number of interacting sites (Sanwal
and Cook, 1966). In most actual cases, however, Hill
coefficients have given an indication of the minimum number
of sites (Atkinson, 1965, 1966; Gerhart and Pardee, 1964;
Stadtman, 1966). In a few examples, the coefficients have
equaled the theoretical number of sites, indicating a very
strong degree of cooperativity (Atkinson et al, 1965; Long
and Pardee, 1967; Long et al, 1970). In many instances the
Hill coefficient is less than the theoretical value. Hemoglobin has a Hill coefficient of 2.8 for a total of 4
oxygen binding sites (Hill, 1910; Wyman, 1948). Fructose-1,
6-diphosphatase from rabbit liver has a Hill coefficient of
1.7 for the binding of 4 equivalents of FDP per mole of
tetrameric enzyme at saturation (Pontremoli and Horecker,
1970).

In considering a schematic model of PP-ribose-P amidotransferase the main observations (pigeon liver enzyme,
chiefly) to be accommodated are (a) tetrameric structure;
(b) kinetic data for binding of at least 3 molecules of
PP-ribose-P; of at least 2 molecules of glutamine; of 2 or
3 molecules of a single type of nucleotide; and of as many
as 5 molecules of mixtures of 6-amino and 6-hydroxypurine
ribonucleotides; (c) catalytic activity and nucleotide
sensitivity of monomer and dimer forms under conditions in
which deliberate attempts to achieve and demonstrate
reaggregation do not indicate a change in molecular weight.

The inhibitor sites would have to be distinct from
substrate sites, i.e., with no shared binding subgroups,
in order to explain protection of even the nucleotide
insensitive enzyme from o-phenanthroline inhibition by AMP
or GMP. Reasonable limiting cases of models extend from 1
to 4 active sites, and from 2 to 8 nucleotide regulatory
sites per tetramer.

A first model envisions 4 identical subunits forming
a tetramer with four catalytic sites and eight regulatory
sites. Each monomer has one site for PP-ribose-P, glutamine, 6-amino- and 6-hydroxypurine ribonucleotides,
respectively. Theoretical values of n or n' of 4 for each
ligand, or of 8 for nonhomologous ribonucleotide inhibitors
acting together, are not observed because the strengths of

the cooperative interactions between sites are not sufficiently great under the conditions of study.

A second model envisions two catalytic sites per tetrameric enzyme of 200,000 MW. Kinetic data do not indicate binding of more than two glutamine molecules by any PP-ribose-P amidotransferase studied thus far. Furthermore, affinity labeling studies show maximal binding of only 1 mole of ^{14}C-labeled DON per 198,000 gm of enzyme (Hartman, 1963b). In this model two PP-ribose-P sites function as catalytic sites and two as modifiers. In addition, the enzyme has 8 nucleotide regulatory sites, as proposed above. Either catalytic site model may be altered to allow for 4 overlapping rather than 8 discrete nucleotide sites. Such a model would allow for cooperative effects between homologous inhibitor sites, and also for synergistic effects of binding of nonhomologous inhibitors, if it is assumed that binding energies for the two types of inhibitors vary with changes of conformational states, and that the tetramer may exist in several different hybrid forms such that one subunit binds a 6-aminopurine preferentially while another binds a 6-hydroxypurine. This model predicts that the maximal value of n' is 4; the one observed value of 4.8 would then presumably represent an actual value of 4 rather than 5. These several schematic possibilities are shown in Fig. 6.

On the basis of these models, the experimental observations with respect to glutamine sites could be explained in two ways, which are not mutually exclusive: (a) The potential glutamine sites are distorted in the inactive form of the enzyme; ligand-induced conformational changes brought about by PP-ribose-P and Mg^{2+} are required to generate four glutamine-binding sites in the tetramer; cooperatively is only moderately strong as indicated by $\underline{n}$ values of 2. (b) The tetramer possesses four glutamine-binding sites but only two are available for the catalytic process and only one can be filled by DON.

CTP-synthetase of E. coli is a tetrameric enzyme (Long et al, 1970) with reactive sites for glutamine, UTP, and ATP. The glutamine sites show negative cooperativity (Levitzki and Koshland, 1969; Long and Pardee, 1967), but values of $\underline{n} > 2$ are required to explain the biphasic substrate-velocity curve of glutamine (Teipel and Koshland, 1969). The maximal $\underline{n}$ values of 3.4 for UTP, and of 3.8 for

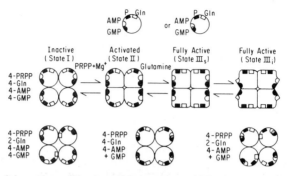

FIG. 6 Schematic models of possible structures of PP-ribose-P amidotransferase.
At the top are models of two types of monomers. In the one on the left two
discrete nucleotide binding sites are shown; in the one on the right overlapping
sites are shown for 6-amino and 6-hydroxypurine ribonucleotides. Open and solid
rectangles indicate substrate binding sites (P = PP-ribose-P, Gln = glutamine).
In the center are shown various hypothetical forms of the tetramer, illustrating the
changes of conformational state brought about by interaction with substrates, and
also indicating that additional conformational variations may determine whether
the enzyme is nucleotide sensitive (s) or insensitive (i). In the lower portion
of the figure, variations of tetramer arrangements are shown, which place different
limits on maximal binding values of substrates and inhibitors. Each of these
models can be substituted for the one at the left on the line above, and carried
through the same sequence of hypothetical conformational changes.

ATP (Long and Pardee, 1967) strongly suggest that four
binding sites exist for each nucleoside triphosphate, and
by inference for glutamine as well (Long et al, 1970).
Nevertheless, only 2 of the 4 probable glutamine sites can
be filled by ^{14}C-labeled DON in affinity labeling experiments. Thus, DON is not a valid probe of the number of
glutamine-binding sites of CTP-synthetase; perhaps it is
also invalid in the case of PP-ribose-P amidotransferase.
Explanations such as strong negative cooperativity, or
steric hindrance by the additional methylene group of DON
have been proposed (Long et al, 1970). Substrate and
inhibitor binding data will be required to discriminate
among the models of PP-ribose-P amidotransferase structure;
the first model (four catalytic and eight nucleotide
regulatory sites) appears to be the most plausible at the
present time.

The regulatory properties of glutamine PP-ribose-P
amidotransferase have been viewed from the standpoint of
the symmetry model of Monod et al (1965) by Nagy (1970) in
the case of the enzyme from S. pombe; and from the standpoint of the ligand-induced or sequential model of

Koshland et al (1966) by Rowe et al (1970) in the case of the enzyme from pigeon liver. The slowly progressive positive homotropic effects of substrates in the activation of pigeon liver enzyme appear strongly in favor of the sequential model. Koshland's (1969, 1970) concepts of enzyme flexibility, of enzyme-substrate interaction in establishment of the catalytically active site, and of enzyme-inhibitor interaction in modifying the activity of the catalytic site would appear to form the best point of departure for considering the regulatory behavior of PP-ribose-P amidotransferase.

References

Atkinson, D. E. (1965). Science 150, 851.

Atkinson, D. E. (1966). Annu. Rev. Biochem. 35, 85.

Atkinson, D. E., Hathaway, J. A., and Smith, E. C. (1965). J. Biol. Chem. 240, 2682.

Buchanan, J. M. Hartman, S. C., Herrmann, R. L., and Day, R. A. (1959). J. Cell. Comp. Physiol. 54, Suppl. 1, 139.

Caskey, C. T., Ashton, D. M., and Wyngaarden, J. B. (1964). J. Biol. Chem. 239, 2570.

Gerhart, J. C., and Pardee, A. B. (1964). Fed. Proc., Fed. Amer. Soc. Exp. Biol. 23, 727.

Goldthwait, D. A. (1956). J. Biol. Chem. 232, 1051.

Goldthwait, D. A., Greenberg, G. R., and Peabody, R. A. (1955). Biochim. Biophys. Acta 18, 148.

Hartman, S. C. (1963). J. Biol. Chem. 238, 3024.

Hartman, S. C. (1963). J. Biol. Chem. 238, 3036.

Hartman, S. C., and Buchanan, J. M. (1957). J. Biol. Chem. 233, 451.

Hartman, S. C., Levenberg, B., and Buchanan, J. M. (1956). J. Biol. Chem. 221, 1057.

Henderson, J. F. (1962). J. Biol. Chem. 237, 2631.

Henderson, J. F. (1963). Biochem. Pharmacol. 12, 551.

Henderson, J. F., Caldwell, I. C., and Paterson, A. R. P. (1967). Cancer Res. 27, 1773.

Henderson, J. F., and Khoo, M. K. Y. (1965). J. Biol. Chem. 240, 3104.

Hill, A. V. (1910). J. Physiol. (London) 40, iv.

Hill, A. V. (1913). Biochem. J. 7, 471.

Hill, D. L., and Bennett, L. L. (1969). Biochemistry 8, 122.

Holmes, E. W., McDonald, J. A., McCord, J. M., Wyngaarden, J. B., and Kelley, W. N. (1972) in press. J. Biol. Chem.

Jha, KK. (1972). Genetics Soc. America, abstract.

Johnson, F. H., Eyring, H., and Williams, R. W. (1942). J. Cell. Comp. Physiol. 20, 247.

Koshland, D. E., Jr., (1963). Cold Spring Harbor 28, 473.

Koshland, D. E., Jr. (1969). Curr. Top. Cell. Regul. 1, 1.

Koshland, D. E., Jr. (1970). In "The Enzymes" (P. Boyer, ed.), Vol. 1, p. 341. Academic Press, New York.

Koshland, D. E., Jr., Némethy, G., and Filmer, D. (1966). Biochemistry 5, 365.

Levitzki, A., and Koshland, D. E., Jr. (1969). Proc. Nat. Acad. Sci. U. S. 62, 1121.

Li, H. C., and Buchanan, J. M. (1971). J. Biol. Chem. 246, 4713.

Long, C. W., and Pardee, A. B. (1967). J. Biol. Chem. 242, 4715.

Long, C. W., Levitzki, A., and Koshland, D. E., Jr. (1970). J. Biol. Chem. 245, 80.

Martin, R. G. (1963). J. Biol. Chem. 238, 257.

McCollister, R. J., Gilbert, W. R., Jr., Ashton, D. M., and Wyngaarden, J. B. (1964). J. Biol Chem. 239, 1560.

Mizobuchi, K., and Buchanan, J. M. (1968). J. Biol Chem. 243, 4853.

Momose, H., Nishikawa, H., and Katsuya, N. (1965). J. Gen. Appl. Microbiol. 2, 211.

Monod, J., Changeux, J.-P., and Jacob, F. (1963). J. Mol. Biol. 6, 306.

Monod, J., Wyman, J., and Changeux, J.-P. (1965). J. Mol. Biol. 12, 88.

Nagy, M. (1970). Biochim. Biophys. Acta 198, 471.

Nierlich, D. P., and Magasanik, B. (1963). Fed. Proc., Fed. Amer. Soc. Exp. Biol. 22, 476.

Nierlich, D. P., and Magasanik, B. (1965). J. Biol. Chem. 240, 358.

Pontremoli, S., and Horecker, B. L. (1970). Curr. Top. Cell. Regul. 2, 173.

Reem, G. H. (1968). J. Clin. Invest. 47, 83a.

Reem, G. H., and Friend, C. (1967). Science 157, 1203.

Remy, C. N., Remy, W. T., and Buchanan, J. M. (1955). J. Biol. Chem. 217, 885.

Rottman, F., and Guarino, A. J. (1964). Biochim. Biophys. Acta 89, 465.

Rowe, P. B., Coleman, M. D., and Wyngaarden, J. B. (1970). Biochemistry 9, 1948.

Rowe, P. B., and Wyngaarden, J. B. (1968). J. Biol. Chem. 243, 6373.

Sanwal, B. D., and Cook, R. A. (1966). Biochemistry 5, 886.

Shiio, I., and Ishii, K. (1969). J. Biochem. 66, 175.

Stadtman, E. R. (1966). Advan. Enzymol. 28, 41.

Taketa, K., and Pogell, B. M. (1965). J. Biol. Chem. 240, 651.

Teipel, J., and Koshland, D. E., Jr. (1969). Biochemistry 8, 4656.

Westby, C. A., and Gots, J. S. (1969). J. Biol. Chem. 244, 2095.

Wood, A. W., and Seegmiller, J. E. (1971). Fed. Proc., Fed. Amer. Soc. Exp. Biol. 30, 1113 (abstract).

Wyman, J. (1948). Advan. Protein Chem. 4, 407.

Wyngaarden, J. B. (1972). Curr. Top. Cell. Regul. 5, 135.

Wyngaarden, J. B., and Ashton, D. M. (1959). J. Biol. Chem. 234, 1492.

FORMYLGLYCINAMIDE RIBONUCLEOTIDE AMIDOTRANSFERASE.*

John M. Buchanan

Department of Biology

Massachusetts Institute of Technology

Cambridge, Massachusetts 02139

The recognition that the amide nitrogen of glutamine is a specific precursor of two nitrogen atoms of the purine ring (Sonne et al., 1953, 1954, Levenberg et al., 1956) and the amino group of glucosamine (Leloir and Cardini, 1953) has led to a wide exploration of this nitrogen donor in metabolic reactions concerned with the formation of carbon-to-nitrogen bonds.

The reaction of purine nucleotide synthesis catalyzed by the enzyme, formylglycinamide ribonucleotide amidotransferase Ѱ, is shown in equation 1 (Melnick and Buchanan, 1957), Mizobuchi and Buchanan, 1968a)

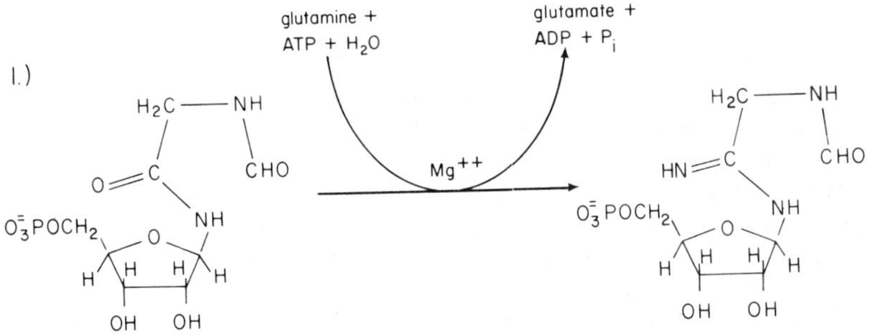

The substrate, formylglycinamide ribonucleotide (FGAR), undergoes amination by the nitrogen donor, glutamine, to yield formylglycinamidine ribonucleotide (FGAM) and glutamic acid. The energy required for the formation of this carbon-to-nitrogen bond is supplied at the expense

of the cleavage of a phosphodiester bond of ATP to ADP
and inorganic phosphate. A mole of H_2O is consumed in
the reaction, which also requires the divalent cation,
Mg^{2+}, Co^{2+} and Mn^{2+} can replace Mg^{2+} (Li and Buchanan,
1972). NH_3 can also replace glutamine as the amino donor
but with a considerably diminished rate of reaction at
pH 8, the optimal pH for reaction of both substrates
(Mizobuchi and Buchanan, 1968b, Mizobuchi et al., 1968).

Assay of the Enzymatic Activity: The most general
assay of the enzyme particularly in impure preparations
is accomplished by coupling the reaction described by
equation 1 with a second reaction shown in equation 2
(Melnick and Buchanan, 1957, French et al., 1963a,
Mizobuchi and Buchanan, 1968a)

2.)

5-aminoimidazole ribo-
nucleotide (AIR)

The enzyme catalyzing reaction 2 is AIR synthetase. The
aminoimidazole product can be diazotized and coupled
with N-(1-naphthyl)ethylenediamine to give a salmon
colored product with an absorption maximum of 500 nm.
In the presence of an excess of the second enzyme, FGAM
is quantitatively converted to AIR and the over-all reaction
is proportional within limits to the concentration of the
amidotransferase and to the duration of incubation at 38°
and pH 8.0. The assay can also be performed with
NH_4Cl ($K_m = 9 \times 10^{-2}$) as the amino donor in place of
glutamine ($K_m = 2 \times 10^{-4}$), and is particularly useful for
determination of the enzymatic activity when the glutamine
site has been blocked by an alkylating agent or glutamine
analog. In spite of its simplicity, however, this assay

388

suffers the disadvantage that certain reagents may inactivate the indicator enzyme and thereby interfere with
the measurement of FGAR amidotransferase activity.

In a second assay, which can only be used with highly
purified preparations free of adenylate kinase or ATPase,
the formation of ADP from ATP is measured. The assay
can be performed with radioactive ATP with measurement
of radioactive ADP separated from ATP by resin chromatography (Mizobuchi and Buchanan, 1968a). In a second
spectrophotometric version of this assay, ADP may be
measured by coupling its formation with the reduction of
NAD in a system composed of pyruvate kinase, lactic
dehydrogenase, phosphoenolpyruvate and NAD (Kornberg
and Pricer, 1951).

A third useful assay involves the measurement of the
formation of glutamate by its oxidation in the presence of
the 3-acetyl pyridine analog of NAD and the indicator
enzyme, glutamate dehydrogenase (Wyngaarden and Ashton,
1959, Li and Buchanan, 1971a). This assay is most
accurately applied when the production and oxidation of
glutamate are measured in a stepwise rather than coupled
system.

Purification of the Enzyme: The enzyme has been
isolated in highly purified form from Salmonella typhimurium (French et al., 1963a), and chicken liver
(Mizobuchi and Buchanan, 1968a). It has also been obtained in a relatively impure state from pigeon liver (Melnick
and Buchanan, 1957 and Ehrlich ascites tumor cells
(Chu and Henderson, 1972a).

The molecular weight of the enzyme either from
Salmonella or chicken liver is 133,000-135,000. From the
molecular weight and specific activity of the enzyme a
molecular activity (defined as the moles of product formed
by 1 mole of enzyme at 38° for 1 min) of 100 has been calculated. Electrophoresis of highly purified chicken liver
enzyme in the presence of a reducing agent and sodium
dodecyl sulfate yields only one component (Frere et al.,
1971). Thus, the enzyme is a single protein and does not
appear to contain subunits.

When enzyme is allowed to stand in the absence of a
reducing agent, such as dithiothreitol or 2-mercapto-
ethanol, it undergoes polymerization, which is reversible
if reductant is supplied at a concentration of 10^{-2} M. It is
believed that polymerization occurs by oxidation of
sulfhydryl groups not associated with the active centers of
the enzyme. The specific activity of the enzyme is not
affected by polymerization (Frère et al., 1971).

Although FGAR amidotransferase exhibits an optimal
activity at pH 8 it is most stable when stored at pH 6.5 in
the presence of the substrates of the reaction, particularly glutamine (Hermann et al., 1959, French et al., 1963a,
Mizobuchi and Buchanan, 1968a). The activity of the
enzyme is stimulated by potassium salts up to a concentration of 0.1 M. At higher concentrations of potassium
salts and at all concentrations of sodium salts the activity
is inhibited (Li and Buchanan, 1971a).

The purified enzyme from chicken liver catalyzes an
exchange reaction between ATP and ADP that is dependent
on Mg^{2+} alone (Mizobuchi and Buchanan, 1968a). FGAR or
glutamine does not affect the rate of the exchange. This
observation indicates that phosphorylation of the enzyme
is significant in the overall reaction.

In contrast, no exchange occurs between ^{32}P-inorganic
phosphate and ATP nor between ^{14}C-glutamate and
glutamine under all conditions tested. The over-all
reaction, is therefore, irreversible.

Inhibitors of FGAR Amidotransferase: O-diazoacetyl-
L-serine (azaserine) and 6-diazo 5-oxo-L-norleucine
(DON): Azaserine, which has been isolated from cultures
of Streptomyces (Bartz et al., 1954) causes a diminution
of the incorporation of either radioactive formate or
glycine into the purines of nucleic acids of several tissues
when injected into tumor bearing mice (Skipper et al.,
1954). This report led to an examination in vitro of the
action of azaserine and a closely related derivative,
DON (Dewald and Moore, 1956, Westland et al., 1956,
Dion et al., 1956). In 1955, Hartman, Levenberg and
Buchanan demonstrated that azaserine, when added to an
extract of pigeon liver, arrested the synthesis of inosinic

acid from its elementary precursors and caused the accumulation of formylglycinamide ribonucleotide. The inhibition could be partially overcome by elevation of the concentration of glutamine (Levenberg and Buchanan, 1957). Subsequently, the structural relationship between the antimetabolites and glutamine (Fig. 1) was established and the particular sensitivity of the reaction catalyzed by FGAR amidotransferase to the action of these inhibitors was determined (Levenberg et al., 1957).

$$NH_2-CO-CH_2-CH_2-CHNH_2-COOH$$

Glutamine

$$NH_2-CO-NH-CH_2-CHNH_2-COOH$$

Albizziin

$$\bar{N}=\overset{+}{N}-CH-CO-O-CH_2-CHNH_2-COOH$$

L-Azaserine

$$\bar{N}=\overset{+}{N}-CH-CO-CH_2-CH_2-CHNH_2-COOH$$

6-Diazo-5-oxo-L-norleucine (DON)

Fig. 1. Structural Relationships between Glutamine and Several Analogs.

The steric properties of the diazo antimetabolites in relation to their capacity to function as antimetabolites of glutamine are of particular interest. In the pigeon liver system the K_m values of glutamine and FGAR are 6.2×10^{-4} M, and 6.4×10^{-5} M, respectively. DON is the most effective of the antimetabolites with a K_i of 1.1×10^{-6} M as compared to a K_i of 3.4×10^{-5} M for L-azaserine. The enantiomorph, D-azaserine, as well as the homolog of DON, 5 diazo-4-oxo-L-norvaline, are ineffective as inhibitors (Levenberg et al., 1957). The stereoisomerism and chain length of the inhibitors are obviously important in the fitting of the antimetabolite to the active site.   Even so, it is recognized that DON is not completely isosteric with glutamine (Levitzki et al.,1971).

Both azaserine and DON exhibit a mixed-type inhibition with respect to glutamine. Because of the structural simi- larity of azaserine or DON to glutamine the inhibitor is directed to the glutamine-active site. The chemically active diazoacetyl group then fixes the inhibitor to the enzyme irreversibly by formation of a covalent bond. Azaserine and DON are probably the first examples of a special kind of antimetabolite now commonly referred to as active-site directed reagents (Baker, 1967, Shaw, 1970). The use of these reagents in the study of enzyme sites is called affinity labeling. Azaserine labeled with ^{14}C in one of the carbon atoms of diazoacetyl group has been reacted with highly purified FGAR amidotransferase from Sal- monella typhimurium (French et al., 1963a). In the absence of glutamine the irreversible inactivation of the enzyme is very rapid.   Approximately 0.3 mole of azaserine was bound per mole of enzyme. This value is low but approaches 1 in more recent experiments carried out with the enzyme from chicken liver.

Digestion of the labeled enzyme with papain yielded ^{14}C-peptides of small molecular weight and an amino acid attached to ^{14}C-labeled inhibitor (Dawid et al., 1968). The labeled amino acid was identified as a derivative of cysteine. Isolation and characterization (French et al., 1968b) of these labeled compounds were exceptionally difficult until it was recognized that in alkaline solution O-substituted esters of serine rearranged to the more stable N-substituted derivatives (Fig. 2).

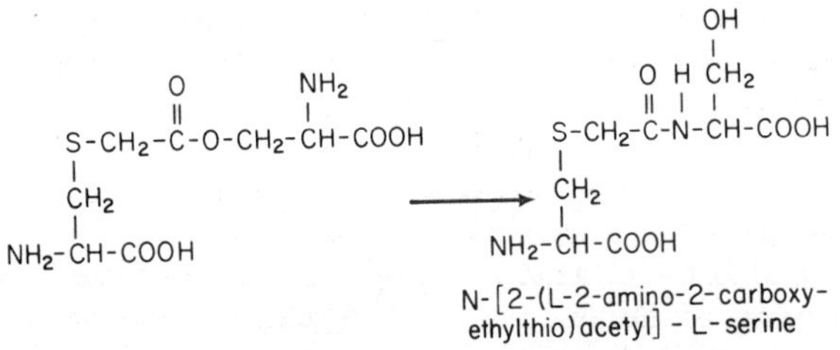

$$\underset{\substack{|\\ CH_2\\ |\\ NH_2-CH-COOH}}{S-CH_2-\overset{\overset{O}{\|}}{C}-O-CH_2-\overset{\overset{NH_2}{|}}{CH}-COOH} \longrightarrow \underset{\substack{|\\ CH_2\\ |\\ NH_2-CH-COOH}}{S-CH_2-\overset{\overset{O}{\|}}{C}-\overset{\overset{H}{|}}{N}-\overset{\overset{CH_2-OH}{|}}{CH}-COOH}$$

N-[2-(L-2-amino-2-carboxy-ethylthio)acetyl] - L-serine

CS

Fig. 2.   Rearrangement of O- to N-substituted derivatives.

Also, the manipulation of sulfur compounds in micro
quantities requires special precautions to prevent oxidation
to sulfones and sulfoxides. Isolation by paper chromatography and electrophoresis were performed routinely in
the presence of thiodiglycol.

Acid hydrolysis of the compounds shown above yielded
S-carboxymethyl cysteine. Other products of papain and
pronase digestion have been identified, namely, the
N-valyl derivative of CS or VCS. Also, labeled derivatives of a tripeptide and pentapeptide with the probable
sequences, gly-val-cys and ala-leu-gly-val-cys have been
isolated. The latter sequence of amino acids thus represents the active site for glutamine on the bacterial enzyme.

Other Inhibitors of the Glutamine-dependent reaction:
FGAR amidotransferase isolated from chicken liver is
inhibited by the action of other reagents, namely iodoacetate, iodoacetamide, L-2-amino-4-oxo-5-chloro-
pentanoic acid, albizziin, cyanate, hydroxylamine and
N-methylhydroxylamine (Mizobuchi et al., 1968a),
Schroeder et al., 1969, Hong and Buchanan, 1972). All of
these compounds in one way or the other inhibit by reaction
at the glutamine site. Indirect evidence obtained by the
differential titration of sulfhydryl groups in the presence
and absence of azaserine or iodoacetate also confirms that
both inhibitors bind to a sulfhydryl group on the enzyme.
The binding of one mole of either iodoacetate or iodoacetamide per mole of enzyme results in the loss of
glutamine-dependent activity. One mole of albizziin also
binds at the same site that combines with azaserine
(Schroeder et al., 1969). The ^{14}C-albizziin residue bound
to the enzyme can be released by reaction with hydroxylamine but does not decompose by treatment with acid or
alkali.

It has been shown that albizziin and probably also
azaserine and DON form saturation complexes reversibly
with the enzyme before combining irreversibly. In contrast, iodoacetate, iodoacetamide and cyanate do not form
saturation complexes but inhibit the enzyme at a rate
proportional to their concentrations.

Both hydroxylamine and N-methylhydroxylamine

exhibit reversible inhibition of the NH_3 reaction but cause
an irreversible loss of the glutamine-dependent activity.
Hill plots of all of the above compounds except the
hydroxylamines indicate that there is one binding site per
molecule of enzyme (Schroeder et al., 1969).

The above mentioned inhibitors of glutamine reaction
have variable effects on the activity of the enzyme when
NH_3 served as the nitrogen substrate. Enzyme treated
with either azaserine or iodoacetamide shows an enhanced
molecular activity with NH_3. On the other hand, enzyme
treated with iodoacetic acid exhibits a lower reaction
and with albizziin and the chloroketone there is no change
in NH_3 activity.

Another interesting comparison concerns the effect of
the other substrates of the reaction on the rate of inhibition
of the enzyme by the various alkylating agents. For
example, when all three components FGAR, ATP and Mg^{2+}
are present, the rate of inactivation of the amidotransferase by iodoacetate and iodoacetamide is accelerated 20 to
30 fold (Mizobuchi, et al., 1968, Schroeder et al., 1969).
These experiments illustrate that conformational changes
in the enzyme induced by these substrates expose the
catalytically reactive sulfhydryl group at the glutamine
site. On the other hand, FGAR, ATP and Mg^{++} have no
effect on the inactivation of the enzyme by azaserine and
actually retard by 2 or 3 fold the rate of inhibition by
albizziin. These apparent inconsistencies may possibly
be explained by the nature of the reactivity of the inhibitors
and their conformational relationships to the glutamine
active site. In the first place azaserine and albizziin,
because of their similarity in structure to glutamine, can
bring about the conformational changes in the enzyme
required for reaction at the active site. However, neither
of the two alkylating agents, iodoacetate or iodoacetamide,
have this potential and hence must rely on FGAR and
MgATP to provide the most suitably reactive form of the
enzyme.

An attractive model is the hypothesis that glutamine
and its structurally similar analogs bind to the enzyme at
the a-carboxyl and amino groups. The γ carbonyl group
and amide nitrogen are then fixed in position for reaction

with the unique sulfhydryl group on the enzyme. In the
case of glutamine these steric relationships are ideally
suited for maximal catalytic activity. The alignment of
all groups are therefore believed to be "in register".
In the case of azaserine or particularly DON the alignment
of atoms nearly approximates those of glutamine, but,
nevertheless, the antimetabolites and glutamine are not
really isosteric. However, azaserine contains an excellent diazo"leaving group" which gives the adjacent
carbon atom a positive charge (Fig. 3). Because of its
favorable position in the active site, the negatively charg-
ed sulfhydryl group of the enzyme readily reacts with the
carbonium carbon of the inhibitor to form a stable covalent
bond. The normal reaction of the sulfhydryl group with
the carbonyl carbon would thus be bypassed (French et al.,
1963). The length of the chain of azaserine or DON
attached to the enzyme site is thus one atom longer than in
the case of glutamine. These compounds are considered
to be " out of register". Iodoacetate and iodoacetamide,
although shorter than DON by several carbon atoms, are
both considered to fall in this same category.

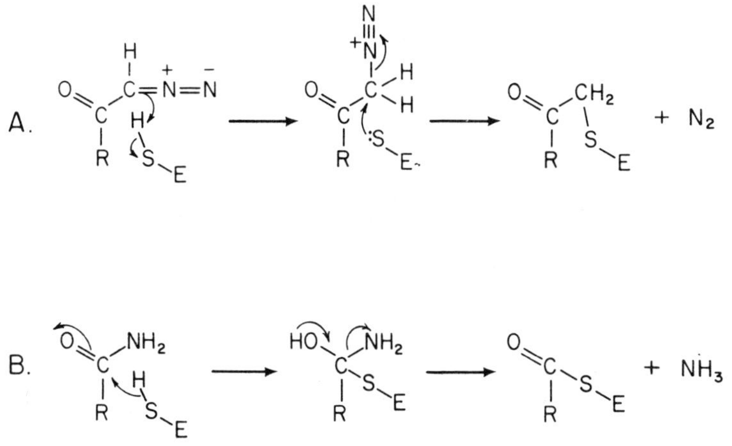

Fig. 3. Scheme for reaction of (A) azaserine and DON
or (B) glutamine with enzyme (E). R stands for remaining portion of structure of glutamine or antimetabolite.

In order to bring further information to bear on this hypothesis, Schroeder et al., (1969) studied the reaction of amidotransferase with albizziin, which binds mole for mole with the enzyme. The mechanism of reaction of albizziin with the enzyme must be quite different from that of azaserine since albizziin contains no group that might supply a potential carbonium ion. Covalent binding of the inhibitor to the enzyme must, therefore, occur by a nucleophilic attack of the sulfhydryl anion of the enzyme on the γ carbonyl carbon with the elimination of NH_3 (Pathway A, Fig. 4). An alternate possibility of formulating the reaction would involve the elimination of 2,3 diamino proprionate (Pathway B, Fig. 4) with the formation of an S-carbamoyl cysteine derivative of the enzyme.

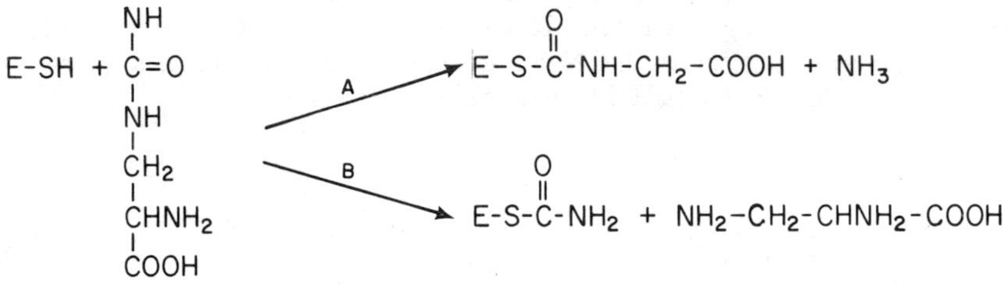

$$E\text{-}SH + \begin{array}{c} NH \\ | \\ C=O \\ | \\ NH \\ | \\ CH_2 \\ | \\ CHNH_2 \\ | \\ COOH \end{array}$$

$$\xrightarrow{A} \quad E\text{-}S\overset{O}{\underset{\|}{C}}\text{-}NH\text{-}CH_2\text{-}COOH + NH_3$$

$$\xrightarrow{B} \quad E\text{-}S\overset{O}{\underset{\|}{C}}\text{-}NH_2 + NH_2\text{-}CH_2\text{-}CHNH_2\text{-}COOH$$

Fig. 4. Alternate pathways of reaction of albizziin with enzyme.

Since S-carbamoyl cysteine is unstable at pH 8 (Stark, 1964) and since the albizziin adduct labeled in the γ carbonyl carbon does not decompose in alkaline solution, we believe that pathway A represents the correct formulation of the reaction.

A stable adduct with albizziin is probably formed because the enzymatic mechanism responsible for the cleavage of the E-glutamyl complex in the normal reaction is not able to catalyze the hydrolysis of the E-albizziin complex. Furthermore, the formation of covalently bound albizziin might occur with more difficulty than the corresponding γ-glutamyl-enzyme complex. As previously mentioned, FGAR, ATP and Mg^{2+} have an adverse effect on the rate of inactivation of the enzyme by albizziin. These compounds may either inhibit the binding of this inhibitor to the enzyme or even more possibly promote its dissociation, once bound.

Although definitive work has not yet been done on the chemistry of the enzyme-albizziin complex, we are probably justified at present in postulating the structure shown in Fig. 4 pathway A. If this structure proves valid, albizziin represents an inhibitor that is "in register" with the active site of the enzyme and closely duplicates the reaction of glutamine.

These experiments support the hypothesis that FGAR amidotransferase catalyzes the transfer of the amide nitrogen of glutamine by a nucleophilic attack of a sulfhydryl group of the enzyme on the γ carbonyl carbon (Fig. 3). In the case of albizziin reaction is interrupted because the enzyme cannot displace the residue attached to the enzyme, and in the case of azaserine because an abortive attachment of inhibitors to the enzyme occurs. Direct evidence for the nucleophilic attack of an enzyme on the carbonyl carbon of DON has been reported by Hartman and McGrath (1972), who have shown that glutaminase can effect the splitting of a carbon-to carbon bond of DON to yield glutamic acid, methanol and nitrogen. At the same time a second reaction between DON and glutaminase occurs to cause its irreversible inactivation probably by the same mechanism described for these compounds on other enzymes.

Examination of Glutamyl Site with γ Glutamyl Amides and Esters: FGAR amidotransferase catalyzes the hydrolysis of a number of γ-glutamyl amides and esters including γ glutamylhydroxamate, γ glutamylmethoxyamide, γ glutamyl hydrazide, γ methyl glutamate and γ ethyl glutamate (Li and Buchanan, 1971d). Although the hydrolytic reactions of glutamyl derivatives were dependent on the presence of FGAR, ATP and Mg^{2+}, they were not accompanied by the stoichiometric formation in

the case of the substituted amides, and there was no ADP
formation when esters and thioesters were used as
substrates. These data suggested that hydrolysis of the
γ glutamyl derivatives did not involve a stoichiometric
transfer of the γ substituent to FGAR with equivalent
formation of ADP as in the case of the glutamine reaction.
In fact, when the products of splitting of γ ethyl glutamate
were measured, it was found that ethanol and glutamate
were formed in equivalent amounts. FGAR and MgATP
were thus needed in the reaction for their effect on the
conformational state of the enzyme rather than as an
acceptor (FGAR) of the γ glutamyl substituent. In support
of this hypothesis it was found that the K_m of FGAR for
several of the reactions is greatly reduced as compared
to the reaction with glutamine. For example, the K_m
of FGAR for the transfer of the amide group of glutamine
is 1×10^{-4} M: for the enzymatic hydrolysis of γ ethyl
glutamate the K_m of FGAR is 3×10^{-6}. In several instances the amount of FGAR present in the reaction vessel
was far less than the total amount of γ glutamyl compound
hydrolyzed.

In the absence of FGAR and MgATP, FGAR amidotransferase splits glutamine at about 0.5% of the rate as
that observed in the complete reaction. The glutaminase
activity is a property of the amidotransferase itself since
the pH optimum is around 8 and the enzyme is irreversibly inhibited by azaserine. The K_m for glutamine in the
glutaminase reaction is about 5×10^{-5} M as compared to
2×10^{-4} M for the amidotransferase reaction.

This residual glutaminase activity seen in several of
the amidotransferases (Li and Buchanan, 1971a, Nagano
et al., 1970) suggests the hypothesis that amidotransferases
are actually composed of two catalytic activities, a
glutaminase, which might be developed from a common
primordal enzyme, and a "transferase", which is
specifically designed for an energy source, ATP, and for
different acceptors. In the absence of the acceptor and
ATP, the hydrolytic activity of the glutaminase is min-
imized by a restriction imposed by the transferase
component of the enzyme in order to prevent unnecessary
hydrolysis of glutamine. When the specific acceptor and
ATP are bound to the transferase component of the enzyme,

the restriction on the glutaminase activity is removed and
the hydrolysis starts to operate for the generation of
"activated" ammonia, which is efficiently transferred via
the ammonia site to the specific acceptor.

Formation of Enzyme-substrate Complexes: The
substrates of the FGAR amidotransferase reaction are
capable of forming stable complexes with the enzyme.
Glutamine forms a complex presumably at one site
(Mizobuchi and Buchanan, 1968b), whereas FGAR, ATP
and Mg^{2+} together form a complex at a second site
(Mizobuchi et al., 1968). These two complexes are stable
enough to permit their isolation by chromatography on
Sephadex G-50. Both complexes have a characteristic
half-life. That of the glutamyl-enzyme complex is 125
min at 2°C. Approximately one mole of glutamine binds
per mole of enzyme. Azaserine prevents formation of
the complex and ^{14}C-γ-glutamyl-enzyme readily exchanges with ^{12}C-glutamine. None of the other substrates
of the reaction are needed for formation of the complex.
In the presence of all substrates the amount of ^{14}C-
glutamine bound to the enzyme is lowered (0.45 mole per
mole enzyme) but is not reduced to zero. Glutamate is
the product of the degraded complex. Also, glutamate
cannot form a stable complex with the enzyme as does
glutamine.

Although ATP and FGAR can interact individually with
the enzyme, they cannot form a stable complex unless
both substrates plus Mg^{2+} are present. By labeling the
γ phosphate of ATP with ^{32}P and the adenine with ^{14}C,
we have shown that all portions of ATP are a part of the
complex. In other words the complex is not formed by
the elimination of ADP or the terminal phosphate of ATP.
The binding ratio of either FGAR or ATP to the enzyme
expressed in molar quantity is approximately 0.7. FGAR
and ATP are therefore present at the enzyme site in the
same proportion.

The half-life of the enzyme-FGAR-Mg^{2+}-ATP complex
is approximately 62 min at 0° C. This half-life was
greatly affected by reaction of glutamine or its analogs at
the other site. In the presence of glutamine the products
of the reaction, FGAM, ADP and P_i, are formed and

dissociate from the enzyme so rapidly that no radioactive compounds can be found bound to the $FGAR-Mg^{2+}-ATP$ site. In the presence of azaserine to the half-life of the $FGAR-Mg^{2+}-ATP$ complex is reduced to 18-20 min. On the other hand, enzyme treated with iodoacetate forms an $FGAR-Mg^{2+}-ATP$ complex with a half-life of 340 min. It is thus seen that conformational changes brought about by the addition of ligands to the glutamine site may affect either positively or negatively the dissociation of substrates at the other site.

Kinetics of Addition of Substrates to FGAR Amido-transferase: From the application of steady-state kinetics to an enzymatic reaction it is possible to gain information about the order of combination of enzymes with substrates and products. Initial rate equations for several possible mechanisms for enzyme reactions involving three substrates have been derived by various authors (Dalziel, 1969, Frieden, 1959, Cleland, 1963). The use of competitive inhibitors to distinguish among several mechanisms have been discussed by Fromm (1967).

In the first of our kinetic experiments with FGAR amidotransferase (Li and Buchanan, 1971b) , one substrate concentration was varied while the other two were maintained at a fixed level in the general concentration range of their Michaelis constants. The experiment was then repeated but at a different concentration of the final substrates while the ratio concentrations remained constant. All three sets of data yielded straight lines that intercepted to the left of the $1/V$ axis on the Lineweaver-Burk plot. Plots with these characteristics indicate that substrates reacted with the enzyme to yield a quaternary complex prior to reaction of any one of the substrates to yield a dissociable product.

In order to classify the FGAR amidotransferase reaction according to one of the subclasses (ordered, partly ordered or random) a second experimental procedure, first reported by Frieden (1959) for a three substrate system, was used. This second procedure involves varying the concentration of one substrate at several levels of a second substrate while holding the

400

third substrate at a constant and high level. A partially
ordered mechanism is indicated when one set of converg-
ent lines and two sets of parallel lines are obtained. The
parallel lines should occur when the last two substrates,
B and C, are held at a constant and high level. Kinetic
data of this kind was obtained for the reaction catalyzed
by FGAR amidotransferase.

Finally the pattern of inhibition observed when
competitive inhibitors (Table 1) are employed confirm the
finding by the independent procedure cited above that the
addition of substrates to the enzyme to form the quaternary complex occurs by the partially compulsory order
mechanism.

Table I
Inhibition of FGAR-amidotransferase by competitive
and product inhibitors

Inhibitor	Type of inhibition relative to		
	Glutamine	ATP	FGAR
Albizziin	Competitive	Mixed	Mixed
β- γ-5' adenylyl methylene di- phosphonate	Uncompet- itive	Compet- itive	Noncompet- itive
ADP	Uncompet- itive	Compet- itive	Mixed

The interpretation that best fit these data is that A
(glutamine) is the obligatory first substrate whereas B
(ATP)and C (FGAR) can add randomly to the enzyme
(Equation 3).

3.)
$$E \rightleftharpoons EA \underset{EAC}{\overset{EAB}{<>}} EABC \longrightarrow EPQR \longrightarrow Products$$

E = Enzyme   A = Glutamine   B = ATP   C = FGAR

The interpretation of the foregoing experiments is summarized in the scheme of reactions shown in Fig. 5.

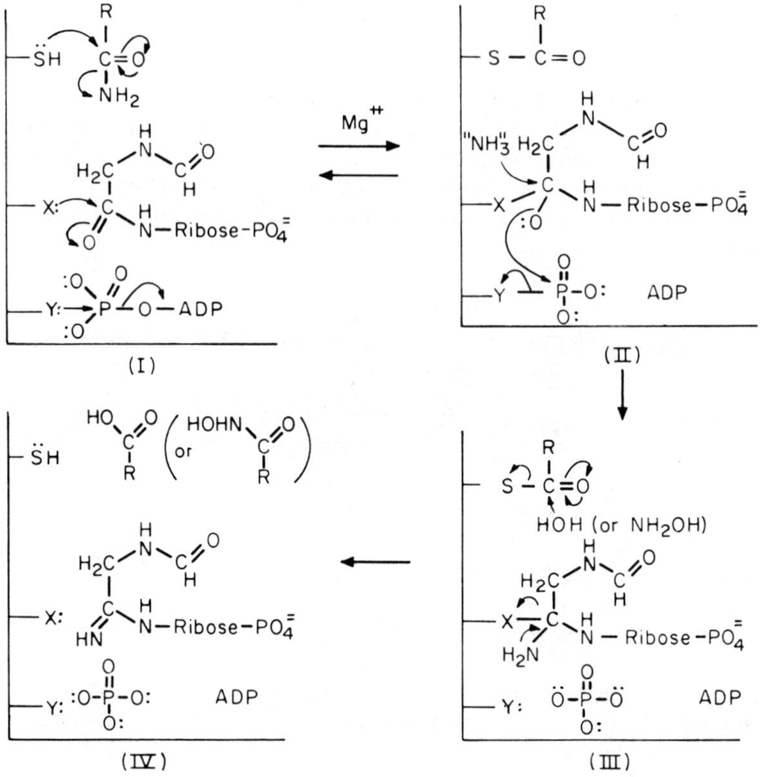

Fig. 5. Proposed mechanism for the reaction of FGAR amidotransferase. X and Y represent hypothetical binding sites for FGAR and ATP on the enzyme, respectively.

Glutamine reacts first at its active site to yield a γ glutamyl thioester. The liberated ammonia does not dissociate but is transferred to a second site on the enzyme. ATP or FGAR with Mg^{2+} then bind randomly to the enzyme. The enzyme is phosphorylated by ATP but retains ADP in bound form. When all substrates are fixed on the enzyme in proper position, the nitrogen of "nascent" ammonia attacks the carbonyl carbon of FGAR, which transfers its oxygen to the phosphate of the enzyme releasing it as inorganic phosphate. The product of the reaction, FGAM, ADP and inorganic phosphate, are released from the enzyme. The γ-glutamyl thioester is then hydrolyzed in the normal reaction to yield glutamate, or in the presence of hydroxylamine reacts to form γ glutamylhydroxamate. In our present scheme the reaction is considered to be concerted with the amide nitrogen of glutamine exerting a nucleophilic attack on the carbonyl carbon of FGAR at the same time that the phosphorus of phosphorylated enzyme exhibits an electrophilic attraction for the carbonyl-oxygen.

The postulation of phosphorylated enzyme at first seems a contradiction to the kinetic analysis indicating that the substrates react with the enzyme to form a quaternary complex. One might have expected to observe ping-pong kinetics, a kinetic pattern generally seen for enzymes whose reactions involve covalent enzyme-substrate intermediates. This analysis requires, however, that the unbound product of the first reaction dissociates from the enzyme before reaction of the second substrate. Otherwise, ping-pong kinetics will not be obtained. Our experiments on binding of FGAR and ATP to the enzyme site show that all parts of ATP are contained in the complex.

The ordering of the reaction of the substrates with the enzyme with glutamine binding first seems to be in contradiction to the fact that ATP and FGAR can form a complex with the enzyme in the absence of glutamine. This seeming contradiction is resolved with the knowledge that the affinity of the enzyme for glutamine is far greater than that of ATP or FGAR. Thus, although glutamine and the pair, FGAR-ATP, can react with the enzyme randomly, only the pathway in which glutamine reacts first will be kinetically significant.

Chu and Henderson (1972b) have shown that the reaction catalyzed by FGAR amidotransferase isolated from Erhlich ascites tumor cells involves a fully ping-pong mechanism. In their system glutamine binds to the free enzyme and glutamate is released before the addition of ATP. ADP is released and FGAR then binds. The liberation of inorganic phosphate is rapid and FGAM is the last product released from the enzyme.

There are other differences between the two enzyme systems, chicken liver and tumor, that are presently puzzling. In the reaction with the tumor enzyme there is an FGAR-dependent exchange between ATP and P_i. An exchange between ADP and ATP has not been reported. A major problem is that the enzyme has not been extensively purified and extraneous exchange reactions are difficult to rule out. Their method of enzyme preparation does not separate FGAR amidotransferase and AIR synthetase, for example.

Nevertheless, if all of the present uncertainties between the two enzyme systems are satisfactorily resolved, the considerable differences in the mechanism of the reactions should prove very interesting.

FOOTNOTES

* The investigations reported from the author's laboratory were supported by grants-in-aid from the Damon Runyon Memorial Fund, the National Science Foundation and the National Cancer Institute, National Institutes of Health (grant CA02015).

ᐯ formylglycinamide ribonucleotide amidotransferase, 2-formamido-N-ribosylacetamide 5'-phosphate: L-glutamine amido ligase (adenosine diphosphate), E.C. 6.3.5.3

The author wishes to express his disagreement with the policy of the International Committee on Nomenclature in designating trivial names for the group of enzymes catalyzing the transfer of the amide group of glutamine to a number of acceptors. With the exception of 5-phosphoribosyl pyrophosphate (PRPP) amidotransferase the term "synthetase" has been generally adopted in devising names for a number of this group of enzymes. The trivial name of an enzyme should aid in its ready recognition and indicate its function. In the opinion of the author the term "synthetase" should be used only when all other means of classification have failed. Therefore, the name "amidotransferase" used in conjunction with the substrate undergoing reaction seems a more informative method of designating specific members of this group of enzymes.

[2] The abbreviations used are: FGAR, 2-formamido-N-ribosylacetamide 5'-phosphate (formylglycinamide ribonucleotide); FGAM, 2-formamido-N-ribosyl-acetamidine 5'-phosphate (formylglycinamidine ribonucleotide); AIR, 5-amino-1-ribosylimidazole 5'-phosphate (5-aminoimidazole ribonucleotide).

REFERENCES

Baker, B. R. (1967). Design of active-site directed irreversible enzyme inhibitors, p 15. John Wiley and Sons, Inc., New York

Bartz, Q. R., Elder, C. C., Frohardt, R. P., Fusari, S. A., Haskell, T. H., Johannessen, D. W., and Ryder, A. (1954). Nature, 173, 72.

Chu, S. Y., and Henderson, J. F. (1972a). Can. J. Biochem. 50, 484.

Chu, S. Y., and Henderson, J. F. (1972b). Can. J. Biochem. 50, 490.

Cleland, W. W. (1963). Biochim. Biophys. Acta 67, 104.

Dalziel, K. (1969) Biochem. J. 114, 547.

Dawid, I. B., French, T. C., and Buchanan, J. M. (1963). J. Biol. Chem. 238, 2178.

Dion, H. W., Fusari, S. A., Jakubowski, Z. L., Zora, J. G., and Bartz, Q. R. (1956) Abstracts of the American Chemical Society Meeting, Dallas, April, 1956, p. 13 M.

Dewald, H. A., and Moore, A. M. (1956) Abstracts of the American Chemical Society Meeting, Dallas, April, 1956, p. 13 M.

French, T. C., Dawid, I. B., Day, R. A., and Buchanan, J. M., (1963a). J. Biol. Chem. 238, 2171.

French, T. C., Dawid, I. B., and Buchanan, J. M. (1963b). J. Biol. Chem. 238, 2171.

Frère, J-M., Schroeder, D. D. and Buchanan, J. M. (1971). J. Biol. Chem. 246, 4727.

Frieden, C. (1959). J. Biol. Chem. 234, 2891.

Fromm, H. J., (1967). Biochim. Biophys. Acta 139, 221

Hartman, S. C., and McGrath, T. F. (1972) in preparation.

Hartman, S. C., Levenberg, B., and Buchanan, J. M. (1955). J. Am. Chem. Soc. 77, 501.

Hartman, S. C., Levenberg, B., and Buchanan, J. M. (1956). J. Biol. Chem. 221, 1057.

Hermann, R. L., Day, R. A., and Buchanan, J. M. (1959). Abstracts of the American Chemical Society Meeting, Boston, April, 1959, p. 45 C.

Hong, B-S., and Buchanan, J. M. (1972) in preparation.

Kornberg, A., and Pricer, W. E., Jr. (1951). J. Biol. Chem. 193, 481.

Leloir, L. F., and Cardini, C. E. (1953). Biochim. Biophys. Acta 12, 17.

Levenberg, B., Hartman, S. C., and Buchanan, J. M. (1956). J. Biol. Chem. 220, 379.

Levenberg, B., and Buchanan, J. M. (1957) J. Biol. Chem 224, 1005.

Levenberg, B., Melnick, I., and Buchanan, J. M. (1957) J. Biol. Chem. 225, 163.

Levitzki, A., Stallcup, W. B., and Koshland, D. E., Jr. (1971). Biochemistry 10, 3371.

Li, H-C., and Buchanan, J. M. (1971a). J. Biol. Chem. 246, 4713.

Li, H-C., and Buchanan, J. M. (1971b). J. Biol. Chem. 246, 4720.

Li, H-C., and Buchanan, J. M. (1972). in preparation.

Melnick, I., and Buchanan, J. M. (1957). J. Biol. Chem. 225, 157.

Mizobuchi, K., and Buchanan, J. M. (1968a). J. Biol. Chem. 243, 4842.

Mizobuchi, K., and Buchanan, J. M. (1968b). J. Biol. Chem. 243, 4853.

Mizobuchi, K., Kenyon, G. L., and Buchanan, J. M. (1968). J. Biol. Chem. 243, 4863.

Nagano, H., Zalkin, H., and Henderson, E. J. (1970). J. Biol. Chem. 245, 3810.

Shaw, E. (1970). Physiol. Rev. 50, 244.

Skipper, H. E., Bennett, L. L., Jr., and Schabel, F. M. Jr. (1954). Federation Proc. 13, 298.

Sonne, J. C., Lin, I., and Buchanan, J. M. (1953). J. Am. Chem. Soc. 75, 1516.

Sonne, J. C., Lin, I., and Buchanan, J. M. (1954). J. Biol. Chem. 220, 369.

Stark, G. R. (1964). J. Biol. Chem. 239, 1411.

Westland, R. D., Fusari, S. A., and Crooks, H. M., Jr. (1956) Abstracts of the American Chemical Society Meeting, Dallas, April, 1956, p 14 M.

Wyngaarden, J. B., and Ashton, D. M. (1959). J. Biol. Chem. 234, 1492.

GUANOSINE 5'-PHOSPHATE SYNTHETASE

Standish C. Hartman

Department of Chemistry, Boston University
Boston, Massachusetts

and

Stanley Prusiner*

Laboratory of Biochemistry, National Institutes of Health
National Heart and Lung Institute, Bethesda, Maryland

Abstract

This article reviews the literature concerning GMP synthetase, the enzyme which catalyzes amination of xanthosine 5'-phosphate to guanosine 5'-phosphate. Synthetases from various biological sources are considered with respect to amino group donor, mechanism of action, and regulatory properties.

*Present Address: Department of Neurology, University of California, School of Medicine, San Francisco

Introduction

The biosynthetic pathway leading to guanine nucleotides in many living organisms uses glutamine as the donor
of the 2-amino group. In a two-step sequence inosine 5'-
phosphate, the central metabolite of puring nucleotide interconversions, is oxidized to xanthosine 5'-phosphate
(XMP) by action of IMP dehydrogenase. Amination of XMP
yields GMP according to Reactions 1 and 2.

1. $XMP + ATP + glutamine + H_2O \xrightarrow{Mg^{++}} GMP + AMP + PP_i + glutamate$

2. $XMP + ATP + NH_3 \xrightarrow{Mg^{++}} GMP + AMP + PP_i$

Under certain conditions and with certain isolated enzymes
ammonia can serve as substrate in place of glutamine (Reaction 2). The enzymes which catalyze these processes are
classified as xanthosine 5'-phosphate:L-glutamine amidoligase (AMP), EC 6.3.5.2, or as xanthosine 5'-phosphate:ammo-
nia ligase (AMP), EC 6.3.4.1, and are trivially called GMP
synthetase or XMP aminase. The former term will be used in
this review without implication of nitrogen donor.

Nitrogen Donors

GMP synthetases which employ glutamine as the preferred nitrogen donor have been partially purified from
pigeon liver (Lagerkvist, 1958 a,b), calf thymus (Abrams
and Bentley, 1959), and Escherichia coli NTCC 7020 (Mardashev and Iarovaia, 1965; Mardashev, et al., 1967). Ammonia
is much less effectively utilized by these enzymes (see
Table I). From the pH dependence of the reaction with ammonia it may be concluded that this substance and not the

Table I

Properties of Various GMP Synthetases

	Calf Thymus	Pigeon Liver	E. coli 7020	A. aerogenes PD-1	E. coli B-24-1 Extract	E. coli B-24-1 Pure
K_M (mM)						
Glutamine	0.05	0.025	3	no reaction	<1	--
NH_4^+	39	50	no reaction	0.9(as NH_3)	2	25
XMP	--	0.03	<1	--	--	0.029
ATP	0.2	<1	<1	--	--	0.53
Relative Rate, Glutamine:NH_3	--	7	∞	0	0.56	0.065
Inhibitors	DON, azaserine	PCMB	DON, azaserine, mercapto-ethanol, phosphate	hydroxyl-amine		hydroxyl-amine, psico-furanine
References	a	b	c	d	e	f

a. Abrams and Bentley (1959); b. Lagerkvist (1958b); c. Mardashev and Iarovaia (1965); d. Moyed and Magasanik (1957); e. Brevet, et al. (1967); f. Udaka and Moyed (1963).

ammonium ion is the actual substrate. The very low concentrations of free ammonia expected in cells at physiological
pH, together with the low K_M of glutamine compared to that
of ammonium, makes it highly likely that glutamine is the
preferred donor of the 2-amino group of guanine in these
systems.

As is generally the case with glutamine-requiring enzymes, these systems are irreversibly inhibited by low levels of 6-diazo-5-oxonorleucine and azaserine. Iarovaia,
et al. (1967) report that these compounds are competitive
with respect to glutamine, and in addition, that the analogues, S-methylcarbamylcysteine, S-carbamylcysteine, and
S-carbamylhomocysteine, also inhibit but in an apparently
non-competitive manner. S-methylcarbamylcysteine irreversibly inactivates the enzyme in a process promoted by ATP,
Mg^{++}, and XMP, while glutamine affords protection.

The substrate specificities of essentially homogeneous
GMP synthetases isolated from the purine-requiring auxotrophs Aerobacter aerogenes PD-1 (Moyed and Magasanik,
1957) and E. coli B-24-1 (Fukuyama and Moyed, 1964) are in
marked contrast to those described above: only ammonia and
not glutamine serves as the nitrogen donor. These strains
are convenient sources of the synthetase because elevated
amounts of the enzyme are produced when they are grown with
limiting guanine supplies. While crude fractions from the
A. aerogenes and E. coli mutants utilize glutamine to some
extent, this activity is completely lost upon purification
(Moyed and Magasanik, 1957; H.S. Moyed, personal communication).

The completely opposite behavior of the two enzymes
isolated from strains of E. coli is particularly puzzling.
It may be, in the one case, that the ability to use glutamine as substrate was lost during purification (e.g., by
dissociation or inactivation of a glutamine specific subunit). Alternatively, the two strains may contain intrin-
sically different GMP synthetases with different substrate
specificities. Brevet, et al. (1967), who have reexamined
the behavior of E. coli B-24-1 with this question in mind,
come to the second conclusion. In agreement with the previously described results, they report that unfractionated
cell extracts utilize both glutamine and ammonia, that

fractionation with ammonium sulfate yields components with
variable substrate preferences, and that after chromatography on DEAE-Sephadex, a preparation is obtained which almost exclusively employs ammonia as nitrogen donor. Further, cell extracts show different pH-activity profiles depending upon the substrate: in the presence of glutamine,
two peaks are observed, at pH 6.5 and 9, while, with ammonium ions, only one peak at pH 9 is seen. Glutamine satu-
rates the system at lower concentrations than does ammonium, but the maximum rate with glutamine cannot be explained
on the assumption that it is first hydrolyzed to ammonia as
the true substrate. They argue that E. coli B-24-1 thus
contains two GMP synthetases, one a glutamine amidoligase
and one an ammonia ligase.

This interpretation must be viewed in the light of
genetic analyses in E. coli (described below) from which it
may reasonably be concluded that only one gene for GMP synthetase exists in these cells. Also, as a result of a single mutation, the related organism, A. aerobacter P-14,
genotypically is GMP synthetase negative and phenotypically
guanine requiring.

An explanation superficially consistent with both the
metabolic and genetic observations is based on the assumption that the XMP aminating and glutamine transferring
functions of E. coli GMP synthetase are properties of enzyme components specified by different operons. Derepression of the aminase (by 20 to 50-fold in the IMP dehydrogenase deficient mutant, B-24-1) may not be accompanied by
correspondingly elevated levels of the glutamine subunit,
with the result that the enzyme exists in two forms in the
cells: a complete enzyme, characteristic of the wild
strain, which uses glutamine, and an incomplete enzyme,
which uses only ammonia. The physical properties of the
two forms would presumably be sufficiently different to account for their separation during purification. One would
conclude, therefore, that the GMP synthetase in wild-type
E. coli should preferentially employ glutamine as nitrogen
donor, as is found to be the case by Mardashev and Iarovaia
(1965).

A precedent for the existence of two types of components and their genetic separation exists in Bacillus

413

subtilis: the gene for a glutamine subunit common to anthranilate synthetase and p-aminobenzoate synthetase is separated on the chromosome from those for the aminating components. In this case, however, both the glutamine subunit
and anthranilate synthetase are under genetic control by the
end product, tryptophan (Kane, et al. (1972).

Mechanism of Action

Despite the differences in respect to their preferred
nitrogen donors, GMP synthetases appear to follow similar
pathways in certain other aspects. Both the pigeon liver
and calf thymus enzymes effect transfer of the oxygen atom
of XMP (as ^{18}O) to the phosphate of AMP, a result most directly interpretable in terms of an adenylyl-XMP intermediate:

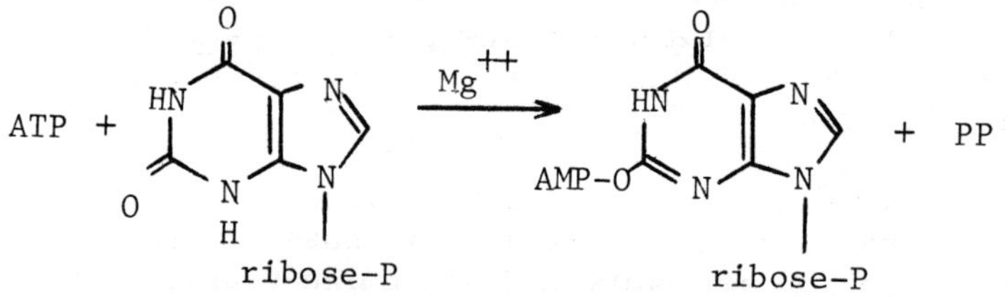

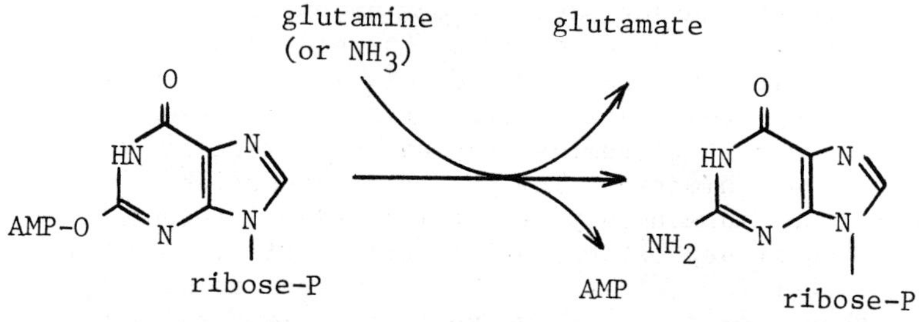

Although this formulation would imply the occurrence of iso-
tope exchange between $^{32}PP_i$ and ATP in the presence of XMP,
in analogy to related processes in which ATP is cleaved to

414

AMP and PP_i, such an exchange was either not detected or not directly correlated with action of the synthetase. It could be argued that pyrophosphate is unable to dissociate from the enzyme until completion of the catalytic cycle, or alternatively, that the amination reaction proceeds in a "concerted" manner without formation of adenylyl-XMP as intermediate having a significant lifetime (Abrams and Bentley, 1959).

With the aid of stoichiometric amounts of GMP synthetase from E. coli B-24 direct evidence for the formation of adenylyl-XMP as an enzyme-bound intermediate was obtained (Fukuyama, 1966). A compound deriving labeled carbon from both ^{14}C-XMP and ^{14}C-ATP was detected when these substrates were reacted in the absence of amine donor (ammonia). The amounts of the new compound approached equivalency with the enzyme used. This substance, together with comparable amounts of pyrophosphate, was released from the enzyme upon acidification and separated electrophoretically. After isolation the intermediate could be enzymatically converted to GMP in the presence of ammonia, or, more slowly, hydrolyzed to XMP and AMP in its absence.

Other than the observations described in the previous section concerning the inhibitory action of certain glutamine analogues, a more detailed examination of the mechanistic aspects of glutamine utilization has not been reported. The fact that ATP, Mg^{++}, and XMP promote the inactivation produced by S-methylcarbamylcysteine is circumstantial evidence that binding of glutamine is preceded by binding of the three components required for the formation of the intermediate, adenylyl-XMP.

The binding of labeled substrates to the E. coli B enzyme has been studied by Fukuyama and Moyed (1964). Both XMP and ATP bind independently of each other and in amounts corresponding to one mole per mole of enzyme (M = 140,000). Thus there would appear to be one catalytic site per enzyme molecule.

Hydroxylamine is a potent inhibitor of GMP synthetase. When this substance replaces ammonia as substrate an abnormal product is formed, presumably N-hydroxy-GMP, which does not dissociate from the enzyme (Fukuyama and Donovan, 1968).

415

Regulation of Enzyme Synthesis and Activity

Most of the information concerning control of GMP biosynthesis comes from studies with bacterial systems. IMP dehydrogenase and GMP synthetase, the two enzymes responsible for GMP formation from IMP, appear to be components of the same operon in E. coli judging from the fact that they are coordinately controlled (Kuramitsu, Udaka, and Moyed, 1964; Nijkamp and de Haan, 1967). Guanine derivatives repress formation of these enzymes while adenine compounds reverse the repression.

The relevant genetic region has been mapped in E. coli K 12 by phage transduction, which locates the gua A and gua B genes (for the synthetase and dehydrogenase, respectively) close together and between tyr and ade G markers (Nijkamp and de Haan, 1967). This corresponds to about 48 minutes on the conjugation map of the K 12 chromosome (Taylor, 1970). E. coli strains with polar gua mutations also provide evidence that gua A and gua B are in a single operon with the operator on the gua A side (Nijkamp and Oskamp, 1968). The gene for GMP reductase, which enzyme catalyzes reductive deamination of GMP to IMP, is separate and found to be between ade K and thr markers by conjugation experiments.

Since IMP dehydrogenase catalyzes the first committed reaction in the pathway leading to GMP, this enzyme would be the expected site of feedback control. The observation that GMP inhibits this reaction is in accord with this expectation (Magasanik and Karibian, 1960; Hampton and Nomura, 1967). In bacteria the pools of guanine nucleotides are also regulated by control of GMP reductase, ATP acting as an inhibitor of this "degradative" reaction (Mager and Magasanik, 1960). These workers have pointed out that the complex set of controls, at both the gene and metabolite levels, act to coordinate the steady state concentrations of guanine and adenine nucleotides.

Studies with bacterial GMP synthetase suggest that this enzyme may also be a site of feedback regulation. Mardashev, et al. (1967) report that GMP synthetase of E. coli 7020 is susceptible to end product inhibition. Thus GDP, GMP, and GTP inhibit with K_I's of 2.2, 3.2, and 5.1

mM, respectively, values somewhat larger than the K_M for XMP, 1.7 mM. (There appears to be an inconsistancy in the values recorded in this manuscript, being described in one place as µmole/ml (i.e., mM) and in another as µM. In view of the probable sensitivity of the assay method and in comparison to K_M values reported by Mardashev and Iarovaia (1965), the former interpretation seems correct.) In each case the inhibition by guanine nucleotides appears to be non-competitive with respect to XMP, although possible competition with the substrate, ATP, was not examined. Also, the liklihood that the added nucleotides, especially GTP and GDP, might chelate and thus remove the required divalent cation, Mg^{++}, was not ruled out. The cellular regulation of this enzyme may involve binding of divalent metal ions by nucleoside triphosphates, as has been shown in other cases (Stadtman, 1970; Prusiner, 1973).

Studies with the E. coli B system by Moyed and his collaborators have identified quite a different form of metabolic regulation. The adenine containing compounds, 5'-AMP, adenosine, and psicofuranine (9-β-D-psicofuranosyl-adenine, an antibiotic analogue of adenosine) inhibit the ammonia-dependent GMP synthetase with effectiveness increasing in that order. Psicofuranine inhibits growth of E. coli by specifically blocking GMP biosynthesis at this point (Udaka and Moyed, 1963). Mutants of E. coli B which are relatively resistant to psicofuranine are of two types: (a) Those having a defective IMP dehydrogenase, with resulting guanine deficiency. They are therefore derepressed in the guanine operon and GMP synthetase levels are elevated. (b) Those having an altered GMP synthetase with reduced sensitivity to the inhibitor. The latter mutants do not exhibit guanine deficiency and are not derepressed for the enzyme. In fact, they contain only about one-third the levels of synthetase found in wild-type cells. The mutant GMP synthetase exhibits an even greater reduction in sensitivity toward the natural metabolities, adenosine and AMP, than it does toward psicofuranine.

These results, and others mentioned below, are interpreted to indicate that GMP synthetase possesses an allosteric regulatory site for adenine compounds and that, in the wild strain, the enzyme exists in a partially inhibited state. The synthetase is hyper-repressed in the mutant

417

strain, presumably because of increased production of GMP by the uncontrolled enzyme. While these studies present a strong case for the regulation of guanine nucleotide bio-synthesis by adenine derivatives, its metabolic function is not clear as it seems to work in the reverse sense from all other controls in purine nucleotide interconversions. Conceivably, it is a response to the "energy charge" of the adenosine phosphate pool: ATP is a required substrate and AMP is an allosteric inhibitor. A decrease in the energy charge would sharply decrease the synthesis of guanine compounds required for nucleic acid and protein synthesis. At the same time, a decrease in ATP would activate GMP reductase, the effect of which should also be to restrict guanine nucleotide levels.

The mechanism by which psicofuranine inhibits GMP synthetase has been extensively examined. A rapidly reversible mode of inhibition occurs in the presence of the product, and Mg^{++}. When the substrate, XMP, is present in addition a time-dependent "irreversible" inhibition is observed, characterized by an extremely tight binding of psicofuranine, XMP, and pyrophosphate to the enzyme in stoichiometric amounts (one mole of each per mole of enzyme) (Udaka and Moyed, 1963; Fukuyama and Moyed, 1964). Removal of the inhibitor from the "irreversibly" inhibited enzyme allows recovery of activity, but only after many hours. ATP does not reduce inhibitor binding. The psicofuranine-resistant form of GMP synthetase undergoes the reversible stage of inhibition but not the XMP-induced irreversible mode; the mutant enzyme is not obviously altered in any of its other substrate-binding parameters. These results support the conclusion that the inhibitor binding site is distinct from the sites for the substrates. Furthermore, various chemical treatments can differentially alter the catalytic and inhibitory properties of the enzyme (Kuramitsu and Moyed, 1966). The evidence indicates that a conformational change leading to disorientation of the catalytic site occurs when psicofuranine is bound. This binding does not interfere with the ability to bind the substrates, XMP and ATP, but prevents their reaction to form the adenylyl-XMP intermediate (Fukuyama, 1966).

Summary

GMP synthetases have been isolated from animal and bacterial sources which show a strong preference for glutamine over ammonium ions as the nitrogen donor. It is highly likely that glutamine is the principal reactant under conditions in vivo. The activity of GMP synthetase in certain purine-requiring bacterial mutants is very much elevated through derepression. Two enzymes appear in these strains, one apparently reactive with glutamine and one with ammonia. It is argued that the ammonia-dependent activity may be an artifact of mutation, resulting from limited production of a glutamine-specific subunit.

ATP and XMP react to produce an enzyme-bound intermediate, adenylyl-XMP, which is subsequently aminated to form GMP and AMP as products. GMP synthetase is a site of metabolic regulation in the synthesis of guanine nucleotides. Its formation is regulated at the gene level by guanine compounds and its activity is inhibited both by guanosine phosphates and by the adenine-containing derivatives, adenosine, AMP, and psicofuranine, an analogue of adenosine.

References

Abrams, R. and Bentley, M. (1959) Arch. Biochem. Biophys., 79, 91.

Brevet, A., Hedegaard, J., and Roche, J. (1967) Compte Rend. Soc. Biol., 161, 1938.

Fukuyama, T.T. (1966) J. Biol. Chem., 241, 4745.

Fukuyama, T.T. and Donovan, K.L. (1968) J. Biol. Chem., 243, 5798.

Fukuyama, T.T. and Moyed, H.S. (1964) Biochemistry, 3, 1488.

Hampton, A. and Nomura, A. (1967) Biochemistry, 6, 679.

Iarovaia, L.M., Mardashev, S.R., and Debov, S.S. (1967) Vop. Med. Khim., 13, 176.

Kane, J.F., Holmes, W.M., and Jensen, R.A. (1972) J. Biol. Chem., 247, 1587.

Kuramitsu, H. and Moyed, H.S. (1966) J. Biol. Chem., 241, 1596.

Kuramitsu, H., Udaka, S., and Moyed, H.S. (1964) J. Biol. Chem., 239, 3425.

Lagerkvist, U. (1958a) J. Biol. Chem., 238, 138.

Lagerkvist, U. (1958b) J. Biol. Chem., 238, 143.

Magasanik, B. and Karibian, D. (1960) J. Biol. Chem., 235, 2672.

Mager, J. and Magasanik, B. (1960) J. Biol. Chem., 235, 1474.

Mardashev, S.R. and Iarovaia, L.M. (1965) Ukr. Biokhim. Zh., 37, 751.

Mardashev, S.R., Iarovaia, L.M., and Debov, S.S. (1967) Dokl. Akad. Nauk SSSR, 174, 484.

Moyed, H.S. and Magasanik, B. (1957) J. Biol. Chem., 226, 351.

Nijkamp, H.J. and de Haan, P.G. (1967) Biochim. Biophys. Acta, 145, 31.

Nijkamp, H.J. and Oskamp, A. (1968) J. Mol. Biol., 35, 103.

Prusiner, S. (1973) In "The Enzymes of Glutamine Metabolism" (S. Prusiner and E.R. Stadtman, eds.), Academic Press, New York, p.

Stadtman, E.R. (1970) In "The Enzymes" (P.D. Boyer, ed.), Vol. II, 3rd Edition, Academic Press, N.Y., p. 398.

Taylor, A. (1970) Bact. Rev., 34, 155.

Udaka, S. and Moyed, H.S. (1963) J. Biol. Chem., 238, 2797.

NAD SYNTHETASE

Bernard Witholt*

Department of Chemistry
University of California, San Diego
La Jolla, California 92037

NAD (nicotinamide adenine dinucleotide) can be synthesized from several precursors, as shown in Figure 1. In the absence of nicotinic acid or nicotinamide, NAD is synthesized de novo from quinolinic acid, via NaMN (nicotinic acid mononucleotide). Eucaryotes synthesize quinolinic acid from tryptophan (Kaplan, 1961), while Escherichia coli synthesizes quinolinic acid from aspartate and glycerol (Ogasawara et al., 1967). The last reaction in de novo synthesis of NAD is catalyzed by NAD synthetase, which converts NaAD to NAD by amidating the nicotinic acid carboxyl group of NaAD. This reaction requires ATP and glutamine or ammonia as the amide donor.

In the presence of nicotinic acid or nicotinamide, NAD can be synthesized via the salvage or Preiss-Handler pathway (Preiss and Handler, 1958b). Nicotinic acid is converted to NaMN by NaMN pyrophosphorylase; this reaction requires ATP and PRPP (5-phosphorylribose-1-pyrophosphate). The further conversion of NaMN to NAD is identical to that observed in de novo NAD synthesis. Nicotinamide can be deamidated to nicotinic acid which is then converted to NAD as outlined above. Alternatively, nicotinamide can be converted directly to NMN (nicotinamide mononucleotide) which is then converted to NAD, in a set of reactions which parallel the corresponding synthesis of NaAD from nicotinic acid. The direct synthesis of NAD from nicotinamide via NMN provides the only known pathway for NAD biosynthesis which does not utilize NAD synthetase.

Table 1 lists a number of experiments performed over the last twenty years, which indicate how widespread the Preiss-Handler pathway is.

NaMN and NaAD were synthesized from nicotinic acid in human erythrocytes (Preiss and Handler, 1958a), mouse

---

*Present address: Biochemistry Laboratory, University of Groningen, Groningen, Netherlands.

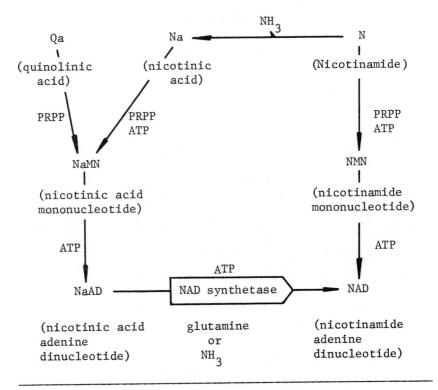

Figure 1.   Pathways of NAD biosynthesis.

liver (Ijichi et al., 1966; Ichiyama et al., 1967; Purko
and Stewart, 1967), rat liver (Preiss and Handler, 1958a,b)
Ehrlich ascites tumor cells (Purko and Stewart, 1967),
yeast (Preiss and Handler, 1958b), E. coli (Imsande, 1961),
and barley leaves (Ryrie and Scott, 1969).

The deamidation of nicotinamide to nicotinic acid, and
the subsequent appearance of NaMN and NaAD have also been
observed frequently. NaMN and NaAD were found in mouse
liver (Purko and Stewart, 1967), cat liver (Pallini and
Ricci, 1965), and barley leaves (Ryrie and Scott, 1969)
after administration of nicotinamide. These compounds were
also produced by Ehrlich ascites tumor cells (Purko and
Stewart, 1967) in the presence of nicotinamide. It is
likely that NAD synthetase was involved in the salvage
synthesis of NAD from nicotinic acid and nicotinamide in
each of these experiments. Thus, NAD synthetase is

TABLE 1

Synthesis of NAD and Its Precursors in Different Tissues

Tissue	Precursors*	Products	Ref**
Erythrocytes			
Rabbit (hemolysate)	N ATP	NAD	a
Human	Na	NaMN NaAD	b
Liver			
Mouse	N	NAD	c
	N	NaMN NaAD NAD	d
	Na	NaMN NaAD NAD	d,e,f
Rat	N	NAD	g
	Na or NaMN	NAD	h
	Na	NaMN NaAD NAD	b,i
Rat (perfused liver)	N or Na	NAD	j
Rat (crude extracts)	Qa and	NaMN	k,l
	Qa precursors	NaMN NaAD	
Cat	N	Na NaMN NaAD NAD	m
<u>Brain</u>, Rat	N NMN or NAD	NAD	h
Adipose tissue			
Rat (perfused)	N or Na	NAD	n
Polymorphonuclear leu-<u>kocytes</u> (guinea pig)	N Na or NaMN	NAD	o
<u>Ehrlich Ascites</u> <u>tumor cells</u>	N or Na	NaMN NaAD NAD	d
<u>Yeast</u>	Na	NaMN NaAD NAD	i
<u>E. coli</u>	Na	NaMN	NAD	p
<u>Barley leaves</u>	N Na or Qa	NaMN NaAD NAD	q

*Abbreviations: N = nicotinamide; Na = nicotinic acid; Qa = quinolinic acid. See text for other abbreviations.

**References: a - Hofmann, 1955; b - Preiss and Handler, 1958a; c - Kaplan et al., 1957; d - Purko and Stewart, 1967; e - Ijichi et al., 1966; f - Ichiyama et al., 1967; g - Greengard et al., 1967; h - Minard and Hahn, 1963; i - Preiss and Handler, 1958b; j - Hagino et al., 1968; k - Nishizuka and Hayaishi, 1963a; l - Nishizuka and Hayaishi, 1963b; m - Pallini and Ricci, 1965; n - Hanson and Ziporin, 1967; o - Flechner et al., 1970; p - Imsande 1961; q - Ryrie and Scott, 1969.

probably ubiquitous in nature, even though it has been detected directly in only a few sources.

Properties of NAD Synthetase

NAD synthetase has been purified from yeast (Preiss and Handler, 1958b; Yu and Dietrich, 1972), E. coli (Imsande, 1961; Spencer and Preiss, 1967), and, to a limited extent, from rat liver (Preiss and Handler, 1958b). It catalyzes the following reaction:

NaAD + ATP + glutamine (or NH_3) ⟶

NAD + AMP + PPi + glutamic acid (or H_2O)

The assay for NAD synthetase is based on the formation of NAD, in the presence of NaAD, ATP and the appropriate amide donor. The NAD formed is assayed using the alcohol dehydrogenase procedure of Racker (1950). Alternatively, the conversion of ^{14}C-NaAD to ^{14}C-NAD is followed by paper chromatography of the reaction mixture (Preiss and Handler, 1958b; Imsande, 1961).

Table 2 summarizes some of the characteristics of NAD synthetase from yeast and E. coli. The enzyme has been purified to various extents; from 35 fold for E. coli NAD synthetase (Imsande, 1961) to 2,000 fold for yeast NAD synthetase (Yu and Dietrich, 1972). The K_m's for NaAD and ATP fall in the range of 10^{-4} M and 2 to 6 x 10^{-4} M, respectively for both the yeast and the E. coli enzyme. Glutamine and ammonia are approximately equally good amide donors for yeast NAD synthetase, with K_m's of approximately 5 x 10^{-3} M. The pH optimum of the enzyme in the presence of glutamine was broad, from 6.2 to 8.4, while the pH optimum in the presence of ammonia was rather narrow, centering at pH 8.5.

E. coli NAD synthetase exhibited a strong preference for ammonia rather than glutamine, as shown in Table 2. Mg^{++} was required for optimal activity of both the yeast and the E. coli enzymes, at a concentration of about 5 mM. K^+ was required by yeast NAD synthetase.

Yu and Dietrich (1972) have examined the highly puri-fied yeast NAD synthetase, and have concluded that it is composed of non-identical subunits of 65,000 and 80,000 daltons. The molecular weight of the "native" enzyme

TABLE 2

Characteristics of NAD Synthetase From Yeast and $\underline{E}$. $\underline{coli}$

	Source			
	Yeast[a]	Yeast[b]	$\underline{E}$. $\underline{coli}$[c]	$\underline{E}$. $\underline{coli}$[d]
Purification	106 x	2000 x	35 x	450 x
K_m's				
NaAD	1.4×10^{-4}	1.9×10^{-4}	2×10^{-5}	1.1×10^{-4}
ATP	6×10^{-4}	1.7×10^{-4}	4×10^{-4}	2×10^{-4}
Glutamine	3.5×10^{-3}	5×10^{-3}		1.6×10^{-2}
NH$_3$	1.4×10^{-3}	6.4×10^{-3}	10^{-5}	6.5×10^{-5}
Mg^{++}	1.3×10^{-3}			
Optimal con-centrations				
Mg^{++}	5×10^{-3}	required	4×10^{-3}	$>3 \times 10^{-3}$
Mn^{++}				1.5×10^{-3}
K$^+$	56×10^{-3}	required		
pH Optima				
Glutamine	6.2-7.4	6.2-8.4		
NH$_3$	8.4	8.4-8.8	8.5	8.4

[a]Preiss and Handler, 1958b; [b]Yu and Dietrich, 1972; [c]Imsande, 1961; [d]Spencer and Preiss, 1967.

appeared to be 630,000 daltons.

A number of compounds inhibit NAD synthetase although here again there are differences in the yeast and $\underline{E}$. $\underline{coli}$ enzymes. As is shown in Table 3, psicofuranine and adenosine inhibit $\underline{E}$. $\underline{coli}$ NAD synthetase (Spencer and Preiss, 1967) while azaserine, which has little effect on the $\underline{E}$. $\underline{coli}$ enzyme, inhibits yeast NAD synthetase (Preiss and Handler, 1958b). Mor and Lichtenstein (1970) have found that several carbobenzoxy amino acids inhibit rat liver NAD synthetase.

TABLE 3

Inhibitors of NAD Synthetase From Several Sources

Source	Inhibitor	Degree of Inhibition
E. coli B[a]	psicofuranine	$K_I = 6 \times 10^{-4}$
	adenosine	$K_I = 1.5 \times 10^{-3}$
Yeast[b]	azaserine	$K_I = 1.3 \times 10^{-3}$ against glutamine
		$K_I = 2.7 \times 10^{-3}$ against NH_4Cl
Rat Liver[c]	carbobenzoxy-L-glutamine	44% inhibition at 16.7 mM
	carbobenzoxy-L-phenylalanine	85% inhibition at 16.7 mM
	carbobenzoxy S-benzyl-L-cysteine	95% inhibition at 16.7 mM

[a]Spencer and Preiss, 1967; [b]Preiss and Handler, 1958b; [c]Mor and Lichtenstein, 1970.

Control and Regulation of NAD Synthetase

E. coli produces 50-100 times more NAD synthetase than is needed to maintain the supply of NAD in dividing cells (Imsande, 1961); this amount of enzyme is produced constitutively (Imsande and Pardee, 1962). Thus, the synthesis of NAD synthetase appears not to be regulated, at least in E. coli. This is in keeping with the fact that the regulation of NAD synthesis occurs at the level of NaMN synthesis in E. coli (Saxton et al., 1968).

It is unlikely that NAD synthetase is regulated by the inhibitors listed in Table 3, since these inhibitors are generally not found intracellularly and only affect NAD synthetase at rather large concentrations. Even if one of these inhibitors were present intracellularly at a concentration sufficient to inhibit NAD synthetase, NAD synthesis would probably not be affected significantly, since NAD synthetase is present in such large amounts, relative to the other enzymes of NAD synthesis (Imsande and

Pardee, 1962).

One possible regulatory factor for NAD synthetase may be the available supply of glutamine. Kapoor and Bray (1968) have found that NAD inhibits glutamine synthetase in Neurospora. An increasing supply of NAD would therefore inhibit glutamine synthetase, reducing the concentration of glutamine available to NAD synthetase. Such a control mechanism would function only in organisms in which ammonia is not an effective amide donor for NAD synthetase.

SUMMARY

NAD synthetase is present in a variety of tissues. It has been purified extensively from yeast and E. coli. There are significant differences in the NAD synthetase isolated from these two sources, with respect to the amide donor, and with respect to inhibitor specificity. There appears to be no regulation of this enzyme at the genetic level, at least in E. coli. Little is known about in vivo allosteric effectors.

REFERENCES

Flechner, I., Amar, A., and Bekierkunst, A. (1970). Biochim. Biophys. Acta 222, 320.

Greengard, P., Petrack, B., and Kalinsky, H. (1967). J. Biol. Chem. 242, 152.

Hagino, Y., Lan, S. J., Ng, C. Y., and Henderson, L. M. (1968). J. Biol. Chem. 243, 4980.

Hanson, R. W. and Ziporin, Z. Z. (1967). J. Lipid Res. 8, 30.

Hofmann, E. C. G. (1955). Biochem. Z. 327, 273.

Ichiyama, A., Nakamura, S., and Nishizuka, Y. (1967). Arzneim. Forsch. 17, 1525.

Ijichi, H., Ichiyama, A., and Hayaishi, O. (1966). J. Biol. Chem. 241, 3701.

Imsande, J. (1961). J. Biol. Chem. 236, 1494.

Imsande, J., and Pardee, A. B. (1962). J. Biol. Chem. 237, 1305.

Kaplan, N. O. (1961). Metab. Pathways 2, 627.

Kaplan, N. O., Goldin, A., Humphreys, S. R., and Stolzenbach F. E. (1957). J. Biol. Chem. 226, 365.

Kapoor, M., and Bray, D. (1968). Biochemistry 7, 3583.

Minard, F. N., and Hahn, C. H. (1963). J. Biol. Chem. 238, 2474.

Mor, G., and Lichtenstein, N. (1970). FEBS Lett. 9, 183.

Nishizuka, Y., and Hayaishi, O. (1963a). J. Biol. Chem. 238, PC 483.

Nishizuka, Y., and Hayaishi, O. (1963b). J. Biol. Chem. 238, 3369.

Ogasawara, N., Chandler, J. L. R., Gholson, R. K., Rosser, R. J., and Andreoli, A. J. (1967). Biochim. Biophys. Acta 141, 199.

Pallini, V., and Ricci, C. (1965). Arch. Biochem. Biophys. 112, 282.

Preiss, J., and Handler, P. (1958a). J. Biol. Chem. 233, 488.

Preiss, J., and Handler, P. (1958b). J. Biol. Chem. 233, 493.

Purko, J., and Stewart, H. B. (1967). Can. J. Biochem. 45, 179.

Racker, E. (1950). J. Biol. Chem. 184, 313.

Ryrie, I. J., and Scott, K. J. (1969). Biochem. J. 115, 679.

Saxton, R. E., Rocha, V., Rosser, R. J., Andreoli, A. J., Shimoyama, M., Kosaka, A., Chandler, J. L. R., and

Gholson, R. K. (1968). *Biochim. Biophys. Acta* 156, 77.

Spencer, R. L., and Preiss, J. (1967). *J. Biol. Chem.* 242, 385.

Yu, C. K., and Dietrich, L. S. (1972). *J. Biol. Chem.* (In press).

STRUCTURE-FUNCTION RELATIONSHIPS IN GLUTAMINE-DEPENDENT CARBAMYL PHOSPHATE SYNTHETASE

Paul P. Trotta, Lawrence M. Pinkus, Vaira P. Wellner, Leonard Estis, Rudy H. Haschemeyer, and Alton Meister

Department of Biochemistry
Cornell University Medical College
New York, N.Y. 10021

ABSTRACT

This paper reviews studies carried out in this laboratory on the subunit structure of the glutamine-dependent carbamyl phosphate synthetase from E. coli, its ability to catalyze several partial reactions, its allosteric regulation by ammonia, ornithine, UMP, and IMP, the specific inhibition of the glutamine-dependent activity by L-2-amino-4-oxo-5-chloropentanoic acid, the specific reaction of this compound with a sulfhydryl group on the light subunit, and the remarkable enhancement of the glutaminase activity of the enzyme by treatment with sulfhydryl reagents. The monomer is composed of a heavy subunit which can catalyze the synthesis of carbamyl phosphate from ammonia but not from glutamine, and a light subunit designed for the binding and hydrolysis of glutamine.

INTRODUCTION

Carbamyl phosphate is a key intermediate in the biosynthesis of pyrimidines and of arginine, and is thus essential for the synthesis of both nucleic acids and proteins. It is therefore not surprising that carbamyl phosphate synthetases are widely distributed in nature, being found in plants, animal tissues, and bacteria, and that they are subject to allosteric regulation by a number of effectors. Two main classes of carbamyl phosphate synthetases, designated I (ammonia-dependent) and II (glutamine-dependent), have been described.

431

Carbamyl phosphate synthetase I is found in liver mito-chondria of ureotelic vertebrates (Marshall et al., 1958; 1961). It can utilize only ammonia (even at low concentrations of NH_4^+ + NH_3) as the nitrogen source, and exhibits a requirement for N-acetyl-L-glutamate:

$$2 \text{ ATP} + HCO_3^- + NH_4^+ \xrightarrow[Mg^{++}]{\text{N-acetyl-glutamate}} NH_2CO_2PO_3^{-2} + 2 \text{ ADP} + Pi$$

The general properties of synthetase I, which has been isolated in the most stable form and highest yield from frog liver mitochondria, have been the subject of much investigation; the reader is referred to reviews by Cohen (1962) and Jones (1965). A highly purified, stable preparation can also be obtained from rat liver, provided that 20% glycerol is added during isolation (Guthohrlein and Knappe, 1968). The subunit interactions and molecular weights of the various forms of this enzyme have been investigated (Virden, 1972); the data obtained show a number of striking similarities to those found in our laboratory and reported here for E. coli carbamyl phosphate synthetase II.

For a number of years it was generally thought that the N-acetylglutamate-dependent ammonia bicarbonate-utilizing enzyme and carbamate kinase were the major physiologically important sources of carbamyl phosphate. Carbamate kinase, first described in Streptococcus faecalis (Jones and Lipmann, 1960), catalyzes a freely reversible reaction utilizing non-enzymatically formed carbamate:

$$(1) \quad NH_4^+ + HCO_3^- \underset{\xleftarrow{\hspace{1cm}}}{\overset{\text{(nonenzymatic)}}{\xrightarrow{\hspace{1cm}}}} NH_2COO^- + H_3O^+$$

$$(2) \quad NH_2COO^- + ATP \underset{\xleftarrow{\hspace{0.5cm}}}{\overset{Mg^{++}}{\xrightarrow{\hspace{0.5cm}}}} NH_2CO_2PO_3^{-2} + ADP$$

However, the role of this enzyme in catalyzing the production of significant amounts of carbamyl phosphate in vivo is questionable since the equilibrium of the reaction markedly favors ATP synthesis. Furthermore, mutants of microorganisms were isolated which retained carbamate kinase activity but yet lacked ability to synthesize carbamyl phosphate (Kanazir et al., 1959). It was also recognized that vertebrate tissues which are capable of synthesizing pyrimidines lack carbamyl phosphate synthetase I (Jones, 1963). Evidence had previously been obtained that in Lactobacillus arabinosus glutamine is essential for the synthesis of arginine (Ory et al., 1954; Lyman et al., 1954) and that in liver homogenates glutamine can serve as a nitrogen donor for urea biosynthesis (Leuthardt, 1938; Bach and Smith, 1956).

Levenberg (1962) reported the partial purification of an enzyme from the common mushroom Agaricus bisporus which catalyzed the incorporation of [^{14}C] bicarbonate into citrulline in the presence of ornithine, ATP, magnesium ions, and L-glutamine; strong evidence was obtained that carbamyl phosphate is an intermediate in this reaction in which glutamine was much more active than was ammonia. In 1964, Pierard and Wiame found evidence for a similar carbamyl phosphate synthetase in cell free extracts of E. coli; thus, they demonstrated a glutamine dependent synthesis of citrulline (Pierard and Wiame, 1964). The finding by these investigators of a one-step E. coli mutant simultaneously auxotrophic for arginine and uracil and lacking the glutamine-dependent carbamyl phosphate synthetase indicated that a single enzyme catalyzes the synthesis of carbamyl phosphate that is utilized for both uracil and arginine biosynthesis. Kalman and associates (1965) subsequently achieved a partial purification of this carbamyl phosphate synthetase and concluded that it was probably identical with carbamate kinase. However, independent investigations in this laboratory (Anderson and Meister, 1965a; 1965b) led to the isolation of very highly purified carbamyl phosphate synthetase (about 300-fold), and studies with this enzyme preparation established the following stoichiometry and metal ion requirements:

PAUL P. TROTTA *et al.*

$$2\ \text{ATP} + \text{HCO}_3^- + \text{L-Glutamine} \xrightarrow[\text{K}^+]{\text{Mg}^{++}} \text{NH}_2\text{CO}_2\text{PO}_3^{-2} + 2\ \text{ADP}$$
$$+ \text{L-Glutamate} + \text{Pi}$$

The stoichiometry of the reaction catalyzed by carbamyl phosphate synthetase II was thus shown to be identical to that of the reaction catalyzed by the liver N-acetylglutamate-dependent enzyme rather than that catalyzed by carbamate kinase which utilizes one mole of ATP for each mole of carbamyl phosphate formed. This observation was later confirmed in independent studies by Kalman and associates (1966). Also in distinction to the reaction catalyzed by carbamate kinase, that catalyzed by the E. coli glutamine-dependent synthetase was shown to be essentially irreversible (Anderson and Meister, 1965b).

Like other glutamine amidotransferases this enzyme can utilize ammonia (at relatively high concentrations of $\text{NH}_4^+ + \text{NH}_3$) as well as glutamine. It is notable, however, that the apparent Km values at pH 7.8 for glutamine and un-ionized ammonia are of about the same order of magnitude.

Carbamyl phosphate synthetase II is present in many tissues and microorganisms while carbamyl phosphate synthetase I is found chiefly in the liver mitochondria. Since ornithine transcarbamylase is also present in the mitochondrion, the carbamyl phosphate synthesized by synthetase I is generally considered to be used mainly for arginine biosynthesis and urea production. In animal cells synthetase II is found in the cytosol, as is aspartate transcarbamylase, and is therefore thought to be associated mainly with the orotic acid pathway for pyrimidine biosynthesis. However, in microorganisms the carbamyl phosphate produced by synthetase II is available for arginine biosynthesis as well. Synthetase II from various sources other than E. coli is briefly discussed in a later section.

An interesting mitochondrial enzyme exhibiting a combination of the properties of the two classes of synthetases has

434

been recently described in the hepatopancreas of the land snail
Strophocheilus oblongus (Tramell and Campbell, 1970). This
system which utilizes glutamine as nitrogen source but also
requires N-acetylglutamate may constitute still another class of
synthetase.

E. coli GLUTAMINE-DEPENDENT CARBAMYL PHOSPHATE SYNTHETASE

Enzyme Purification

The enzyme was purified from E. coli B (three-quarter
log phase; minimal medium enriched with arginine) essentially
by the method described by Anderson and associates (1970). This
procedure consists of a sonication, followed by heating in the
presence of glutamine, treatment with protamine sulfate, DEAE-
Sephadex chromatography, and finally gel filtration on Sephadex
G-200. Subsequent electrophoretic studies of the purified preparation in the presence of sodium dodecyl sulfate demonstrated at
least two minor contaminant bands which represented about 5-10%
of the total protein. Rechromatography on Sephadex G-200 (150
mM potassium phosphate, 20 mM ammonium chloride; pH 7.8) was
carried out to remove the lower molecular weight impurities; the
higher molecular weight impurities were removed by fractionation
with ammonium sulfate. The final preparation is homogeneous by a
number of independent criteria including acrylamide gel electrophoresis carried out in a variety of dissociating and non-dissociat-
ing solvents, sedimentation velocity and sedimentation equilibrium.
Foley and coworkers (1971) have used an alternative method for
purification based on successive Sephadex G-200 chromatograph-
ies in UMP (for removal of high molecular weight contaminants)
and ornithine (for removal of low molecular weight contaminants).

Shortly after purification the specific activity of the
enzyme decreases to a value of about 160 micromoles of carbamyl
phosphate per hour per milligram of protein from a value of about
280 initially obtained after the final purification step.

Mechanism Studies and Partial Reactions

Early studies on purified E. coli glutamine-dependent carbamyl phosphate synthetase provided evidence that the enzyme catalyzes the activation of carbon dioxide in the absence of glutamine (Anderson and Meister, 1965). In these experiments, a relatively large amount of enzyme was incubated with [^{14}C] bicarbonate and ATP; then a solution containing both glutamine and a large excess of unlabeled bicarbonate was added and the labeled carbamyl phosphate formed was determined. The amount of labeled carbamyl phosphate formed was much greater than could be accounted for if it was assumed that all of the radioactive bicarbonate had equilibrated with the unlabeled bicarbonate, and that all of the glutamine present was utilized. These findings indicated that labeled bicarbonate is bound to the enzyme in a reaction that requires ATP and that binding permits the bound bicarbonate to be immediately available for reaction with glutamine. Similar experiments in which [^{32}P] ATP was used indicated that the activation of bicarbonate is associated with cleavage of ATP to ADP. These findings suggested the following sequence of steps:

(a) $\text{Enzyme} + \text{ATP} + \text{HCO}_3^- \rightleftharpoons \text{Enzyme-}(\text{HCO}_3\text{PO}_3^=) + $

$$\text{ADP}$$

(b) $\text{Enzyme-}(\text{HCO}_3\text{PO}_3^=) + \text{L-Glutamine} \rightleftharpoons \text{Enzyme-}$

$$(\text{HCO}_3\text{PO}_3^=) (\text{L-Glutamine})$$

(c) $\text{Enzyme} (\text{HCO}_3\text{PO}_3^=) (\text{L-Glutamine}) \longrightarrow \text{Enzyme}$

$$(\text{NH}_2\text{CO}_2^-) + \text{L-Glutamate} + \text{Pi}$$

(d) $\text{Enzyme} (\text{NH}_2\text{CO}_2^-) + \text{ATP} \rightleftharpoons \text{Enzyme} + \text{NH}_2\text{CO}_2\text{PO}_3^{-2}$

$$+ \text{ADP}$$

Additional evidence in support of this sequence of reaction steps came from the discovery that the enzyme could catalyze several partial reactions (Anderson and Meister, 1966a; 1966b). Thus, it was found that the enzyme catalyzes a bicarbonate-dependent cleavage of ATP to ADP in the absence of added glutamine or ammonia:

$$ATP + H_2O \xrightarrow[Mg^{++},\ K^+]{HCO_3^-} ADP + Pi$$

Duffield and associates (1969) demonstrated a related reaction, a bicarbonate-dependent [^{3}H]ADP-[^{3}H]ATP exchange. Thus, step (a) (see p.6) is reversible. The apparent Km value for bicarbonate in the bicarbonate-dependent hydrolysis of ATP is 1 mM; this value is similar to that for bicarbonate in the synthetase reaction. However, the apparent Km value for ATP is 0.7 mM or about 10% of that found for ATP in the synthetase reaction. The bicarbonate-dependent cleavage of ATP presumably reflects the formation and breakdown of the activated carbon dioxide (reaction (a), see p. 6); the rate of this reaction is about 10% of that of the overall synthetase reaction.

Another partial reaction catalyzed by the enzyme is the synthesis of ATP from ADP and carbamyl phosphate:

$$ADP + NH_2CO_2PO_3^{-2} + H_2O \xrightarrow{Mg^{++},\ K^+} NH_4^+ + HCO_3^- + ATP$$

The rate of this reaction is about 20% of that of carbamyl phosphate formation in the synthetase reaction. This step appears to reflect reversal of the phosphorylation step (reaction (d)) and is thus a partial reversal of the overall synthetase reaction. The relation between ADP concentration and reaction velocity is sigmoidal as is the curve for ATP dependence in the synthetase reaction.

The enzyme was also found to catalyze ATP- and bicarbonate-dependent hydrolysis of γ-glutamylhydroxamate; this reaction is not associated with stoichiometric formation of ADP:

$$\gamma\text{-L-Glutamylhydroxamate} + H_2O \xrightarrow[K^+, HCO_3]{ATP, Mg^{++}} \text{L-Glutamate} +$$

Hydroxylamine

The reaction proceeds at about 80-90% of the rate of the overall synthetase reaction. It is notable that only catalytic amounts of ATP and bicarbonate are required to activate this glutaminase-like activity. γ-Glutamylhydroxamate cannot be utilized for the synthesis of the hydroxamate analog of carbamyl phosphate.

In addition, the enzyme catalyzes the hydrolysis of glutamine in the absence of the other substrates and of added metal ions:

$$\text{L-Glutamine} + H_2O \longrightarrow \text{L-Glutamate} + NH_4^+$$

The rate of the reaction is about 2% of the rate of the overall synthetase reaction. This reaction and the ATP and bicarbonate-dependent hydrolysis of γ-glutamylhydroxamate appear to be related to the postulated steps (reactions b, c, p. 6) involving the binding of glutamine and the cleavage of its amide group.

Allosteric Regulation

The activity of carbamyl phosphate synthetase is significantly affected by a number of purine and pyrimidine nucleotides (Anderson and Meister, 1966b). Thus, purine nucleotides (e.g., IMP) stimulate activity, and pyrimidine nucleotides are either inhibitory (e.g., UMP) or have no effect (cytidine nucleotides) (Figure 1). These effects appear to be exerted maximally by the first nucleotide synthesized in each pathway and to decrease as the number of steps required to synthesize the various nucleotides from IMP or UMP, respectively, increases. The data indicate that the allosteric effectors act by altering the affinity of the enzyme for ATP. For example, activation or inhibition

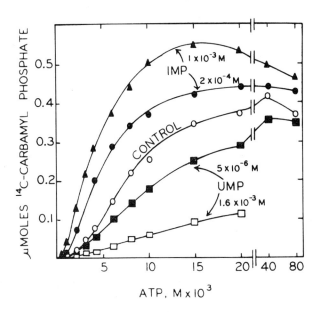

Fig. 1 Effect of IMP and UMP on carbamyl phosphate synthetase activity as a function of ATP concentration. The reaction mixtures contained $MgCl_2$ in concentrations that were equimolar with ATP, $[^{14}C]$-$NaHCO_3$ (20 μmoles, 700,000 cpm), KCl (100 μmoles), L-glutamine (10 μmoles), Tris–HCl buffer (100 μmoles, pH 8.2), IMP and UMP as indicated, and enzyme (0.015 mg) in a final volume of 1 ml. The $[^{14}C]$ carbamyl phosphate synthesized after incubation for 10 min at 37° was determined. (Anderson and Meister, 1966b).

is substantially diminished at relatively high concentrations of ATP. Furthermore, since the bicarbonate-dependent ATP-ase is not affected by the allosteric effectors, it was concluded that step (d) (phosphorylation of carbamate) is allosterically regulated rather than the activation of bicarbonate (step a). This finding suggests that the enzyme has two separate binding sites for ATP. Independent studies by Pierard (1966) showed that L-ornithine counteracts the inhibitions produced by pyrimidine nucleotides while other intermediates in the arginine biosynthetic pathway have no such effect. Since E. coli possesses only one carbamyl phosphate synthetase which therefore must catalyze the formation of carbamyl phosphate for both arginine and pyrimidine biosynthesis, this effect of ornithine appears necessary to insure a supply of carbamyl phosphate for the arginine pathway when excess UMP is present. Subsequent studies by Anderson and Marvin (1968) showed that ornithine alone (in the absence of UMP) is a potent allosteric activator.

The allosteric regulation of carbamyl phosphate synthetase in E. coli by various metabolites from the pyrimidine, purine and arginine biosynthetic pathways as well as by ammonia (see later section) is summarized in Figure 2.

The different conformational states produced by the effectors ornithine and UMP, and the substrate ATP, were shown to influence in different ways the reaction of N-ethylmaleimide or 5,5'-dithiobis (2-nitrobenzoate) with two different sulfhydryl groups which affected catalytic activity (Anderson and Marvin, 1970; Foley et al., 1971; Mathews and Anderson, 1972). One sulfhydryl group was available for reaction with reagent under all conditions and such reaction had no apparent effect on activity; in the presence of ornithine, this was the only reactive sulfhydryl group. Addition of ATP, magnesium ions, and bicarbonate exposed a second sulfhydryl group, whose reaction with N-ethylmaleimide was accompanied by an increase in the apparent Km value for glutamine and a 50% loss in the glutamine-dependent synthetase reaction and in the synthesis of ATP from carbamyl phosphate and ADP. A third sulfhydryl group reacted at higher concentrations of reagent; UMP accelerated this reaction, while ornithine and a mixture of ATP, MgCl2, and bicarbonate apparently protected this

440

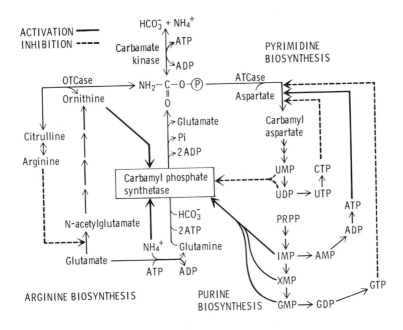

Fig. 2 Schematic summary of the role of the glutamine-dependent carbamyl phosphate synthetase of E. coli in providing an essential precursor for pyrimidine and arginine biosynthesis and the feed-back regulation of this metabolite from these pathways and by purines and ammonia.

group. Reaction of this group with the sulfhydryl reagent was accompanied by a doubling of the rate of ATP synthesis and a complete loss of the overall synthetase reaction. A scheme summarizing the conditions for the availability for the various sulfhydryl groups of the enzyme is given in Figure 3. Mathews and Anderson (1972) reacted   the enzyme with [^{14}C] N-ethylmaleimide followed by dissociation in 6 M guanidinium hydrochloride and demonstrated that the first and third sulfhydryl groups were located on the heavy subunit while the second sulf-hydryl group was located on the light subunit (see next section for a discussion of the non-identical subunits). However, a role for these groups in the catalytic or regulatory functions of the enzyme

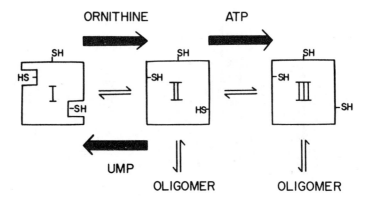

Fig. 3 Scheme summarizing the conditions under which the SH groups described in the text are available for reaction with DTNB or NEM. (Foley et al., 1971).

has not yet been established.

Subunit Structure and Function

Trotta and associates (1971a; 1971b) and Anderson and associates (Anderson et al., 1970; Mathews and Anderson, 1972) independently demonstrated that the enzyme could be dissociated in sodium dodecyl sulfate into two non-identical polypeptide chains: a heavy chain (molecular weight $\sim$ 130,000) and a light chain (molecular weight $\sim$ 42,000). A similar dissociation could be obtained under a variety of other conditions of strong denaturation, including high concentrations of urea or guanidinium hydrochloride and treatment with maleic or succinic anhydrides (Anderson et al., 1970; Trotta et al., 1971b).

A major breakthrough toward understanding the relationship between the structure and function of this enzyme was the finding by Trotta and associates (1971b) that a relatively mild solvent perturbation; i.e., treatment with 1 M potassium thiocyanate promoted a reversible dissociation into non-identical subunits with retention of biological activity. Figure 4 summarizes the various activities associated with the heavy and light subunits and the reconstituted enzyme.

The heavy subunit was found to catalyze carbamyl phosphate production from ammonia but not from glutamine. Consistent with this observation, the heavy subunit could also catalyze bicarbonate-dependent ATPase and ATP synthesis from ADP and carbamyl phosphate. The only activity associated with the light subunit is the capacity to catalyze the hydrolysis of glutamine. When the light and heavy subunits were recombined, the glutamine dependent synthetase reaction was restored. These observations led to the suggestion that the enzyme may represent the evolutionary combination of a more primitive ammonia-dependent synthetase and a glutaminase.

A schematic model based on these findings is presented in Figure 5. The evidence for localizing the allosteric sites on the heavy subunit is summarized in the section entitled: "Enzymatic and Hydrodynamic Properties of the Non-identical Subunits".

Studies on the Glutamine Site of the Enzyme.

    a. Inhibition by the Glutamine Analog L-2-Amino-4-oxo-5-chloropentanoic Acid.

That the enzyme can utilize ammonia in place of glutamine indicates that an enzyme-ammonia complex can form directly without utilization of the complete glutamine binding site. It therefore seemed feasible to design a reagent which could react with the glutamine binding site of the enzyme without interfering with its other catalytic functions; it was thought that such treatment might then convert the enzyme from one which could utilize both glutamine and ammonia to one which could use only ammonia. This goal was achieved through synthesis of a chloroketone analog of

443

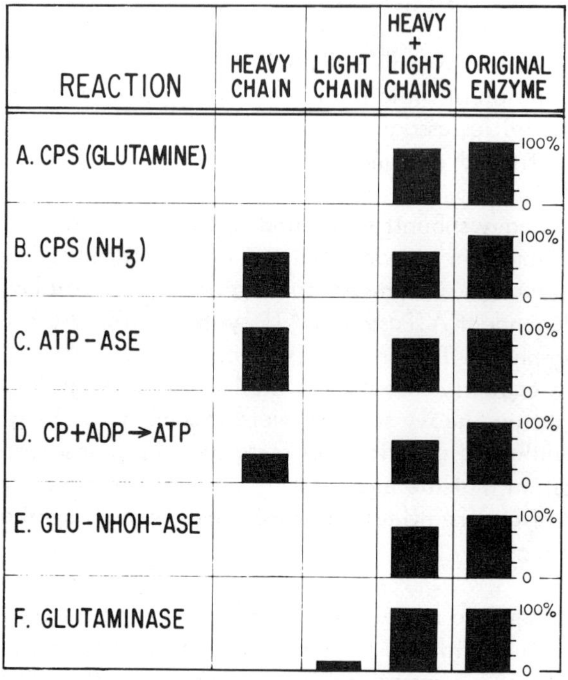

REACTION	HEAVY CHAIN	LIGHT CHAIN	HEAVY + LIGHT CHAINS	ORIGINAL ENZYME
A. CPS (GLUTAMINE)			■	■ 100% / 0
B. CPS (NH₃)	■		■	■ 100% / 0
C. ATP-ASE	■		■	■ 100% / 0
D. CP+ADP→ATP	■		■	■ 100% / 0
E. GLU-NHOH-ASE			■	■ 100% / 0
F. GLUTAMINASE		▪	■	■ 100% / 0

Fig. 4 The enzymatic activities exhibited by the heavy and light subunits, the native enzyme, and the reconstituted enzyme. The heights of the bars represent the percentages recovered of the specific activities of the native enzyme. The specific activities used to determine the heights of the bars in the column labeled "heavy + light chains" are expressed in terms of nmoles of heavy subunit, except for glutaminase which is expressed in terms of nmoles of light subunit. The various activities were determined as described in Trotta et al. (1971b), except for the glutaminase, which was assayed as described in Figure 20.

---

glutamine, L-2-amino-4-oxo-5-chloropentanoic acid (Figure 6) (Khedouri et al., 1966).

444

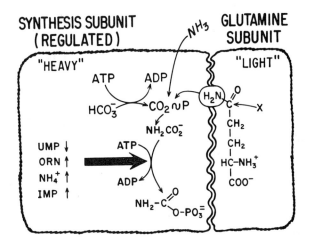

Fig. 5 Schematic representation of the functional role of the two subunits of carbamyl phosphate synthetase. "X" represents the postulated nucleophile. (Trotta et al., 1971b).

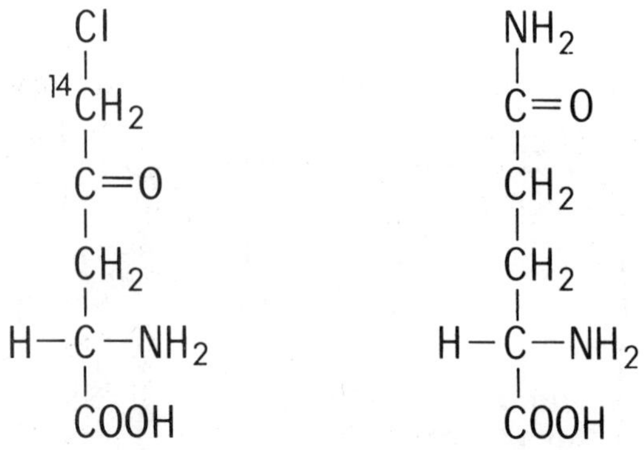

$$CI$$
$$|$$
$$^{14}CH_2$$
$$|$$
$$C{=}O$$
$$|$$
$$CH_2$$
$$|$$
$$H{-}C{-}NH_2$$
$$|$$
$$COOH$$

L-[5-^{14}C] 2-amino-4-oxo-
5-chloropentanoic acid
(chloroketone)

$$NH_2$$
$$|$$
$$C{=}O$$
$$|$$
$$CH_2$$
$$|$$
$$CH_2$$
$$|$$
$$H{-}C{-}NH_2$$
$$|$$
$$COOH$$

L-glutamine

Fig. 6 Structures of L-2-amino-4-oxo-5-chloro[5-^{14}C] pentanoic acid and L-glutamine.

Thus, as shown in Figure 7 when the enzyme was incubated with the chloroketone, its ability to utilize glutamine as the nitrogen donor disappeared.

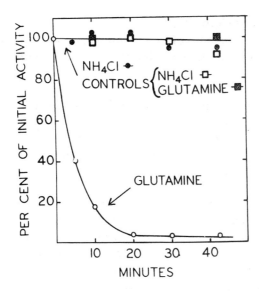

Fig. 7 Effect of chloroketone as a function of time on the activity
of carbamyl phosphate synthetase with NH_4Cl or glutamine as
amino donor. A solution (0.5 ml) containing enzyme (0.3 mg),
potassium phosphate buffer (100 μmoles, pH 6.9), potassium
chloride (100 μmoles), and EDTA (0.25 μmole) was added to 0.5
ml of a solution containing chloroketone (0.25 μmole) at 0°. The
mixture was then warmed rapidly at 37° and maintained at this
temperature. At the indicated intervals 0.05 ml aliquots were
added to separate test tubes and frozen in a dry ice–acetone bath.
The activity of the enzyme samples thus removed was then determined with NH_4Cl or glutamine by adding 0.95 ml of a mixture
containing ATP (20 μmoles), $MgCl_2$ (20 μmoles), [^{14}C] $NaHCO_3$
(20 μmoles, 800,000 cpm), potassium phosphate buffer (80 μmoles,
pH 7.8), and L-glutamine (10 μmoles) or potassium phosphate
buffer (40 μmoles, pH 7.8) and NH_4Cl (100 μmoles). The [^{14}C]
carbamyl phosphate formed after incubation for 10 min at 37° was
determined. The control experiments were carried out in the same

way, except that chloroketone was omitted. The initial activities
were 0.28 and 0.49 μmole of $[^{14}C]$ carbamyl phosphate formed,
respectively, with NH_4Cl and glutamine. (Khedouri et al., 1966).

Inhibition by the chloroketone was prevented by L-glutamine, L-γ-glutamylhydroxamate, and to a much smaller extent by
a mixture of ATP, magnesium ions and bicarbonate. Albizziin, a
glutamine analog, and glutamate have also been shown to protect
against the chloroketone (Pinkus and Meister, 1972).

The ability of the enzyme to catalyze carbamyl phosphate
synthesis from ammonia was not impaired by treatment with the
chloroketone, nor was the activity responsible for the synthesis of
ATP from ADP and carbamyl phosphate affected. However, as
expected, the chloroketone-treated enzyme was unable to catalyze
the hydrolysis of γ-glutamyl hydroxamate or of glutamine (Wellner
et al., 1972). It was concluded that the chloroketone selectively
reacts with the enzyme site that normally accepts glutamine and
that it probably alkylates a specific nucleophilic group of the
enzyme.

It was observed that the bicarbonate-dependent ATPase
of the chloroketone-treated enzyme was about three times that
of the untreated enzyme. This is of particular interest since the
ATPase site and the residue reacting with the chloroketone are on
different subunits (see below).

b. Identification of a Reactive Cysteine Residue at the
Glutamine Binding Site

The evidence reviewed above on the subunit structure of
the enzyme indicates that in the native enzyme glutamine binds
to the light subunit and that its amide nitrogen atom is then transferred to the heavy subunit where it is used for the synthesis of
carbamyl phosphate. Additional evidence for this idea has been
provided by recent studies in this laboratory in which L-2-amino-
4-oxo-5-chloro-$[5-^{14}C]$ pentanoic acid was synthesized and the
enzyme was incubated with this labeled chloroketone preparation

448

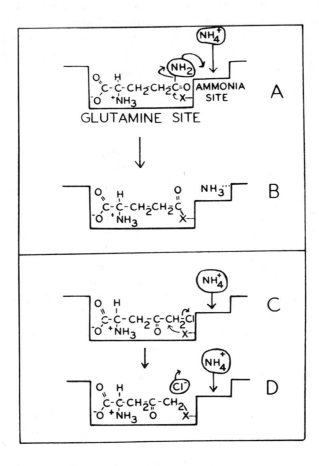

Fig. 8 Schematic diagram of the reaction of the enzyme with glutamine and chloroketone (Khedouri et al., 1966).

(Pinkus and Meister, 1972). The data indicate that close to 1 mole of 4-oxo-norvaline becomes attached to the light subunit of the enzyme after it is treated with chloroketone (Figure 9) and that such binding is prevented by glutamine or albizziin. Although treatment of the enzyme with chloroketone is accompanied by some binding to the heavy subunit and slight additional binding to the light subunit, this seems to reflect the ability of the chloroketone to function as a general alkylating agent similar to (but less reactive than) iodoacetate.

The findings also indicate that virtually all of the chloroketone bound at the site protected by glutamine bound to a cysteine residue. This was established by studies in which the labeled enzyme was treated with mild performic acid and then subjected to acid hydrolysis. According to the pathways indicated in Figure 10, a Baeyer-Villiger type of rearrangement (Hassall, 1967) followed by acid hydrolysis might be expected to yield a [^{14}C] carboxymethyl amino acid and unlabeled serine (Scheme 1) or a [^{14}C] hydroxymethyl amino acid and aspartate (Scheme 2). The hydroxymethyl amino acid would be expected to decompose during acid hydrolysis releasing the radioactivity in a volatile form ([^{14}C] formaldehyde) but the postulated carboxymethyl amino acid would be expected to be stable. [^{14}C] S-Carboxymethylcysteine was isolated in good yield from such hydrolysates of the enzyme and of the isolated labeled light subunit.

These studies with the chloroketone provided the first clear-cut evidence that the glutamine-dependent and ammonia-dependent activities of a glutamine amidotransferase can be separately affected. Certain earlier observations in which unexpected variability was found in the relative rates of reaction with ammonia and glutamine (see Meister, 1962) may now probably be explained in terms of effects on the glutamine binding site or subunit.

Properties of the Glutaminase Activity of the Enzyme

When the enzyme is incubated with [^{14}C] L-glutamine in the absence of other substrates, a labeled enzyme complex is

450

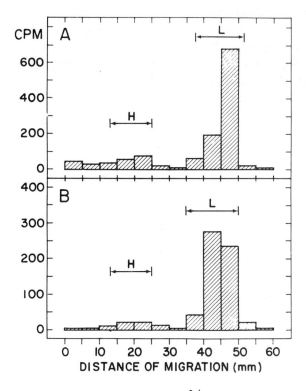

Fig. 9 Separation of the subunits of the [^{14}C]-labeled enzyme by gel electrophoresis in 0.1% SDS-100 mM Tris-acetate (pH 7.0). A, the enzyme was labeled with 0.1 mM chloroketone. The binding of chloroketone was 1.8 moles per monomer; the enzyme was 85% inhibited (glutamine-dependent activity). The inhibited enzyme (130 ug) was then applied to the gel in 50 ul of 100 mM sodium phosphate buffer (pH 7.0), and electrophoresis was carried out for 2.5 hours at 4 ma per gel (volume, 1.2 ml) and 75 volts until the dye marker emerged. The total radioactivity applied to the gel was 1570 cpm; 1240 cpm (79%) were recovered from the gel. 1050 cpm were in the regions which stained for protein; i.e., those marked H and L. Distribution of protein bound radioactivity: light subunit (L), 87%; heavy subunit (H), 13%. B, the enzyme was labeled as above but in the presence of ATP, Mg^{++}, and HCO$_3^-$. The binding of chloroketone was 0.9 mole per monomer; the enzyme

was 55% inhibited. The conditions of electrophoresis were the same as in A. The total radioactivity applied to the gel was 1060 cpm; 765 cpm (72%) were recovered. Distribution of protein bound radioactivity: light subunit (L), 88%; heavy subunit (H), 12%. (Pinkus and Meister, 1972).

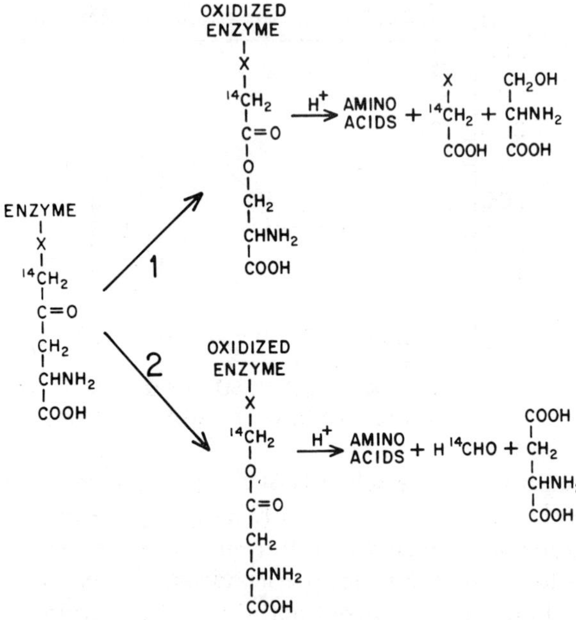

Fig. 10 Possible pathways for the reaction of the chloroketone-treated enzyme with performic acid (see the text). (Pinkus and Meister, 1972).

formed, which contains the equivalent of one mole of glutamine per mole of enzyme. The bound label does not exchange at an appreciable rate with added unlabeled glutamine, and it reacts readily to yield glutamate when ATP, Mg^{++}, and bicarbonate are added (Wellner et al., 1972). Denaturation of the labeled enzyme with trichloroacetic acid leads to release of the radioactivity as glutamate and glutamine. The available data are consistent with the intermediate formation of a γ-glutamyl-enzyme. As stated previously, the enzyme catalyzes glutamine hydrolysis at about 2% of the rate for glutamine-dependent carbamyl phosphate synthesis, and enzyme treated with the chloroketone does not exhibit any glutaminase activity.

In the course of studies on the glutaminase activity of the enzyme the very interesting observation was made that storage of the enzyme at 4° at pH 9 led to a very substantial increase (about 40-fold) in its glutaminase activity. It was subsequently found that treatment of the enzyme with 5,5'-dithiobis (2-nitrobenzoate) or N-ethylmaleimide also produced striking activation of the glutaminase activity (about 200-fold; Figure 11); this was accompanied by loss of the synthetase activity. Since N-ethylmaleimide and p-hydroxymercuribenzoate are potent inhibitors of the synthesis reaction, we sought to learn whether these compounds could interact with the cysteine residue at the glutamine binding site. It was found that glutamine did not protect against inhibition by these reagents whereas as described above glutamine does protect the enzyme completely against inhibition by the chloroketone.

The enhanced glutaminase activity of the N-ethylmaleimide-treated enzyme is markedly inhibited by treatment with the chloroketone, and this effect of the chloroketone can also be diminished by addition of glutamine. Studies with [^{14}C] chloroketone showed that close to 1 mole of chloroketone binds to the N-ethylmaleimide-treated enzyme, and that more than 90% of the radioactivity is bound to the light subunit (Pinkus and Meister, 1972). These studies indicate that N-ethylmaleimide does not inhibit the glutaminase activity, nor does it bind to the sulfhydryl group at the glutamine binding site; this group may therefore be considered to be buried. These studies indicate that sulfhydryl group modified by reaction with the chloroketone is of special significance in the binding and utilization of glutamine; as discussed elsewhere (Li and Buchanan, 1971; Pinkus and Meister, 1972),

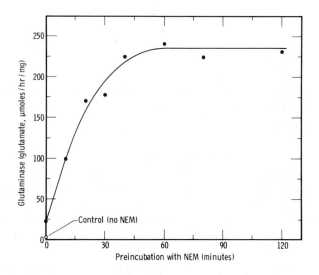

Fig. 11 Effect of N-ethylmaleimide on the glutaminase activity. The enzyme (64 μg/ml) was preincubated at 37° with 10 mM N-ethylmaleimide in 100 mM potassium phosphate-100 mM Tris, pH 7.5. At various intervals glutaminase activity was assayed in 0.1 ml of a solution containing 50 μg/ml enzyme, 10 mM glutamine [^{14}C] (50,000 cpm/μmole), 7.5 mM N-ethylmaleimide, and 100 mM phosphate Tris, pH 7.5; 37°, 15 min. The control point was obtained under the same conditions but in the absence of N-ethylmaleimide. The reaction was stopped by acidification with HCl and the solution was immediately placed on ice. The amount of [^{14}C] glutamate produced was quantitated after separation from [^{14}C] glutamine by paper electrophoresis. (Wellner et al., 1972).

this functional group may be of general significance in the mechanisms of action of the glutamine amidotransferases. The remarkable enhancement of the enzyme's glutaminase activity by treatment with N-ethylmaleimide and other sulfhydryl reagents

454

(Pinkus et al., 1972) may probably be ascribed to conformational changes triggered by the modification of the other sulfhydryl groups in the enzyme. The result is a selective inactivation of the heavy subunit, while the glutamine binding site on the light subunit is maintained. The relationship between the heavy and light subunits may be altered in such a manner as to facilitate the reaction of water with the enzyme-bound glutamyl moiety.

In the course of these studies it was found that treatment of the enzyme with dithiothreitol led to an enzyme preparation that is in many ways similar to the chloroketone-treated enzyme; thus, the glutaminase activity of carbamyl phosphate synthetase can readily be destroyed by incubation with a low concentration of dithiothreitol (Figure 12). Such sensitivity to dithiothreitol is also

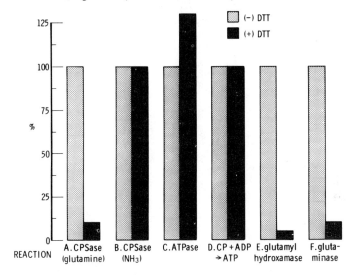

Fig. 12 The effect of prolonged incubation of the enzyme with dithiothreitol on the various reactions catalyzed by carbamyl phosphate synthetase. The enzyme (0.2 mg/ml) was allowed to stand for 16 hours in 150 mM potassium phosphate, 0.5 mM EDTA, pH 7.8 at 23° in the presence and absence of 1.5 mM dithiothreitol. Assays were then performed for the various activities by procedures already described (Trotta et al., 1971b). The activity of the control (no dithiothreitol) is arbitrarily set at 100% for each reaction.

455

characteristic of the two other reactions catalyzed by the enzyme which involve glutamine or a glutamine analog; namely, the glutamine-dependent synthesis of carbamyl phosphate and the hydrolysis of γ-glutamylhydroxamate. On the other hand, the synthesis of carbamyl phosphate from ammonia and the synthesis of ATP from ADP and carbamyl phosphate are unaffected by dithiothreitol while the bicarbonate-dependent ATP-ase is some-what activated. The most straight forward interpretation of these results is that there is a disulfide bond in the light subunit which is critical for maintaining the integrity of the glutamine binding site.

The time course of the dithiothreitol inhibition of the glutamine-dependent carbamyl phosphate synthesis is unusual in that it exhibits two phases: activation followed by inhibition (Figure 13).

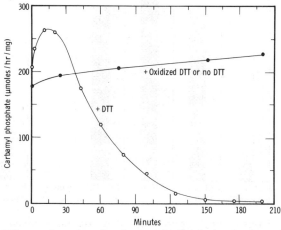

Fig. 13 Effect of preincubation with dithiothreitol on the gluta-mine-dependent carbamyl phosphate synthetase. The enzyme (0.54 mg/ml) was incubated with 1.0 mM dithiothreitol in 150 mM potassium phosphate, 0.5 mM EDTA; pH 7.8 at 37°. 0.02 ml ali-quots were assayed by quantitating the amount of [^{14}C] carbamyl phosphate produced from NaH^{14}CO$_3$ (30,500 cpm/µmole) (Anderson and Meister, 1965b). (Trotta et al., 1972b).

The initial rapid activation suggests that another disulfide bond may be present, whose reduction leads to an increase in activity. It is conceivable that the loss in activity observed shortly after purification of the enzyme may be related to the formation of this disulfide.

Hydrodynamic Properties of the Isolated Enzyme

Carbamyl phosphate synthetase is usually isolated and stored in potassium phosphate buffer. A sedimentation velocity study in this buffer (Figure 14, curve c) showed that the sedimentation coefficient increased with increasing protein concentration, indicating self-association. This equilibrium was rapid compared with the time of the experiment since only one boundary was observed over a protein concentration range of 0.1 to 20 mg/ml. Qualitatively similar observations were made by Anderson and Marvin (1968; 1970) using sucrose gradient centrifugation. The addition of low concentrations of urea (0.2–2.0 M) or of guanidinium hydrochloride (0.05–0.5 M) in potassium phosphate shifts the equilibrium toward monomer. In the absence of potassium phosphate (e.g., in Veronal–sodium chloride or Tris–sodium chloride buffer), a non–self–associating species with a symmetrical boundary is produced which has a sedimentation coefficient that is essentially concentration independent (Figure 14, curve d; $S_{20,w}^{o}$ = 7.3 S). The absence of both potassium and phosphate is required for the production of a non–associating monomer: in the presence of either of these ions alone, there is still a residual tendency toward self–association. Furthermore, in the absence of potassium phosphate, urea (or guanidinium hydrochloride) has no effect on the sedimentation coefficient at the reagent concentrations indicated above.

The addition of low concentrations of ammonia (Figure 14, Curve b) or ornithine (Figure 14, Curve a) to the enzyme in potassium phosphate increases the sedimentation coefficient to a maximum of about 15 S at high protein concentration. The effect of ornithine has also been observed in sucrose density centrifugation by Anderson and Marvin (1968;1970), who also reported that the allosteric activator IMP increased the sedimentation coefficient,

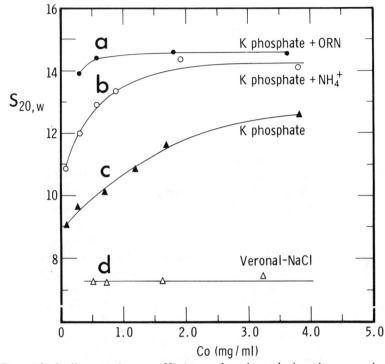

Fig. 14 Sedimentation coefficient of carbamyl phosphate synthetase as a function of protein concentration under various solvent
conditions. Curve a: 150 mM potassium phosphate, 0.5 mM EDTA,
20 mM ornithine; pH 7.8; Curve b: 150 mM potassium phosphate,
0.5 mM EDTA, 20 mM ammonium chloride; pH 7.8; Curve c:
150 mM potassium phosphate, 0.5 mM EDTA; pH 7.8; Curve d:
30 mM Veronal,   0.5 mM EDTA, 100 mM
sodium chloride; pH 7.6. These experiments were performed in a
model E analytical ultracentrifuge equipped with the RTIC unit,
electronic speed control, and photoelectric scanning system. 30-,
12-, and 3-mm path length double-sector cells were used; scanning
was done at 280 nm. Sedimentation coefficients were measured
from the rate of movement of the 50% position of apparent
boundaries with time; these values were in good agreement with
those obtained from weight-average calculation.

while the allosteric inhibitor UMP diminished it. The increased
sedimentation coefficient of carbamyl phosphate synthetase in the
presence of ammonia prompted us to investigate a possible role for
ammonia as an allosteric regulator (Figure 15). The results

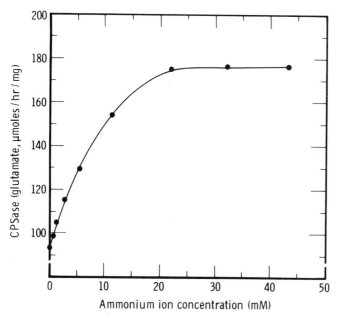

Fig. 15 The effect of ammonium ion on the glutamine-dependent
carbamyl phosphate synthetase. 0.02 ml enzyme (0.21 mg/ml) was
assayed (10 minutes at 37°) in a final volume of 0.3 ml containing
5 mM ATP, 5 mM $MgCl_2$, 20 mM $NaHCO_3$, 20 mM glutamine,
105 mM KCl, 60 mM Tris-HCl; pH 7.7. The reaction was stopped
by adding 0.1 ml of 1 N HCl. The solution was immediately
placed on ice for 10 minutes and then neutralized with 0.1 ml of
1 M Tris. The amount of glutamate formed was determined by the
DPN-glutamate dehydrogenase system   (Bernt and Bergmeyer,
1963). (Trotta et al., 1972a).

indicate that low concentrations of ammonia stimulate the glutamine-dependent synthesis of carbamyl phosphate. Ammonia can
therefore apparently serve dual functions, i.e., both substrate
and effector.

459

In general, it can be stated that those factors which shift the equilibrium toward 7.3 S monomer (UMP, absence of potassium ions, the presence of urea or guanidinium–HCl) inhibit enzymatic activity, while those which favor higher states of association stimulate enzymatic activity (potassium ions, ATP + Mg^{++}, ornithine, IMP, ammonia). The only exception thus far noted to the activation–association relationship is the finding that the absence of phosphate promotes monomer formation but does not produce significant decrease in activity (Trotta et al., 1972). This suggests that phosphate may promote a conformational state which is different from that which exists in the presence of other modifiers of the state of association; such a conformational alteration is supported by the observation that addition of phosphate diminished the extent of allosteric activation by ornithine and actually abolished that due to IMP (Anderson and Marvin, 1970).

Carbamyl phosphate synthetase was characterized by sedimentation equilibrium in a number of different solvent systems. The results confirm and extend the interpretation of sedimentation velocity data cited above. The results of these experiments can be summarized as follows:

(1) In a Veronal–sodium chloride buffer the enzyme is a highly homogeneous monomer by the criterion of sedimentation equilibrium (Figure 16). The weight average molecular weight under these conditions is 163,000 ± 4,000 based upon the assumption that the partial specific volume is 0.734, as calculated from the amino acid composition.

(2) In potassium phosphate the enzyme exists in an equilibrium which is predominantly monomer–dimer. When either potassium or phosphate is removed from the system, the equilibrium still exists but is shifted toward monomer. The removal of phosphate has a greater effect on the equilibrium than the removal of potassium.

(3) The addition of ADP + $MgCl_2$ promotes dimer formation both in the presence and absence of potassium phosphate.

460

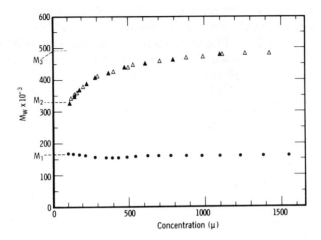

Fig. 16 Sedimentation equilibrium studies on carbamyl phosphate synthetase. The apparent weight average molecular weight as a function of protein concentration (fringe displacement was analyzed from interference data with the aid of the computer program of D. Roark (1971). The data were obtained under the following conditions: ▲ = 150 mM potassium phosphate, pH 7.8, 0.5 mM EDTA, 20 mM ornithine; 21.9°, 12,000 rpm, initial loading concentration: 0.25 mg/ml. △ = 150 mM potassium phosphate, pH 7.8, 0.5 mM EDTA, 20 mM ornithine; 8.4°, 8000 rpm, initial loading concentration: 1.17 mg/ml. ● = 30 mM Veronal, pH 7.6, 0.5 mM EDTA, 100 mM NaCl; 20.4°, 19,978 rpm; initial loading concentration: 0.25 mg/ml. (Trotta et al., 1972a).

Thus, the binding of at least one ADP shows no absolute requirement for potassium phosphate.

(4) The addition of ornithine to the enzyme in potassium phosphate shifts the equilibrium strongly toward oligomeric species

461

(Figure 16). At low concentration (0.1 mg/ml) some monomer is still present, although dimer predominates. The equilibrium must be interpreted by including oligomeric species corresponding to tetramer or higher. Curiously, the concentration dependence of the weight average molecular weight at the two temperatures indicated (8.4° and 21.8°) is identical. The interpretation of these data in terms of thermodynamic parameters must await additional data obtained at other temperatures.

Enzymatic and Hydrodynamic Properties of the Non-Identical Subunits

The 7.3 S monomer can be dissociated into its constituent polypeptide chains by high concentrations of urea (8 M), by sodium dodecylsulfate (SDS), or by succinic or maleic anhydrides. Under all of these conditions two bands were seen on gel electrophoresis; these exhibited different molecular weights and an approximate staining ratio of about 3 to 1 (Figure 17). SDS gel electrophoresis with appropriate molecular weight markers gave estimates of 42,000 and 130,000 for the molecular weights of the light and heavy subunits, respectively. The sum of these molecular weights (172,000) is in reasonable agreement with the molecular weight obtained from sedimentation equilibrium (163,000) for the enzyme in monomer form. Direct evidence for a one to one ratio of heavy to light subunit in the monomer was obtained by crosslinking the enzyme in a Veronal-sodium chloride buffer with the bifunctional reagent dimethylsuberimidate, (Davies and Stark, 1971). SDS gel electrophoresis of the cross-linked species produced a third band in addition to the light and heavy subunits corresponding in molecular weight to about 170,000 (Figure 17); cross-linked aspartate β-decarboxylase was employed as a molecular weight marker. In contrast, cross-linking of the enzyme in the presence of potassium phosphate and the allosteric activator IMP followed by SDS gel electrophoresis produced in addition to the light and heavy chain bands six other bands which correspond to species containing various light and heavy chain combinations derived from a dimer.

Under the conditions given above for dissociation of the

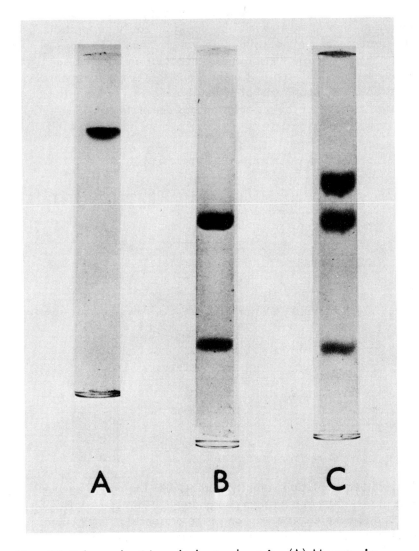

Fig. 17 Polyacrylamide gel electrophoresis. (A) Monomeric carbamyl phosphate synthetase was run in a 6% gel; (B) The SDS-dissociated enzyme was run in a 5% gel that contained 0.1% SDS; (C) Enzyme (1.5 mg/ml) was first cross-linked with dimethyl-suberimidate (4.0 mg/ml), and was then treated as in (B). (Trotta et al., 1971b).

chains no significant reversibility could be easily demonstrated. In
the presence of 1 M potassium thiocyanate, and 0.1 potassium
phosphate (pH 7.6), there was also dissociation into light and
heavy subunits (Figure 18, Curves B and C). Under these conditions however, upon removal of the thiocyanate by dialysis, substantial reversal of the dissociation was demonstrated (Figure 18,
Curve D).

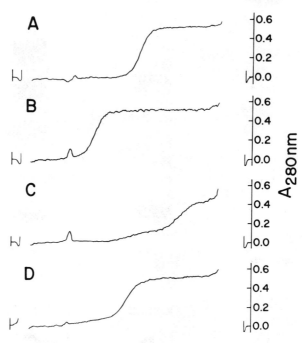

Fig. 18 Sedimentation velocity studies. (A) The native enzyme
was run in 150 mM potassium phosphate-0.5 mM EDTA (pH 7.8);
(B) and (C) The enzyme was run in the presence of 1.0 M potassium
phosphate (pH 7.6); (D) The enzyme was first treated with potassium thiocyanate for 2 hr. at 4° under the buffer conditions given
in (B), and was then dialyzed against the buffer given in (A). The
photoelectric scanning patterns (at 60,000 rpm and 4.8°) were
obtained 44, 42, 238, and 46 min. after reaching speed,
respectively. (Trotta et al., 1971b).

This reversibility in physical state was also accompanied by a sub-stantial restoration of enzymatic activity. There is observed, however, a time-dependent loss in the synthetase activity after exposure of the native enzyme to potassium thiocyanate followed by dialysis to remove it. The specific cause for this loss is not clear, but it may be relevant that the original specific activity can be fully restored from the isolated subunits only when a four-fold molar excess of light subunit is used. This result suggests that during exposure to thiocyanate damage to the light chain may occur. However, it has also been shown that during the dissocia-tion procedure thiocyanate treatment affects the heavy chain as well (see below).

A Sephadex G-200 column equilibrated in 1 M potassium thiocyanate was employed for the separation of the subunits. Typically, three peaks were obtained (Figure 19): an aggregate in the excluded volume, which is composed exclusively of heavy subunit, and two additional peaks corresponding to the separated heavy and light subunit (Figure 20). It was observed that the isolated heavy subunit activities are unstable in the absence of dithiothreitol, while in contrast the light chain glutaminase activity is destroyed by mercaptans. Dithiothreitol was therefore added only to those fractions containing the heavy subunit. Similarly, when thiocyanate was removed by dialysis, the light subunit was dialyzed against potassium phosphate, while the heavy subunit was dialyzed against the same buffer containing dithio-threitol.

Table I summarizes the influence of allosteric effectors on the synthesis of ATP from ADP and carbamyl phosphate catalyzed by the isolated heavy subunit. It is clear that at the low concentration of ADP + $MgCl_2$ employed (3 mM each), ornithine, IMP, and ammonia activate while UMP inhibits. Orni-thine is the most potent of the activators, followed by IMP and then ammonia. This order of effectiveness in enhancing the bind-ing of ADP is the same as that observed in the native enzyme. There is strong evidence therefore that all of the binding sites for the allosteric effectors are located on the heavy subunit.

465

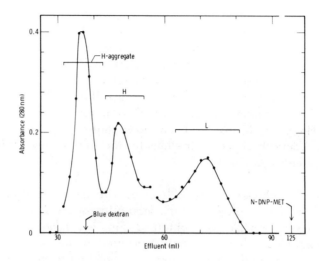

Fig. 19 Sephadex G-200 chromatography of the subunits of carbamyl phosphate synthetase in potassium thiocyanate. The column (60 x 0.9 cm) was equilibrated with 1.0 M potassium thiocyanate, 5 mM EDTA, 100 mM potassium phosphate; pH 7.6 at 4°. 1.0 ml of enzyme (9.0 mg/ml; 150 mM potassium phosphate, 0.5 mM EDTA, pH 7.8) was applied. 0.9 ml fractions were collected; the flow rate was approximately 4 ml/hr.

Fig. 20 SDS-polyacrylamide gel electrophoresis of the (A) heavy and (B) light subunits of carbamyl phosphate synthetase isolated by Sephadex G-200 gel chromatography. (Trotta et al., 1971b).

466

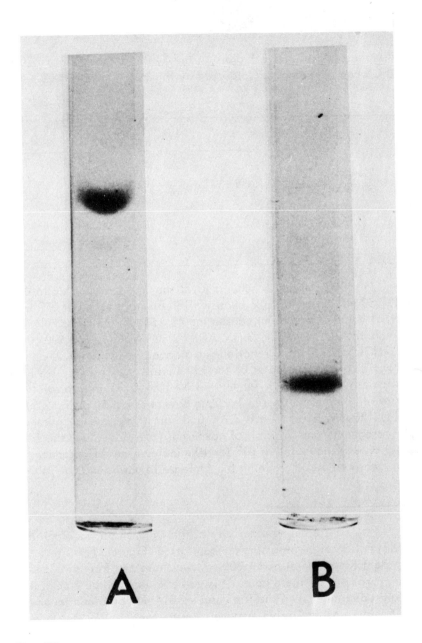

Fig. 20

TABLE I

Influence of Allosteric Effectors on the Synthesis of ATP from ADP and Carbamyl Phosphate Catalyzed by the Heavy Subunit*

Effector Added	(μmoles ATP produced/hr/ mg) $\times 10^2$
None	333
UMP	117
Ammonium ion	576
IMP	1530
Ornithine	2857

*0.01 ml heavy subunit (0.55 mg/ml, 2 mM dithiothreitol, 150 mM potassium phosphate, pH 7.8) was assayed at 37° in a final volume of 0.16 ml containing 40 mM lithium carbamyl phosphate, 100 mM KCl, 3 mM ADP, 3 mM $MgCl_2$, and 100 mM Tris-HCl, pH 7.5. The reaction was stopped by adding 0.05 ml of 1 N HCl; after standing at 0° for 15 minutes, the solution was neutralized by addition of 0.05 ml of 1 M Tris. 0.25 ml of a solution containing 40 mM glucose, 1.25 mg bovine serum albumin, 2 mM TPN, 100 mM Tris-HCl (pH 7.6), 0.8 unit of glucose-6-phosphate dehydrogenase and 0.8 unit of hexokinase was added and the solution was incubated at 37° for 10 minutes. The ATP produced was measured by determining the increase in absorbance at 340 nm.

The glutaminase activity of the light chain was studied as a function of L-glutamine concentration (Figure 21). An apparent Km value of about 300 mM was obtained from the double reciprocal plot of these data. This value is about three orders of magnitude higher than the Km value of 0.4 mM obtained for the glutaminase of the native enzyme (Wellner et al., 1972). Upon

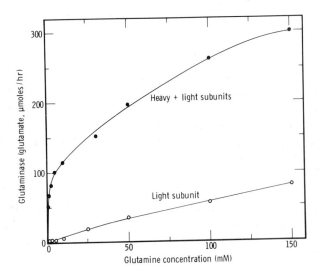

Fig. 21 The glutaminase activity of the isolated light subunit and of the reconstituted enzyme as a function of glutamine concentration. The assay was carried out at 37° in a final volume of 0.12 ml containing 150 mM potassium phosphate, pH 7.8, 0.5 mM EDTA, 0.67 mM dithiothreitol. The reaction was stopped with 0.05 ml of 1 N HCl. The solution was immediately placed on ice for 10 min. and then neutralized with 0.05 ml of 1 M Tris. The amount of glutamate produced was determined by the DPN-glutamate dehydrogenase system. (Bernt and Bergmeyer, 1963).

addition of an excess of heavy subunit to the light subunit, the glutaminase activity was stimulated at all concentrations of glutamine. Two phases are discernible: a sharp rising phase between 0 and 5 mM glutamine followed by a more slowly rising phase. A biphasic response has also been observed with preparations of the isolated enzyme. The glutaminase activity exhibiting the higher Km value may reflect a population of altered light subunits present in the purified enzyme.

When the light subunit was treated with a low concentration of the chloroketone, there was a dramatic inhibition of glutaminase activity, which was partially protectable by high concentrations of albizziin (Figure 22). Failure to demonstrate complete protection may probably be ascribed to the fact that the albizziin concentration was nonsaturating, since the Km for glutamine in the light subunit is very high. About 0.8–1.0 mole of [^{14}C] chloroketone reacted per mole of light subunit (molecular weight:42,000); this result implies a high degree of specificity. Dithiothreitol also inhibited the glutaminase. This inhibition is presumably related to reduction of the disulfide bond which causes the intact enzyme to lose its glutamine related functions (Figure 12).

In the presence of 150 mM potassium phosphate (pH 7.8), the isolated light and heavy subunits exhibit sedimentation coefficients of 2.9 S and 7.6 S, respectively. In the absence of potassium phosphate (e.g., in Veronal–sodium chloride buffer), the sedimentation coefficient of the light subunit was virtually unaffected. The sedimentation coefficient of the heavy subunit, however, was decreased to about 4.9 S. The latter value is similar to that obtained for the heavy chain in the presence of potassium phosphate and potassium thiocyanate (Trotta et al., 1971). From the relationship $(S_1/S_2) = (M_1/M_2)^{2/3}$ where $S_1 = 7.3$ S, $M_1 = 170,000$, and $M_2 = 130,000$ and assuming no change in shape or degree of hydration between the heavy subunit and the monomer,

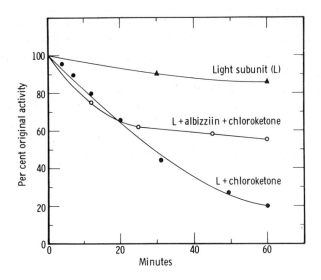

Fig. 22 Effect of L-2-amino-4-oxo-5-chloropentanoic acid on the glutaminase activity of the light subunit. The light subunit (0.53 mg/ml; 150 mM potassium phosphate, 0.5 mM EDTA, pH 7.8) was incubated with 0.1 mM chloroketone at 37° both in the presence and absence of 200 mM albizziin. Light chain alone was also incubated at 37° in the same buffer. At various times 0.02 ml aliquots were withdrawn for glutaminase assay in a final volume of 0.1 ml containing 180 mM glutamine, 150 mM potassium phosphate, 0.5 mM EDTA, pH 7.8. After 40 minutes at 37° the reaction was stopped by adding 0.05 ml of 1 N HCl and immediately placed on ice for 10 minutes. After neutralization with 0.05 ml of 1 M Tris, the amount of glutamate formed was measured with the DPN-glutamate dehydrogenase system (Bernt and Bergmeyer, 1963). The zero time values were determined on aliquots assayed immediately after addition of the chloroketone. These values reflected an inhibition occurring during the assay of 20% for the chloroketone alone and of 44% for the chloroketone plus albizziin. For each of the curves, the data were normalized to the corresponding zero-time activity, which was set at 100%. (Trotta et al., 1972b).

471

the predicted sedimentation coefficient of the non-associated heavy subunit is 6.1 S. The difference between this value and the observed result (4.9 S) implies that the isolated heavy subunit in Veronal-sodium chloride buffer is markedly different in conformation (and/or hydration) from the heavy subunit in the monomer. A sedimentation coefficient of 7.6 S for the heavy subunit in potassium phosphate is consistent with the hypothesis that it is a dimer of the 4.9 S form. The isolated light subunit exhibits a sedimentation behavior expected for a non-associating subunit which has undergone negligible conformational (and/or hydrational) alteration.

DISCUSSION

Although they synthesize carbamyl phosphate for
different biosynthetic pathways, carbamyl phosphate synthetases
I and II exhibit a number of common features; most notably, the
catalysis of an overall synthesis reaction and a number of partial
reactions which exhibit the same stoichiometry and which have
similar substrate requirements. The studies discussed here indicate
additional possible points of similarity in relation to size and
subunit structure. Marshall and co-workers (1961) reported a
value of 315,000 for the molecular weight of frog liver carbamyl
phosphate synthetase I, a value which is virtually identical with
that reported for rat liver carbamyl phosphate synthetase I (Virden,
1972). The dimer molecular weight for the E. coli synthetase II
studied here is about 2 x 163,000 or 326,000. The rat liver enzyme
may also dissociate to a 7.5 S form (molecular weight: 160,000),
a monomer which is in a chemical equilibrium with an 11 S dimer
in the presence of potassium-enriched Tris buffer. Centrifugation
in a sucrose gradient showed a single peak, with protein and
activity coincident, which displayed an increase in sedimentation
coefficient with increasing protein concentration, as is characteristic of rapid equilibrium system. These observations are analogous
to those presented here for E. coli carbamyl phosphate synthetase;
i.e., the finding of a 7.3 S form (molecular weight: 163,000) in
equilibrium with the dimer in potassium phosphate buffer. It is
also of interest that Cohen (1970) cites unpublished evidence for
the presence of non-identical chains in the frog liver synthetase I.
It is tempting to speculate that in the latter enzyme N-acetylglutamate may bind to a separate subunit, distinct from the catalytic
synthetase subunit, which may bear an evolutionary relationship
to the glutamine-binding light subunit reported here.

Genetic and biochemical evidence can be found in the
literature that indicates the presence in other glutamine amidotransferases of a subunit designed for the binding of glutamine. The
most extensively studied enzyme is anthranilate synthetase, which
has been shown in a variety of micoorganisms to consist of two
non-identical polypeptide chains which have been designated as
components I and II (Ito and Yanofsky, 1966). Component I alone

473

can utilize ammonia, but not glutamine, for the synthesis of anthranilate from chorismate; addition of component II restores the capacity to utilize glutamine. In Salmonella typhimurium, 6-diazooxonorleucine (DON) a glutamine analog, irreversibly inhibits the glutamine-dependent anthranilate synthetase, while the ammonia-dependent activity is relatively unaltered; L-glutamine protects against the inactivation (Nagano et al., 1970). When the native enzyme was reacted with [^{14}C]·DON and then dissociated into subunits in 8 M urea, 1 mole of [^{14}C]·DON was incorporated into a sulfhydryl group on each chain of component II. These results strongly indicate the glutamine and ammonia binding sites are located on separate polypeptide chains. It is also relevant that Huang and Gibson (1970) have reported the isolation from two different E. coli mutants of two components with differing molecular weights (48,000 and 9,000, respectively) which are both required for the glutamine-dependent synthesis of 4-aminobenzoate from chorismate; neither species is active alone. Evidence has been presented in work on B. subtilis that a small subunit (molecular weight: 16,000) functions in glutamine binding and is required for both anthranilate synthetase and 4-aminobenzoate synthetase reactions (Kane and Jensen, 1970; Kane et al., 1972). A mutation in trp-X, whose gene product is the glutamine binding subunit, preferentially reduces the level of the glutamine-dependent anthranilate synthetase without affecting the ammonia-dependent reaction. The fact that addition of 4-aminobenzoic acid reversed the tryptophan inhibition (tryptophan inhibited the growth of the mutant by repressing the synthesis of an already deficient glutamine subunit) implicates the subunit in 4-aminobenzoic acid synthesis. It is intriguing to consider the possibility that the same light subunit may be shared in common with a number of glutamine amidotransferases. However, Kane and associates (1972) presented evidence that the glutamine binding subunit coded for by trp-X is not present in cytidine triphosphate synthetase or in the histidine biosynthetic route. It is possible, nevertheless, that these enzymes did possess at one time a separate low molecular weight glutamine binding subunit, and that it was lost as a discrete subunit by the fusion of its gene with that of a primitive ammonia-utilizing enzyme . Evidence for gene fusion has been reported in other enzyme systems

(Kohno and Yourno, 1971; Yourno et al., 1970; Bonner et al., 1965). This hypothesis requires further study; primary sequence comparisons and immunological cross reactivity studies between glutamine binding subunits and various glutamine amidotransferases would clearly be of interest.

If the E. coli carbamyl phosphate synthetase is derived from a combination of the ammonia-dependent synthetase and a glutaminase, the question arises as to whether the latter is the same as, or similar to, the other E. coli glutaminases which have been described. The glutaminase A described by Hartman (1968) has very different properties from the light subunit glutaminase; most notably a specific activity a number of orders of magnitudes higher, a pH optimum of 5.0 (with virtual inactivity at pH values greater than 5.4) and a larger subunit molecular weight (55,000). A second glutaminase, B, active at pH 7 and above, has also been described (Prusiner and Stadtman, 1971). This enzyme, however, is notably different from the light subunit described here in that it exhibits a much lower K_m value for glutamine (2.6 mM) that that of the light subunit (300 mM). Further comparison of the two systems requires more data on their properties.

Although the hydrolysis of glutamine and the utilization of the nitrogen thus released occur on separate polypeptide chains, in carbamyl phosphate synthetase these events are tightly coupled since only 1 mole of glutamate is produced for every mole of carbamyl phosphate formed (Anderson and Meister, 1965b). In addition, substantial evidence exists that the binding of a substrate or substrate analog to one subunit can produce alterations in the other subunit which may reflect functional interrelationships in the synthesis reaction of the native enzyme. Examples of this phenomenon include the following: 1. The addition of ATP + Mg^{++} and bicarbonate dramatically stimulates the hydrolysis of γ-glutamylhydroxamate (about 200 fold) (Anderson and Meister, 1966a). However, it is important to note that catalytic rather than stoichiometric amounts of ATP are hydrolyzed to ADP under these conditions. The formation of "activated carbon dioxide" postulated to be an enzyme-bound mixed anhydride between carbonic and phosphoric acids (Anderson and Meister, 1965b) apparently triggers a conformational alteration which "turns on" the hydrolysis of γ-glutamylhydroxamate on the light subunit. This type of phenomenon is presumably operative in the normal catalytic mechanism with glutamine as substrate since the specific activity for glutamine cleavage in the absence of ATP + Mg^{++} and bicarbonate is only about 2% of the specific activity of the overall synthesis reaction. 2. Modification of the enzyme with the chloroketone, a glutamine analog, stimulates the bicarbonate-dependent ATPase activity about three-fold (Khedouri et al., 1966) and decreases the apparent Km value for ammonia (Pinkus et al., 1972). This effect is apparently due to the specific binding of the chloroketone to the light subunit since it is glutamine-protectable. Likewise, ATP, which binds to the heavy subunit, partially protects against chloroketone inactivation of the gluta-mine site on the light subunit. 3. The presence of ATP-Mg and bicarbonate exposes a sulfhydryl group which reacts with N-ethyl-maleimide to cause a 50% loss in both of the glutamine-dependent carbamyl phosphate synthesis and ATP synthesis (Foley et al., 1971). This group is reported to be located on the light subunit (Matthews and Anderson, 1972).

Another aspect of heavy-light subunit interaction is the observation that the addition of heavy subunit strongly stimulates the light subunit glutaminase. This is apparently associated with decrease of the Km value for glutamine since binding of glutamine to the isolated light subunit is not readily demonstrated by ^{14}C-glutamine binding experiments until the heavy subunit is added (Wellner and Meister, 1972). This increase in substrate affinity can be viewed either as a conformational alteration in the light subunit triggered by interaction with the heavy subunit, or that a portion of the glutamine molecule binds to a site on the heavy subunit.

In relation to the latter possibility, one might speculate that the $-NH_2$ moiety of the amide group of glutamine binds to a region at or near the ammonia binding site on the heavy subunit. This arrangement would facilitate transfer of nitrogen across the intersubunit contact.

476

ACKNOWLEDGEMENT

The authors wish to thank Mrs. Kim Li, Miss Mary Ahland, Mrs. Patricia Leake, and Mr. Nathaniel Burrison for skillful technical assistance.

This research was supported by grants from the Public Health Service, National Institutes of Health, Bethesda, Md.

REFERENCES

Abd-El-Al, A. and Ingraham, J.L. (1969a), J. Biol. Chem., 244, 4033.

Abd-El-Al, A. and Ingraham, J.L. (1969b), J. Biol. Chem., 244, 4039.

Anderson, P.M. and Marvin, S.V. (1968), Biochem. Biophys. Res. Commun., 32, 928.

Anderson, P.M. and Marvin, S.V. (1970), Biochemistry, 9, 171

Anderson, P.M., Mathews, S.L., and Foley, R.E. (1970), Fed. Proc., 29, 400.

Anderson, P.M. and Meister, A. (1965a), Abstracts, 150th National Meeting of the American Chemical Society, p. 35C, Atlantic City, N.J.

Anderson, P.M. and Meister, A. (1965b), Biochemistry, 4, 2803.

Anderson, P.M. and Meister, A. (1966a), Biochemistry, 5, 3157.

Anderson, P.M. and Meister, A. (1966b), Biochemistry, 5, 3164.

Anderson, P.M., Wellner, V.P., Rosenthal, G.A. (1970),

Methods in Enzymol., 17A, 235.

Bach, S.J. and Smith, M. (1956), Biochem. J., 64, 417.

Bernt, E. and Bergmeyer, H.U. (1963), In "Methods of Enzymatic Analysis". (H.U. Bergmeyer, ed.), pp. 384-388. Academic Press, New York.

Bonner, D.M., DeMoss, J.A., and Mills, S.E. (1965), In "Evolving Genes and Proteins". (V. Bryson and H.J. Vogel, eds.), p. 305. Academic Press, New York.

Cohen, P.P. (1962), In "The Enzymes", Second Edition, 6, 477. (P.D. Boyer, H. Lardy, and K. Myrback, eds.), Academic Press, New York.

Cohen, P.P. (1970), Science, 168, 533.

Davies, G.E. and Stark, G.R. (1970), Proc. Natl. Acad. Sci. U.S., 66, 651.

Davis, R.H. (1965), Biochim. Biophys. Acta, 107, 44.

Davis, R.H. (1967), In "Organizational biosynthesis". (H.J. Vogel, J.O. Lampen, and V. Bryson, eds.), pp. 303-322. Academic Press, New York.

Duffield, P.H., Kalman, S.M., and Brauman, J.I. (1969), Biochim. Biophys. Acta, 171, 189.

Foley, R., Poon, J., and Anderson, P.M. (1971), Biochemistry, 10, 4562.

Guthohrlein, G. and Knappe, J. (1969), European J. Biochem., 8, 207.

Hager, S.E. and Jones, M.E. (1967a), J. Biol. Chem., 242, 5667.

Hager, S.E. and Jones, M.E. (1967b), J. Biol. Chem., 242, 5674.

Hartman, S.C. (1968), J. Biol. Chem., 243, 853.

Hassall, C.H. (1967), Org. React., 9, 73.

Hoogenraad, N.J., Levine, R.L., and Kretchmer, N. (1971), Biochem. Biophys. Res. Commun., 44, 981.

Huang, M. and Gibson, F., (1970), J. Bacteriol., 102, 767.

Issaly, I.M., Issaly, A.S., and Reissig, J.L. (1970), Biochim. Biophys. Acta, 198, 482.

Ito, K. and Uchino, H. (1971), J. Biol. Chem., 246, 4060.

Ito, K., Nakanishi, S., Terada, M., and Tatibana, M. (1970), Biochim. Biophys. Acta, 220, 477.

Ito, J. and Yanofsky, C. (1966), J. Biol. Chem., 241, 4112.

Jones, M.E. (1963), Science, 140, 1373.

Jones, M.E. (1965), Ann. Rev. Biochem., 34, 381.

Jones, M.E. and Lipmann, F. (1960), Proc. Natl. Acad. Sci. U.S., 46, 1194.

Kalman, S.M., Duffield, P.H., and Brzozowski, T. (1965), Biochem. Biophys. Res. Commun., 18, 530.

Kalman, S.M., Duffield, P.H., and Brzozowski, T. (1966), J. Biol. Chem., 241, 1871.

Kanazir, D., Barner, H.D., Flaks, J.G., and Cohen, S.S. (1959), Biochim. Biophys. Acta, 34, 341.

Kane, J.F., Holmes, W.H., and Jensen, R.A. (1972), J. Biol. Chem., <u>247</u>, 1587.

Kane, J.F. and Jensen, R.A. (1970), Biochem. Biophys. Res. Commun., <u>41</u>, 328.

Khedouri, E., Anderson, P.M., and Meister, A. (1966), Biochemistry, <u>5</u>, 3552.

Kohno, T. and Yourno, J. (1971), <u>246</u>, 2203.

Lacroute, F., Pierard, A., Grenson, M., and Wiame, J.M. (1965), J. Gen. Microbiol., <u>40</u>, 127.

Leuthardt, F. (1938), Z. Physiol. Chem., <u>252</u>, 238.

Levenberg, B. (1962), J. Biol. Chem., <u>237</u>, 2590.

Li, H.C. and Buchanan, J.M. (1971), J. Biol. Chem., <u>246</u>, 4713.

Lue, P.F. and Kaplan, J.G. (1969), Biochem. Biophys. Res. Commun., <u>34</u>, 426.

Lue, P.F. and Kaplan, J.G. (1970), Biochim. Biophys. Acta, <u>220</u>, 365.

Lyman, C.M., Ory, R.L., and Hood, D.W. (1954), Fed. Proc., <u>13</u>, 256.

Marshall, M., Metzenberg, R.L., and Cohen, P.P. (1958), J. Biol. Chem., <u>233</u>, 102.

Marshall, M., Metzenberg, R.L., and Cohen, P.P. (1961), J. Biol. Chem., <u>236</u>, 2229.

Mathews, S.L. and Anderson, P.M. (1972), Biochemistry, <u>11</u>, 1176.

Meister, A. (1962), In "The Enzymes", Second Edition, 6, 247. (P.D. Boyer, H. Lardy, and K. Myrback, eds.), Academic Press, New York.

Nagano, H., Zalkin, H., and Henderson, E.J. (1970), J. Biol. Chem., 245, 3810.

Nakanishi, S., Ito, K., and Tatibana, M. (1968), Biochem. Biophys. Res. Commun., 33, 774.

O'Neal, D. and Naylor, A.W. (1968), Biochem. Biophys. Res. Commun., 31, 322.

O'Neal, D. and Naylor, A.W. (1969), Biochem. J., 113, 271.

Ory, R.L., Hood, D.W., and Lyman, C.M. (1954), J. Biol. Chem., 207, 267.

Pierard, A. (1966), Science, 154, 1572.

Pierard, A. and Wiame, J.M. (1964), Biochem. Biophys. Res. Commun., 15, 76.

Pinkus, L. and Meister, A. (1972), J. Biol. Chem., in press.

Pinkus, L. M., Wellner, V.P., and Meister, A. (1972), Fed. Proc., 31, 1459.

Prusiner, S. and Stadtman, E.R. (1971), Biochem. Biophys. Res. Commun., 45, 1474.

Roark, D. (1971), Ph.D. dissertation, State University of New York (Buffalo).

Shoaf, W.T. and Jones, M.E. (1970), Biochem. Biophys. Res. Commun., 45, 796.

Tatibana, M. and Ito, K. (1967), Biochem. Biophys. Res. Commun., 26, 221.

Tatibana, M. and Ito, K. (1969), J. Biol. Chem., 244, 5403.

Tatibana, M. and Shigesada, K. (1972), Biochem. Biophys. Res. Commun., 46, 491.

Tramell, P.R., Campbell, J.W. (1970), J. Biol. Chem., 245, 6634.

Trotta, P.P., Haschemeyer, R.H., and Meister, A. (1971a), Fed. Proc., 30, 1058.

Trotta, P.P., Burt, M.E., Haschemeyer, R.H., and Meister, A. (1971b), Proc. Natl. Acad. Sci. U.S., 68, 2599.

Trotta, P.P., Estis, L.F., Haschemeyer, R.H., and Meister, A. (1972a), in preparation.

Trotta, P.P., Pinkus, L.M., and Meister, A. (1972b), in preparation.

Virden, R. (1972), Biochem. J., 127, 503.

Wellner, V.P., Anderson, P.M., and Meister, A. (1972), in preparation.

Wellner, V.P. and Meister, A. (1972), unpublished observation.

Williams, L.G., Bernhardt, S.A., and Davis, R.H. (1971), J. Biol. Chem., 246, 973.

Williams, L.G., and Davis, R.H. (1970), J. Bacteriol., 103, 335.

Yip, M.C.M. and Knox, W.E. (1970), J. Biol. Chem., 245, 2199.

Yourno, J., Kohno, T., and Roth, J.R. (1970), Nature, 228, 820.

A COMPARISON OF THE ORGANIZATION OF CARBAMYL-
PHOSPHATE SYNTHESIS IN SACCHAROMYCES CEREVISIAE
AND ESCHERICHIA COLI, BASED ON GENETICAL AND
BIOCHEMICAL EVIDENCES

A. Piérard, M. Grenson, N. Glansdorff and
J.M. Wiame

Laboratoire de Microbiologie de l'Université
Libre de Bruxelles,
Laboratorium voor Erfelijkheidsleer en
Microbiologie, Vrije Universiteit Brussel,
and Institut de Recherches du C.E.R.I.A.,
Bruxelles, Belgium.

Abstract

Mutations at two distinct loci (cpu and either cpaI
or cpaII) are required to abolish carbamylphosphate (CP)
synthesis in S. cerevisiae. The cpu locus codes for a carbamylphosphate synthetase which is related to the pyrimidine pathway. cpaII is the structural gene for the major
polypeptidic component of a second enzyme which belongs to
the arginine pathway. This component carries out the synthesis of CP from ammonia. A second polypeptide, product
of cpaI, is required for the physiological glutamine-dependent activity. Regulatory mutations affecting the synthesis of the arginine pathway carbamylphosphate synthetase
have been selected and found to delineate a negative control mechanism.
In E. coli, a single carbamylphosphate synthetase
supplies the arginine and pyrimidine pathways with CP
Seventy five mutations impairing the synthesis of CP have
been found to map in a single monocistronic locus. All
these mutations lead to defect in the heavy subunit of the
enzyme which catalyzes the synthesis of CP from ammonia.

————
Supported by grants from the "Fonds de la Recherche
Fondamentale Collective" and the "Nationale Fonds voor
Wetenschappelijke Onderzoek".

The reasons of the failure to select mutants defective in the light glutamine amidotransferase subunit of the enzyme are discussed. Cumulative repression of E. coli carbamyl-phosphate synthetase is found to involve the product of the arginine pathway regulatory gene, argR.

Introduction

Bacterial mutants exhibiting a simultaneous require-ment for arginine and pyrimidine have historical signifi-cance in that they provided the first evidence that the ar-ginine and pyrimidine pathways share a common precursor (Roepke, 1946). This precursor was later identified as carbamylphosphate (CP) and its function as the carbamylat-ing agent in the synthesis of citrulline and carbamylaspar-tate was demonstrated (Jones et al., 1955 ; Lowenstein and Cohen, 1956).

The regulation of transcarbamylation and its rela-tionships with CP biosynthesis have been reviewed recently (Wiame, Stalon, Piérard and Messenguy, 1972). In this paper, we shall discuss genetic data which were critical in our approach to the problem of CP biosynthesis and which complement the biochemical data obtained in this and in other laboratories.

1. The Structural Genes for Carbamylphosphate Synthetases

Whereas it is known since 1946 that a single mutation can lead to the requirement of both arginine and pyrimidine in E. coli, it is clear that in Saccharomyces cerevisiae (Lacroute et al., 1965) as well as in Neurospora (Davis, 1963), two mutations affecting distinct loci must occur in the same cell in order to provoke a double arginine-pyrimi-dine requirement. This simple observation has several im-portant implications with regard to the organization of synthesis and distribution of CP between the arginine and the pyrimidine biosynthetic pathways.

A. Number of genetic loci coding for carbamylphosphate synthetase.

In Saccharomyces cerevisiae : Mutations at three dis-tinct and unlinked genetic loci (cpu, cpaI and cpaII) affect the synthesis of carbamylphosphate synthetase

(Lacroute et al., 1965). None of these mutations alone is able to create a requirement in cells growing on minimal medium, but the combination of cpu with either cpaI or cpaII gives rise to a double arginine-pyrimidine requirement.

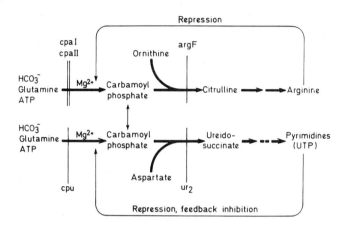

Fig. 1 Regulation of the synthesis of carbamylphosphate in Saccharomyces cerevisiae. Adapted from Lacroute et al., 1965.

Several kinds of evidence support the conclusion that the cpu locus is the structural gene for a carbamylphosphate synthetase (CPUse) integrated in the regulation of the pyrimidine biosynthetic pathway, while the cpa loci code for the two polypeptides forming another carbamylphosphate synthetase (CPAse) linked to the arginine biosynthetic pathway (Fig. 1).

- Although both types of mutants grow perfectly well on minimal medium, the growth of cpu mutants is inhibited by arginine, whereas pyrimidines prevent the growth of cpaI and cpaII mutants.

- Cell free extracts of cpaI and cpaII mutants catalyze the synthesis of CP from glutamine as the nitrogen donor. This activity is sensitive to feedback inhibition by UTP (Lacroute et al., 1965). Feedback insensitive mutants, isolated as fluorouracil resistant, are alleles of the cpu region (Lacroute, 1968). These observations

explain why the growth of the cpa mutants is inhibited by
pyrimidines and provide evidence that the cpu locus is the
structural gene for a pyrimidine pathway related carbamylphosphate synthetase (CPUse). The cpu locus is part of the
ura-2 genetic region coding for an enzyme aggregate which,
in addition to CPUse, contains aspartate carbamyltransferase (ATCase) activity and a UTP binding site. The latter
site controls the feedback inhibition of both enzymatic activities by UTP (Lacroute et al, 1965 ;Lacroute, 1968 ; Lue
and Kaplan, 1969 ; Denis-Duphil and Lacroute, 1971).

   - The CP synthesizing activity which is present in cpu
mutants also uses glutamine as the amino group donor but
does not seem to be sensitive to feedback inhibition. However, the repression of this activity by arginine (Table 1)
gives account of the growth inhibition of cpu mutants by
arginine. This activity depends on the integrity of the
cpaI and cpaII loci which show no close linkage with one
another, neither with cpu. Complementation between these
loci was demonstrated in vitro as well as in vivo (Lacroute
et al., 1965). Accordingly, cpaI and cpaII are clearly
characterized as structural genes coding for two polypeptidic chains of a carbamylphosphate synthetase belonging to
the arginine pathway (CPAse). We shall deal with the role
of these two polypeptides in a subsequent section of this
paper.

   Consequently, in yeast, the independence of the regulation of the arginine and pyrimidine pathways is preserved
by the use of two carbamylphosphate synthetases, each integrated in the regulation of one pathway. These enzymes
supply a single CP pool the size of which is regulated
according to the needs of both pathways. The organization
of CP synthesis in Neurospora is similar to that of yeast.
However, a major difference exists in the distribution of
CP : two distinct CP pools are maintained and channelled to
their respective pathways (Davis, 1967).

   In Escherichia coli : as mentioned above, single E.
coli mutations lead to a simultaneous requirement for arginine and pyrimidine. About fifty such mutations in E. coli
K12 have been localized by reciprocal transduction and
deletion mapping (Fig. 3). All map in the capA locus
which lies in the thr-leu segment of the E. coli chromosome.

It is not linked to any of the pyrimidine or arginine genes
(Piérard et al., 1965 ; Mergeay, 1969 ; Mergeay et al.,
1972). This locus was submitted to functional analysis. Two
types of complementation tests, namely abortive transduction and formation of stable merodiploids, which were performed between mutations located at both extremities of the
capA locus have failed to show more than one functional
unit in this locus (Mergeay et al., 1972).

The occurrence of a single CP generating system in E.
coli, suggested by the existence of the arginine-uracil
requiring mutants, has been confirmed by the data of the
biochemical analysis. These mutants lack a single carbamylphosphate synthetase (CPSase) which uses glutamine as
the nitrogen donor (Piérard and Wiame, 1964 ; Piérard et
al., 1965). The synthesis and the activity of this enzyme
are subject to elaborate control mechanisms which take
into account its double function of providing CP to the
arginine and pyrimidine pathways. The synthesis of CPSase
is repressed in a cumulative manner by arginine and pyrimidines (Piérard and Wiame, 1964) while its activity is under
the compensatory controls of UMP as inhibitor, and ornithine as activator (Piérard et al., 1965 ; Piérard, 1966 ;
Anderson and Meister, 1966b). As the synthesis of ornithine is itself under the control of arginine through feedback

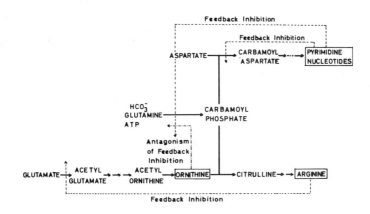

Fig. 2 Control of the enzyme activities in the
arginine and pyrimidine biosynthetic pathways of E. coli.
(Piérard, 1966).

inhibition of acetylglutamate synthetase (Vyas and Maas,
1963), an elegant way of modulating CP synthesis according
to the needs of both pathways is achieved (Fig. 2).In this
way CP synthesis is under the control of feedback inhibition by pyrimidine nucleotides only ; would the arginine
concentration become limiting, ornithine will reverse that
inhibition according to the CP need of the arginine pathway. In addition, IMP activation of CPAse establishes a
link between the purine and pyrimidine pathways (Anderson
and Meister, 1966b).

B. Informations provided by mutations affecting the nitrogen donor recognition sites.

The participation of glutamine as the nitrogen donor
in the synthesis of CP was first discovered in Agaricus
bisporus by Levenberg (1962). The subsequent observation
of the same reaction in S. cerevisiae and E. coli suggested
the generality of this reaction which has since been found
to occur in most types of organisms. Like numerous enzymes which catalyze the transfer of the amide nitrogen of
glutamine (Meister, 1962), glutamine-dependent carbamylphosphate synthetases are able to use either glutamine or
ammonium as substrate.

In Escherichia coli, the Km of CPSase for glutamine is
over 200-fold less than the Km for ammonia (Anderson and
Meister, 1965 ; Kalman et al., 1966). By the study of the
inhibition of the enzyme by a chloroketone analog of glutamine, Khedouri et al. (1966) have obtained evidence that
glutamine binding to the enzyme is followed by transfer of
the glutamine amide nitrogen to a site which can also
accept ammonia. More recently, Trotta et al. (1971) characterized the E. coli B enzyme as an aggregate of two kinds
of subunits : a heavy polypeptide chain (molecular weight
130,000) catalyzes the synthesis of CP from ammonia and
bears the effectors binding sites ; a light subunit (molecular weight 42,000) is required for the glutamine-dependent CP synthesis. The glutamine binding site is located
on this light subunit (Pinkus et al., 1972).

As mentioned in a preceding paragraph, E. coli mutations which create a double requirement for arginine and
pyrimidines map in a single monocistronic locus, capA.
All these mutations have been found to result in the loss

of CPSase activity with either glutamine or ammonium as the
nitrogen donor (Piérard, unpublished), and consequently, it
is suggested that the capA locus codes for the heavy subunit of CPSase. In addition, the size of capA implies a
polypeptide of at least 70,000 daltons (Mergeay, 1969)
which is only compatible with that of the heavy subunit.
A class of mutants which show no growth or slow growth on
minimal medium but are helped by a high ammonia concentration has retained a particular attention. Two of these
mutants, bearing the capA177 and 178 mutations, have been
studied in some detail. Extracts of both mutants exhibit
a low CP synthesizing activity with glutamine but an increased ammonia dependent activity with respect to the
wild-type strain. The affinities of the CPSase of these
mutants for the nitrogen donor and, particularly for glutamine, are relatively little affected. On the contrary, an
inversion of the ratio of maximal velocities attained with
ammonia and glutamine is observed (Piérard and Mergeay, unpublished). An alteration of the transfer of the amide nitrogen from the glutamine site to the ammonia site, as
might result from an incorrect aggregation of the two subunits, is proposed to explain the properties of these enzymes. Such modifications might be produced by mutational
alterations of any of the two subunits and, consequently,
are compatible with the mapping of those ammonia responding
mutations into the capA locus.

In Saccharomyces cerevisiae : the subunits of CPAse.
Although CPAse, in vitro, catalyzes the synthesis of CP
from glutamine or ammonia (Lacroute et al., 1965), the Km
of the enzyme for glutamine (1.1x10^{-3}M) is 35-fold less
than the Km for ammonia (3.9x10^{-2}M) and the maximal velocity attained with glutamine is 10-fold higher than for ammonia (Piérard, Graas and Wiame, unpublished). The values
of these parameters designate glutamine as the probable
physiological nitrogen donor. This is further demonstrated
by the observation that, whereas cpaII mutations abolish
the in vitro CP synthesis from both nitrogen substrates
(table 1), mutations in the cpaI locus result in the loss
of the glutamine-dependent activity but leave the ammonia
activity intact (Piérard, Graas and Wiame, unpublished).
The molecular weight of CPAse is 140,000 as determined
by sucrose gradient centrifugation (Piérard, Schröter and
Wiame, unpublished). The two polypeptide components of
CPAse can be separated by passage through a column of

489

TABLE 1

Influence of the cpaI and cpaII mutations on the activity of the arginine
pathway carbamylphosphate synthetase (CPAse) of Saccharomyces cerevisiae
(Piérard, Graas and Wiame, unpublished)

| Strains | Genotype | Growth Medium | CPAse activity with (substrate) | |
			Glutamine 0.01 M	NH_4^+ 0.1 M
4031c	cpu-2	M.am + ura	0.330	0.030
		M.am + ura + arg	0.054	0.016
6028b	cpu-2, cpaI-3	M.am + ura + arg. lim.	0.003	0.028
MG701	cpu-2, cpaII-3	M.am + ura + arg. lim.	0.003	0.001

Minimal medium (M.am) is supplemented, when indicated, with uracil 25µg/ml (ura), with
arginine 1000µg/ml (arg), with arginine 5µg/ml (arg. lim.). Mutants 6028b and MG701
are harvested at the "plateau" after exhaustion of a limiting amount of arginine.
Activities are in µM CP/H/mg protein.

DEAE-Sephadex and identified by in vitro complementation
with extracts of cpaI and cpaII mutants. Ammonium-dependent CP synthesis was found associated with the component
identified as the product of the cpaII locus (Piérard,
Thuriaux, Graas and Wiame, unpublished). Further characterization of the CPAse structure has been prevented by the
poor stability of the enzyme and its subunits.

These various observations support the conclusion that
cpaII codes for a fundamental catalytic unit of the enzyme
which catalyzes all the steps of CP synthesis from ammonia
whereas the cpaI locus codes for a second subunit which is
required for the expression of the physiological glutamine-
dependent activity. The recent observation of a stimulatory effect by glycine of the ammonia-dependent activity of
CPAse has provided additional knowledge of these two subunits (Piérard, Schröter and Wiame, unpublished). This effect of glycine, fortuitously noticed during attempts at
stabilizing CPAse, consists mainly in an increase, by a
factor of four, of the rate of CP synthesis from ammonium
ions ; the Km for ammonium ions is relatively little affected. Out of a number of amino-acids tested only L-α-
alanine shares this property of glycine. The activation
by glycine was found to depend on the integrity of the
"glutamine" subunit and is absent in the cpaI mutants. This
led to investigate the effects of glutamine analogues on
the ammonia-dependent activity of the native enzyme. No
stimulation was observed in the presence of azaserine,
asparagine, glutamate or acetyl-glutamate ; however, O-carbamylserine exhibited an effect comparable to that of glycine and alanine. In addition, 2-amino-4-oxo-5-chloropen-
tanoic acid which binds irreversibly·at the glutamine site
of E. coli CPSase (Khedouri et al., 1966), was found to
stimulate the ammonia-dependent activity while abolishing
the glutamine activity of the yeast enzyme. The effective
protection against that agent which was achieved with both
glutamine and glycine suggests that all four activators of
the ammonia activity, glycine, alanine, O-carbamylserine
and 2-amino-4-oxo-5-chloropentanoic acid bind at the glutamine site of the enzyme. These results support the view of
separate sites for glutamine and ammonium ion but also show
that interactions exist between these sites, the occupation
of the glutamine site by glutamine analogues, and probably
also by glutamine itself, resulting in a considerable modification of the properties of the NH_4^+ site. These various

observations have been repeated with E. coli enzyme (Piérard, Schröter and Wiame, unpublished) and consequently,
underline the great homology which exists between these
enzymes. A similar behaviour of formylglycinamide ribonucleotide amidotransferase from chicken liver with respect
to azaserine has been reported (Mizobuchi et al., 1968).

C. The regulatory sites of E. coli carbamylphosphate synthetase.

The reaction catalyzed by carbamylphosphate synthetase
is complex. It consists in a sequence of three enzymatic
steps and requires three substrates, HCO_3^-, glutamine and
ATP which is involved in two of these steps (Anderson and
Meister, 1966a).   In addition, various effectors (UMP,
ornithine and IMP) control the activity of the enzyme. The
control of the activity is mediated through modifications
in the distribution of the enzyme between at least three
conformational states and results in changes in the sigmoidal saturation function of the enzyme by one of its substrates, namely ATP-Mg (Anderson and Meister, 1966b; Anderson and Marvin, 1968, 1970).   The physiological importance
of the control of CPSase activity is probably best attested
by the diversity of phenotypes which are encountered among
E. coli mutants affected in CPSase.
      In addition to mutations which lead to an absolute
requirement for arginine and uracil, a large proportion of
mutations at the capA locus lead to perturbed growth behaviours following the addition of arginine or pyrimidines to
the growth medium ; optimal growth rate is restored when
both end-products are present (Mergeay, 1969 ; Mergeay et
al., 1972). The most common of these phenotypes is uracil-
sensitivity (Novick and Maas, 1961 ; Gorini and Kalman,
1963 ; Piérard et al., 1965). These mutants are prototrophic in minimal medium but inhibited by uracil, an inhibition which is specifically reversed by arginine. It is the
behaviour of these mutants which led to discover the feedback inhibition of CPSase by UMP (Piérard et al., 1965).
Another phenotype, arginine-sensitivity, is symmetrical to
the preceding one as far as growth behaviour is concerned.
A strikingly different phenotype with respect to arginine
is that of the "arginine-less like mutants" ; these mutants
grow faster on minimal medium plus arginine than on minimal
medium. The study of mutants exhibiting these various

phenotypes suggests that these mutational alterations often
consist in subtle changes of the balance between the
various conformational states of the enzyme. These modifications result in complex impairments of the regulatory
properties of the enzyme which are generally difficult to
relate with the behaviour of the corresponding mutant (Piérard, Mergeay and Leclercq, unpublished). A fourth phenotype, in which growth is partially restored when the medium
contains a high concentration of ammonium ions, has been
discussed in more details under B, in relation with the
specificity of the enzyme for the nitrogen donor.

    Thirty mutations at the capA locus which do not result
in a double requirement for arginine and pyrimidine have
been precisely localized on the map of the locus (Fig. 3).
Some phenotypes exhibit a tendency towards clustering. For
instance, mutations leading to arginine-sensitivity map in
the right half of the locus whereas a class of mutations
which are expressed by a uracil-sensitivity that can be
released by CO_2 are located in the central part of the
locus. Nevertheless, on the whole length of the capA
locus, mutations which are expressed by non-auxotrophic
phenotypes are distributed at random among mutations leading to complete auxotrophy for arginine and uracil (Mergeay
et al., 1972). This observation is in agreement with the
monocistronic nature of the locus, as already established
by the functional analysis. Consequently, the phenotypic
diversity among mutations affecting capA does not result
from the existence of adjacent cistrons specifying polypeptide chains with differing functional properties. This
eliminates the confusion which had initially occurred when
the observation of phenotypes such as uracil-sensitivity
was taken as evidence for the existence of distinct and
specialized enzymes. It is clear that the causes of uracil
sensitivity in yeast and E. coli are completely different.
The fact that mutations leading to various perturbations of
the regulatory properties of CPSase map in capA provides
additional support to the view that this locus codes for the
heavy subunit of the enzyme (Trotta et.al., 1971).

    Frameshift mutations are randomly distributed along
the whole length of the locus (Fig. 3). Six presumably
frameshift mutations which are expressed by non-auxotrophic
phenotypes map at the left extremity of the locus. The
growth on minimal medium of mutants bearing these mutations
increases from left to right. A counter-clockwise polarity

493

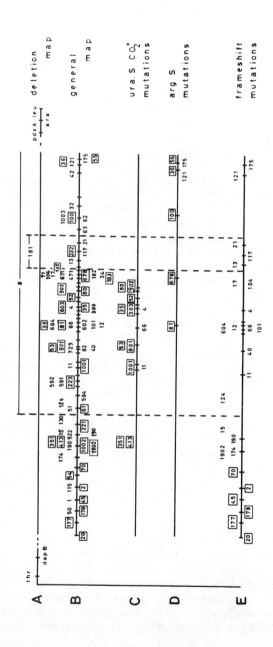

Fig. 3 Genetic map of the capA locus.

Mutations which are not expressed by a complete auxotrophy for arginine and pyrimidines are indicated by numbers surrounded by a square. Other numbers are strict double-auxotrophic mutations (Mergeay et al., 1972).

of the capA locus is suggested ; being operator distal,
these mutations can still allow the synthesis of polypeptide chains which are correct for 85 to 95 % of their
length. A few mutations, cap222, 522, 190, 70, 45, 178,
and 177, bring about a reduction of the glutamine activity.
Six of these mutations map in the extreme left part of
capA, five of them being frameshift (Fig. 3). As discussed under B, the properties of the CPSases of mutants capA
177 and 178 are compatible with mutations impairing the
heavy subunit of the enzyme. It is suggested that mutations at the left end of capA are preferentially expressed
by alterations in the transfer of the amide nitrogen of
glutamine to the ammonia site.

D. The organization of CP synthesis in other microorganisms.

    As was already underlined, the study of CP synthesis
has progressed simultaneously in Neurospora and yeast and
great similarities have been observed in these two organisms. However, they differ by their organization of CP
distribution. Mutants of Neurospora and also of Coprinus
radiatus (Cabet et al., 1967) affected in CP synthesis are
auxotrophic for arginine or pyrimidines, thus suggesting
the existence of arginine and pyrimidines specific CP generating systems. These auxotrophic phenotypes were taken as
evidence for the maintenance of two discrete CP pools in
this organism. Metabolic "channelling" in the form of con-
finement of CP to enzyme aggregates was initially proposed
to explain this segregation of CP (Davis,1967). However,
recent histochemical studies (Bernhardt and Davis, 1972)
have suggested that localization of the pyrimidines and
arginine specific metabolisms of CP within different cellular organelles might also account, wholly or in part, for
that channelling.
    Whereas fungi, in general, seem to have two carbamyl-
phosphate synthetases, the occurrence of a single enzyme
appears as rule in bacteria. Mutants exhibiting a double
requirement for arginine and pyrimidines have been obtained
in many bacteria : Pseudomonas (Loutit, 1952), Salmonella
typhimurium (Yan and Demerec, 1965), Proteus mirabilis
(Prozesky and Coetzee, 1966), Streptomyces coelicolor
(Hopwood, 1967) and Bacillus subtilis (Issaly et al., 1970).
The best studied of these bacteria is S. typhimurium which

495

is very similar to E. coli with respect to phenotypic diversity of mutations affecting CP synthesis as well as to the
properties of CPSase (Yan and Demerec, 1965 ; Eisenstark,
1967 ; Abd-El-Al et al., 1969 ; Abd-El-Al and Ingraham,
1969a, b). These properties are probably shared by CPSase
from Pseudomonas (Stalon and Legrain, unpublished). A somewhat different mechanism exists in B. subtilis (Issaly et
al., 1970). Single mutational events lead to requirement
for arginine and uracil and to the simultaneous loss of an
anabolic CPSase and of carbamate kinase. CPSase is cumula-
tively inhibited by arginine and UTP ; its synthesis is
subject to concerted repression by arginine and uracil.

2. Regulatory Loci controlling the Formation
of Carbamylphosphate Synthetase

A. Cumulative repression of E. coli carbamylphosphate
synthetase.

The synthesis of CPSase is partially repressed in the
presence of either arginine or pyrimidines (Table 2), but
almost totally abolished in the presence of both end-products (Piérard and Wiame, 1964). The question which arises
concerning the mechanism of cumulative repression is
whether the effects of molecules which are structurally so
unrelated as arginine and pyrimidines are mediated through
the action of a single repressor molecule bearing two sets
of specific binding sites, or, on the contrary, through the
combined actions of two specific repressors which might
possibly aggregate to exert a maximal repressive capacity.
So far the attempts at selecting regulatory mutants which
specifically affect the regulation of the synthesis of
CPSase have failed. However, the observation was made recently that mutations at the arginine pathway regulatory
gene, argR mutations (Maas, 1961 ; Gorini et al., 1961),
reduce the repressibility of CPSase by arginine (Piérard et
al., 1972). This effect of argR mutations which is shown
in table 2, suggests that the same regulatory circuit controls the repression by arginine of CPSase and of the entire arginine pathway. It would be interesting to determine
whether mutations which affect the control of the synthesis
of the pyrimidine pathway enzymes, have a similar effect on
the synthesis of CPSase. Such mutations have been recently
identified in Salmonella typhimurium (O'Donovan and Gerhart

496

1972) but their effect on the synthesis of CPSase has not been determined.

TABLE 2

Influence of the argR mutation on cumulative repression of E. coli K12 glutamino-carbamyl-phosphate synthetase (Adapted from Piérard et al., 1972)

Growth Medium	CPSase activity (μmoles/H/mg protein)	
	wild-type strain (P4X)	argR strain (P4XB2)
Minimal	1.04	1.93
Minimal + arginine	0.52	1.71
Minimal + uracil	0.38	1.65
Minimal + arginine + uracil	0.05	1.10

Assay conditions : 0.1 M phosphate buffer (pH 7.5), 0.01 M glutamine, 0.01 M H^{14}CO$_3$, 0.01 M ATP, 0.01 M MgCl$_2$, 0.005 M ornithine, Incubation : 10 min at 37°C. ^{14}CP formed was converted into ^{14}C-citrulline by coupling with ornithine carbamyltransferase (Piérard, 1966) and samples counted in a liquid scintillation spectrometer.

B. Repression of carbamylphosphate synthetase belonging to the arginine biosynthetic pathway in S. cerevisiae.

As mentioned above, cpu mutations are expressed by a phenotype of sensitivity towards arginine which has been explained by the repression of CPSase by arginine. Advantage has been taken of this behaviour for the selection of regulatory mutations affecting the synthesis of CPAse. Two classes of mutations which exhibit a reduced repressibility of CPAse have been obtained among mutants resistant to arginine (Thuriaux et al., 1972). The mutations of the first class, designated as cpaIO, are strongly linked to the cpaI gene and are cis-dominant (Table 3). Consequently, these mutations are of constitutive-operator type.

497

TABLE 3

Regulatory mutations affecting the control of the synthesis of the arginine pathway carbamylphosphate synthetase of S. cerevisiae (Adapted from Thuriaux et al., 1972).

Strain	Genotype	Growth on M.am + arg	Glutamine-dependent CPAse activity after growth on	
			M.am + ura	M.am + arg + ura
4031c	cpu-2	slow	0.65	0.07
6028b	cpu-2, cpaI-3	-	-	0.004
9177b	cpu-2, cpaIO-8	+	0.48	0.33
Diploids				
9177b x 6028b	$\frac{\text{cpu-2, + cpaIO-8}}{\text{cpu-2, cpaI-3 +}}$	+	0.38	0.23
4031c x 6920c	$\frac{\text{cpu-2, +, +}}{\text{cpu-2, cpaI-3, cpaIO-8}}$	slow	0.35	0.04
10601c	cpu-2, cpaR-2	+	0.87	0.23
Diploid				
10601c x 4021b	$\frac{\text{cpu-2, cpaR-2}}{\text{cpu-2, +}}$	slow	0.52	0.06

Media compositions are indicated in legend of Table 1.

The mutations of the second class, cpaR mutations, are re-
cessive towards the wild-type allele (Table 3) and do not
show close linkage with the cpaI or cpaII loci. These mutations have been interpreted as impairing the formation of
an active repressor of CPAse.

While a control of negative type for the synthesis of
the CPAI polypeptide is clearly defined by the existence of
the cpaIO and cpaR mutations, little is known concerning
the regulation of the synthesis of CPAII. As shown in
table 1, the ammonia-dependent CP synthesizing activity,
which corresponds to the CPAII polypeptide, is partially
repressible by arginine. Nevertheless, the operator constitutive mutations cpaIO can achieve effective derepression
of the overall glutamine-dependent activity. As the cpaII
gene is not linked to cpaI and its operator, it must be
concluded either that another coordination mechanism links
the synthesis of their products or that the product of
cpaII is always synthesized in large excess over that of
cpaI. Recent experiments involving the construction of
series of tetraploid yeast cells bearing variable doses of
active cpaI and cpaII genes do not suggest that such an
excess of the product of cpaII exists (Hilger, Piérard,
Grenson and Wiame, unpublished).

Conclusions

Two extremely different ways of controlling the
supply of a common precursor of two biosynthetic sequences
are disclosed by the study of CP synthesis in E. coli and
S. cerevisiae. The single E. coli CP generating system
requires an extremely precise regulation which takes the CP
needs of both pathways into consideration. In S. cerevisiae, the duplication of carbamylphosphate synthetase
allows much simpler control mechanisms acting on each
enzyme independently. This is illustrated, for instance,
by the arginine pathway enzyme which is efficiently regulated by the sole arginine repression mechanism ; no control of its activity has been observed.

This yeast enzyme has not been submitted to a thorough
biochemical study which is prevented by its great lability.
Nevertheless, enough is known of the role of its two subunits to suggest a great homology with the E. coli carbamylphosphate synthetase. One component of the yeast enzyme
alone can carry out the synthesis of CP with ammonia as the

amino donor just as is known for the purified E. coli heavy
subunit. However, clear evidence that ammonia is not used
efficiently as an amino group donor for CP synthesis in
vivo, has been obtained by the study of mutants affected in
the other subunit.

Such observations are interesting as, due to the homology which exists between properties of the enzymes from E.
coli and yeast, similar behaviours with respect to the nitrogen group donor are to be expected in vivo. Nevertheless,
all the E. coli mutants which, so far, have been selected
for being impaired in the synthesis of CP, appear to be
defective in the heavy subunit. Mutations in the light
subunit may be expected to be rarer than mutations in the
heavy subunit (in yeast, cpaI mutations are less frequent
than cpaII mutations). However, this assumption is incom-
patible with the failure to select even one single mutation
affecting the light subunit whereas more than seventy five
mutations at capA are presently identified. An alternative
explanation might be that the lack of mutations affecting
the light subunit is due to a second essential function of
this subunit, the loss of which would not be compensated by
the presence of arginine and uracil in the growth medium.

Authorizations to reproduce published material from
Journal of Microbiology (Fig. 1), Science (Fig. 2), Molecular and General Genetics (Table 2) and Journal of Molecular Biology (Table 3) are acknowledged.

REFERENCES

Abd-El-Al, A. and Ingraham, J.L. (1969a). J. Biol. Chem.
    244, 4033.

Abd-El-Al, A. and Ingraham, J.L. (1969b). J. Biol. Chem.
    244, 4039.

Abd-El-Al, A., Kessler, D.P. and Ingraham, J.L. (1969).
    J. Bacteriol. 97, 466.

Anderson, P.M. and Meister, A. (1965). Biochemistry 4, 2803.

Anderson, P.M. and Meister, A. (1966a). Biochemistry 5, 3157.

Anderson, P.M. and Meister, A. (1966b). Biochemistry 5, 3164.

Anderson, P.M. and Marvin, S.V. (1968). Biochem. Biophys. Res. Commun. 32, 928.

Anderson, P.M. and Marvin, S.V. (1970). Biochemistry 9, 171.

Bernhardt, S.A. and Davis, R.H. (1972). Proc. Nat. Acad. Sci. US. 69, 1868.

Cabet, D., Gans, M., Motta, R. and Prévost, G. (1967). Bull. Soc. Chim. Biol. 49, 1537.

Davis, R.H. (1963). Science 142, 1652.

Davis, R.H. (1967) In "Organizational Biosynthesis" (H.J. Vogel, J.O. Lampen and V. Bryzon, eds.), p.303-322, Academic Press, New York.

Denis-Duphil, M. and Lacroute, F. (1971). Mol. Gen. Genet. 119, 354.

Eisenstark, A. (1967). Nature 213, 1263.

Gorini, L. and Kalman, S.M. (1963). Biochim. Biophys. Acta 69, 355.

Gorini, L., Gundersen, W. and Burger, M. (1961). Cold Spring Harbor Symp. Quant. Biol. 26, 173.

Hopwood, D.A. (1969). Bact. Rev. 31, 373.

Issaly, I.M., Issaly, A.S. and Reissig J.L. (1970). Biochim. Biophys. Acta 198, 482.

Jones, M.E., Spector, L. and Lipman, F. (1955). J. Am. Chem. Soc. 77, 819.

Kalman, S.M., Duffield, P.H. and Brzozowski, T. (1966). J. Biol. Chem.241, 1871.

Khedouri, E., Anderson, P.M. and Meister, A. (1966). Biochemistry 5, 3552.

Lacroute, F.(1968). J. Bacteriol. 95, 824.

Lacroute, F., Piérard, A., Grenson, M., and Wiame, J.M. (1965). J. Gen. Microbiol. 40, 127.

Levenberg, B. (1962). J. Biol. Chem. 237, 2590.

Loutit, J.S. (1952). Austral. J. Exp. Biol. Med. Sci. 30, 287.

Lowenstein, J.M. and Cohen, P.P. (1956). J. Biol. Chem. 220, 57.

Lue, P.F. and Kaplan, J.G. (1969). Biochem. Biophys. Res. Commun. 34, 426.

Maas, W. (1961). Cold Spring Harbor Symp. Quant. Biol. 26, 183.

Meister, A. (1962). In "The Enzymes" (P.D. Boyer, H. Lardy and K. Myrbäck, eds.), vol. 6, p. 247-266, Academic Press, New York.

Mergeay, M. (1969). Thèse de doctorat, Université Libre de Bruxelles.

Mergeay, M., Beckmann, J., Glansdorff, N., and Piérard, A. (1972). In preparation.

Mizobuchi, K., Kenyon, G.L. and Buchanan, J.M. (1968). J. Biol. Chem. 243, 4863.

Novick, R.P. and Maas, W. (1961) J. Bacteriol. 81, 236.

O'Donovan, G.A. and Gerhardt, J.C. (1972). J. Bacteriol. 109, 1085.

Piérard, A. (1966). Science 154, 1572.

Piérard, A. and Wiame, J.M. (1964). Biochem. Biophys. Res. Commun. 15, 76.

Piérard, A., Glansdorff, N., Mergeay, M. and Wiame, J.M. (1965). J. Mol. Biol. 14, 23.

Piérard, A., Glansdorff, N. and Yashphe, J. (1972), Mol. Gen. Genet. In press.

Pinkus, L.M., Wellner, V.P. and Meister, A. (1972). Feder. Proceed. 31, 474.

Prozesky, O.W. and Coetzee, J.N. (1966). Nature 209, 1262.

Roepke, R.R. (1946). Quoted by Tatum, E.L., Cold Spring Harbor Symp. Quant. Biol. 11, 278.

Thuriaux, P., Ramos, F., Piérard, A., Grenson, M. and Wiame, J.M. (1972). J. Mol. Biol. 67, 277.

Trotta, P., Burt, M.E., Haschemeyer, R.H. and Meister, A. (1971). Proc. Nat. Acad. Sci. US. 68, 2599.

Vyas, S., and Maas, W.K. (1963). Arch. Biochem. Biophys. 100, 542.

Yan, Y. and Demerec, M. (1965). Genetics 52, 643.

Wiame, J.M., Stalon, V., Piérard, A. and Messenguy, F. (1972). Symp. Soc. Exp. Biol. In preparation.

THE ALLOSTERIC CONTROL OF CTP SYNTHETASE

Alexander Levitzki

Department of Biophysics, The Weizmann Institute of Science, Rehovot, Israel, and Department of Biochemistry, University of California, Berkeley, California, 94720.

The discovery of the molecular basis of feedback inhibition in aspartate transcarbamylase by CTP (Yates and Pardee, 1956) has revolutionized the thinking on enzymes in general and regulatory enzymes in particular. It became evident that specific ligands, dissimilar to either the substrates or the products of a certain enzymic reaction, can "switch on" or "switch off" the reaction by binding to the enzyme catalyzing that reaction. These regulatory molecules were called allosteric effectors (Monod et al., 1963) or simply effector molecules. It soon became apparent that regulatory enzymes are only one set of biological systems in which the biological activity can be modulated by the binding of specific ligands. Other systems in which the biological activity can be modulated specifically are for example: the peeling of a repressor molecule from DNA by the binding of an inducer molecule, the change in ion fluxes through the nerve membrane by the binding of acetylcholine and the binding of the drug chlorpromazine to brain glutamic dehydrogenase from mammals (Fahien and Shemisa, 1970). It is therefore of great interest to investigate the mechanism of action of an allosteric effector on a system on which many molecular details are known. A system of this kind is CTP synthetase.

CTP SYNTHETASE FROM *E. coli* B.

CTP synthetase was first described by Lieberman (1955, 1956), who demonstrated the synthesis of CTP from UTP, ammonia and ATP (equation 1):

$$UTP+NH_3+ATP \xrightarrow{Mg^{2+}} CTP+ADP+P: \qquad (1)$$

Later (Chakraborty and Hurlbert, 1961) it was shown that L-glutamine can also function as the nitrogen donor in the presence of a guanosine nucleotide. Long and Pardee (1967) obtained a highly purified preparation of CTP-synthetase and demonstrated that the glutamine reaction is absolutely dependent on the presence of GTP as the positive allosteric effector (equation 2).

$$UTP+Glu-NH_2+ATP \xrightarrow[GTP]{Mg^{2+}} CTP+Glu-OH+ADP+Pi \qquad (2)$$

CTP synthetase when isolated from *E. coli* is a dimer of molecular weight of 108,000 (Long et al., 1970). Using electrophoretic techniques and isoelectric focussing in 8M urea only one kind of subunit could be identified. Furthermore, direct binding studies revealed one site per subunit for each ligand involved in the CTP synthetase reaction: ATP, UTP, Glutamine and GTP (Levitzki et al., 1971; Levitzki and Koshland, 1972a). Thus one is led to the conclusion that the enzyme dimer is composed of two identical subunits. Another interesting feature of the enzyme is that the enzyme dimer dimerizes to form a tetramer in the presence of either ATP or UTP (Fig. 1). The ATP and UTP effects are synergistic and detailed studies have shown that in the presence of saturating ATP and UTP the enzyme remains in the tetramer form even at concentrations as low as 10^{-9} M.

The Mechanism of the CTP Synthetase Reaction.

In order to reveal the mechanism of GTP activation it became essential to identify all the elementary steps comprising the synthetic pathway of CTP.

A phosphorylated UTP intermediate: The amination of UTP to form CTP is ATP-dependent, the ATP being cleaved to ADP and Pi concomitantly with the amination. Three possible mechanisms (Fig. 2) can be formulated, by which the 4'-OH on the uracil moiety is replaced by an amino group via a phosphorylated intermediate. It was therefore necessary to check whether a phosphorylated intermediate exists. For this purpose the $4'-O^{18}$ UTP was prepared by treating CTP with $NaNO_2$ in H_2O^{18} at pH 4.0 as described elsewhere (Levitzki and Koshland, 1971). When O^{18}-UTP was incorporated into either the NH_3 reaction (eq. 1) or the glutamine reaction (eq. 2) one mole of O^{18} labeled phosphate per mole of ATP

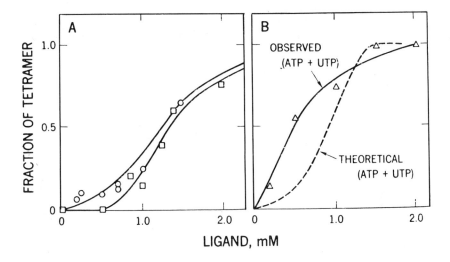

Figure 1: The Dimerization of CTP Synthetase to the Active
Tetramer

The ATP and UTP induced dimerization of the CTP synthetase
dimer (108,000 molecular weight) to the functionally active
tetramer (216,000) was studied using Sephadex G-200 columns.
The columns (2.5x100 cm) were run in the presence of 10 mM
$MgCl_2$, 20 mM phosphate buffer pH 7.2, 1 mM EDTA, 70 mM β-
mercaptoethanol and 4 mM glutamine. The columns were equilibrated with ATP or UTP or both, as is indicated in the
figure. Blue dextran, LDH, γ-globulin, BSA, egg albumin and
cytochrome C as standards were applied together with the CTP
synthetase (4-6 units) in each run. The fraction of enzyme
tetramer in each run was computed as described elsewhere
(Levitzki and Koshland, 1972). (A) The fraction of tetramer of a function of ATP (□) or UTP (○) concentrations.
(B) Summation of the ATP and UTP effects from A (- - -)
and the observed (▲) dimerization against [ATP + UTP] where
at each point [ATP] = [UTP].

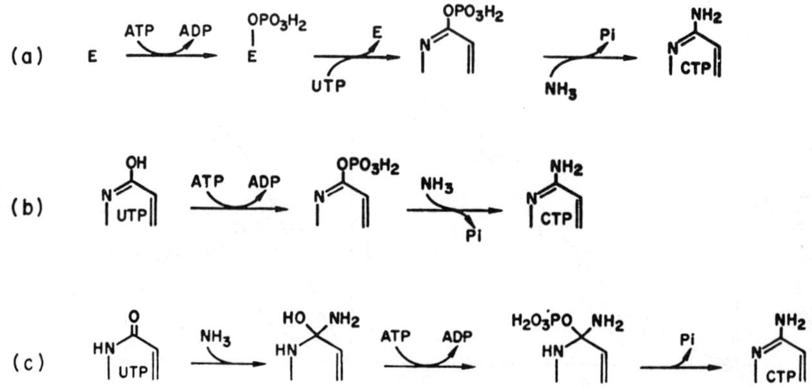

Fig. 2: Alternative Pathways for CTP Synthesis.

cleaved and CTP synthesized was produced. This experiment established the existence of a phosphorylated UTP as an intermediate in the reaction. Still, the mechanism could be either one of the three alternatives suggested in Fig. 2: (a) a phosphorylated enzyme intermediate followed by phosphorylation of the UTP moiety and then displacement by NH_3; (b) direct phosphorylation of UTP by ATP followed by displacement by NH_3, and (c) amination of UTP followed by phosphorylation and elimination of Pi. Mechanism (a) predicts ATP-ADP exchange which is UTP independent; mechanism (b) requires ATP-ADP exchange which is UTP dependent. No ATP-ADP exchange occurs either in the presence or the absence of UTP eliminating the mechanisms described in (a) and (b). ADP-ATP exchange could be demonstrated in the presence of ammonia or glutamine; however the measured exchange rate does not represent the true exchange rate since the reaction progresses to completion within a few minutes and thus the absolute amount of H^3-ADP exchanged to H^3-ATP is small. The exchange measured under these conditions is due to the reversibility of the CTP synthetase reaction prior to the dephosphoryla-

508

tion stage (Fig. 2(c)).

It is thus concluded that both the glutamine reaction and the ammonia reaction proceed via a tetrahedral adduct which is then phosphorylated to form a phorphorylated adduct which readily decomposes to form CTP, as in case (c).

THE ROLE OF GTP

The ammonia reaction and the glutamine reaction proceed via the cleavage of one mole of ATP per mole CTP formed (equations 1 and 2). It appears therefore that the GTP is selectively involved in the activation of glutamine. Indeed, the kinetic parameters K_m and k_{cat} for glutamine change dramatically upon addition of the allosteric activator GTP (Fig. 3). In the absence of GTP, glutamine shows Michaelis-Menten kinetics with a K_m of $1 \times 10^{-3} M$. In the presence of GTP at saturating levels, this K_m drops to $1.6 \times 10^{-4} M$. The addition of GTP also increases the turnover number (k_{cat}) of the enzyme by a factor of 11. One should stress that both the K_m and k_{cat} for the ammonia reaction remain unchanged when GTP is added. The simplest approach to the explanation of the GTP effect is shown in Fig. 3. We assume that the glutamine forms a Michaelis complex which is then converted to a second enzyme glutamine compound E-S. If this conversion (designated k_3) is the one facilitated by GTP it can be easily seen that it will result in an increase of k_{cat} and a decrease in K_m if $k_2 > k_4$ (Fig. 3).

Since glutamine has eventually to yield ammonia it is reasonable to assume that the second intermediate in the glutamine reaction involves a covalent glutamyl-enzyme and enzyme bound ammonia. A search for the glutamyl-enzyme intermediate was therefore initiated. Direct binding studies of C^{14}-glutamine to the enzyme dimer by equilibrium dialysis revealed the existence of 0.9 equivalent of glutamine per subunit of 52,000 molecular weight. The next step was to establish whether part of the bound glutamine is covalently attached. A typical experiment is described in Fig. 4. As is indicated in the figure about 10% of the bound glutamine can be trapped as a glutamyl residue covalently attached to the unfolded CTP synthetase subunit.

509

ROLE OF GTP

Kinetic Parameters for Glutamine

	K_M	k_{cat}
without GTP	1.0×10^{-3} M	14
with GTP	1.6×10^{-4} M	150

$$E + S \underset{k_2}{\overset{k_1}{\rightleftharpoons}} \underset{\substack{\text{Michaelis} \\ \text{complex}}}{ES} \overset{k_3}{\longrightarrow} \underset{\substack{\text{Covalent} \\ \text{intermediate}}}{E - S}$$

$$\downarrow k_4$$

$$E + \text{Products}$$

$$K_M = \frac{k_3 + k_2}{k_1 (1 + k_3/k_4)} \qquad \frac{\partial K_M}{\partial k_3} < 0 \text{ if } k_2 > k_4$$

$$k_{cat} = \frac{k_3}{1 + k_3/k_4} \qquad \frac{\partial k_{cat}}{\partial k_3} > 0$$

Figure 3: The Role of GTP in the Activation of CTP Synthetase

K_m and k_{cat} for the glutamine reaction were measured at saturating ATP, UTP and Mg^{2+} at pH 7.2 and at 38°C. Glutamine concentration was varied between 2×10^{-5}M and 1×10^{-2}M in the absence and in the presence of saturating GTP (1 mM). The standard assay conditions for CTP synthesis were used. (Long et al., 1970).

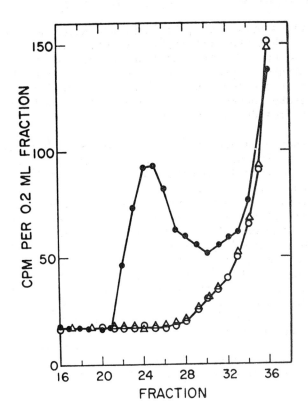

Figure 4: The Formation of a $[^{14}C]$ Glutamyl-Enzyme.

(●) Glutamine-free enzyme (0.2 ml, 2.85 nmoles) in 0.02 M
sodium phosphate buffer containing 1 mM EDTA was incubated
with 10 μl of 0.1 M $[^{14}C]$ glutamine (1050 CPM/nmoles) for
15 minutes at 25°C. Then 10 μl of 0.1 M I_2 (in 5% KI) was
added and the sample cooled to 0°C. The sample was then
applied to a Bio-Gel P-2 column (1.5x10 cm), equilibrated
with 8 M urea in 0.02 M sodium phosphate buffer (pH 7.4),
1 mM EDTA and 70 mM β-mercaptoethanol in the cold. (O)
Same, except enzyme was reacted with I_3 first and then with
^{14}C glutamine. (Δ) Same, except enzyme was treated with
DON first. Fraction of enzyme trapped as covalent glutamyl-
enzyme intermediate in all experiments was 0.08 - 0.12.

The chemical nature of the glutamyl-enzyme intermediate.

Previous work (Long, Levitzki and Koshland, 1970) demonstrated that the affinity label DON(6-diazo-5-oxo-nor-
leucine) which is a structural analog of glutamine reacts
with a sulfhydryl group on the protein. This finding indicates that the covalent intermediate formed by glutamine
is a thioester between the glutamyl residue and the SH of
a cysteine residue at the active site.

Further evidence for the involvement of a sulfhydryl
group in the formation of a glutamyl-enzyme intermediate
was obtained from differential carboxymethylation experiments. When CTP synthetase is carboxymethylated at pH 8.5
in the presence of glutamine 3.9 sulfhydryl groups out of
a total of 5.1 per 52,000 molecular weight are labeled
within 15 minutes. In the absence of glutamine 5 sulfhydryl groups are readily labeled. Taken together with the
results of the DON affinity labeling the results suggest
very strongly that a glutamyl-enzyme thioester intermediate
is formed. It is now possible to suggest a complete scheme
for the glutamine reaction pathway as described in Fig. 5.
On the basis of the GTP effect on the glutamine reaction
(Fig. 3) and the chemistry of the glutamine site we suggest
that GTP affects specifically the glutamine cleavage by
promoting the conversion of the enzyme glutamine Michaelis
complex to the glutamyl enzyme intermediate and bound ammonia (Fig. 6).

*The nascent nature of the ammonia released from glut-
amine.* - The problem at hand now is to resolve whether the
ammonia released from glutamine reacts directly in nascent
form with UTP or equilibrates as free ammonia in solution
before adding to UTP. A second aspect of the glutamine reaction is whether the glutamine and ammonia sites are separate or overlapping, and whether the release of ammonia in
the glutamine reaction is sufficiently rapid to allow the
overall synthesis of CTP. In order to resolve the above
questions we looked for means to block the CTP synthesis
and retain the glutamylation. This was achieved by the use
of ATP analogs which cannot be hydrolyzed. Two compounds
were tested for this role as shown in Table I. The β,γ-NH-
ATP (ADPNP) prepared by Yount et al. (1971) and the β,γ-CH$_2$
-ATP prepared by Moos et al. (1960). It can be readily

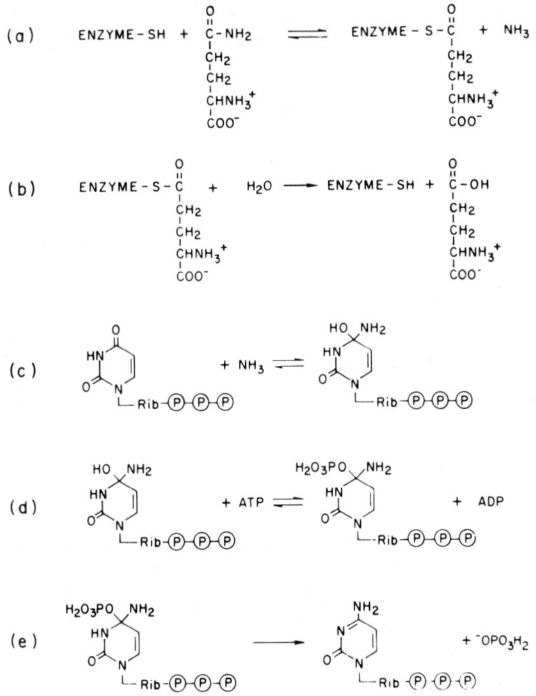

Figure 5:   The Covalent Changes Catalyzed by CTP Synthetase

seen from Table I that the enzyme-ADPNP complex has a dissociation constant equal to that of the enzyme-ATP complex.
When ATP is substituted for ADPNP in the CTP synthetase
reaction mixture no CTP synthesis takes place.   Both the
imido and the methylene analogs acted as competitive inhibitors of ATP. In the presence of ADPNP (or ADPCP) the
enzyme retains its full capacity to cleave glutamine and
is thus converted to a glutaminase.

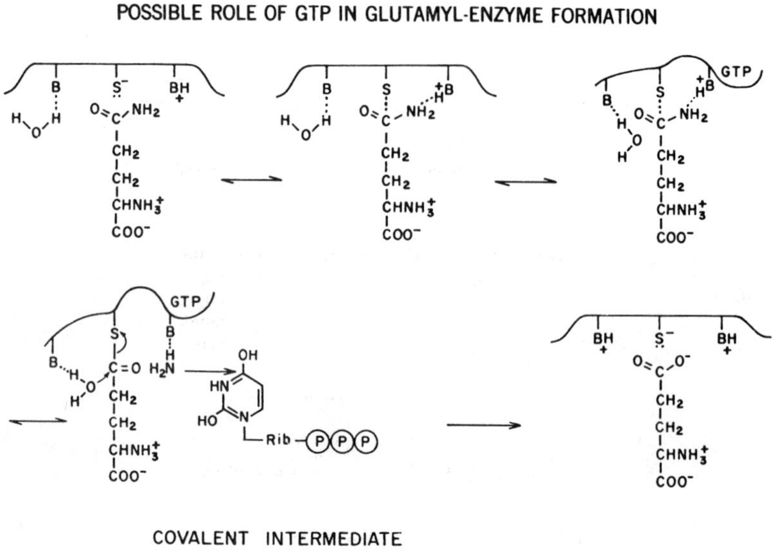

POSSIBLE ROLE OF GTP IN GLUTAMYL-ENZYME FORMATION

COVALENT INTERMEDIATE

Figure 6: The Role of GTP in the Glutamine Reaction.

As is indicated in the text, the role of GTP is to facilitate the conversion of the Michaelis complex to the glutamyl enzyme and enzyme bound ammonia.

Table I: *Effect of ATP analogs on the CTP synthetase reaction*

ATP analog	Dissociation Constant (M) 25°	Rate of glutamine cleavage (moles NH_3 released/mole enzyme sites/min) (k_{cat})	Formation of CTP
β,γ-CH$_2$-ATP	6.6×10^{-4}	145	no
β,γ-NH-ATP	3.3×10^{-4}	150	no
ATP	3.0×10^{-4}	150	yes

The assay mixture contained 0.75 mM UTP, 10 mM MgCl$_2$, 20 mM
phosphate (pH 7.2), 6 mM glutamine, 0.1-0.7 mM ATP and varying the concentration of the ATP analogs. Dissociation

constants for the analogs were computed from double reciprocal plots. In the presence of ATP the rate of glutamine
cleavage was assumed to be equal to the rate of CTP formation. The ATP-enzyme dissociation constant was measured by
direct binding studies (Levitzki and Koshland, 1972). The
nitrogen analog seems to be the best substitute for ATP
since it has the same binding constant towards the enzyme
as ATP (Table I). This identity between ATP and β,γ-NH-ATP
with respect to the interaction with the ATP site probably
reflects the fact that ATP and β,γ-NH-ATP are both isoelec-
tronic and isosteric (Yount et al., 1971). From Table I it
is readily seen that the glutamine hydrolysis in the presence of β,γ-NH-ATP and the other ligands, occurs at a rate
equal to that of CTP synthesis in the presence of ATP. It
is therefore clear that the rate of ammonia production from
glutamine matches the rate of subsequent steps leading to
the final product CTP.

It remains now to be established whether the ammonia
produced from glutamine equilibrates with the solution or
whether it is directly available to UTP. This aspect of
the glutamine reaction was resolved by studying the pH profiles of the ammonia and of the glutamine reactions. (Fig.
7). The glutamine activity in the presence of saturating
GTP reaches a maximum at pH 7.5 and begins to decline at
pH 9.3. The ammonia activity reaches a maximum at pH 10.3
a value at which the enzyme begins to undergo a fast inactivation. The pH profile of the ammonia activity indicates
strongly that the unprotonated form is the reactive species.
At pH 7.2 where 150 moles of CTP are synthesized per minute
per mole active site in the glutamine reaction, only 27
moles of CTP are formed at saturating ammonia $NH_4^+ + NH_3$).
The reaction is measured to the extent of 5% in order to
obtain initial velocities. If the 5% of the glutamine (10^{-2}
M) decompose instantaneously 5×10^{-4} M ammonia will be formed.
This concentration of ammonia at pH 7.2 will yield a rate
of $27 \times 5 \times 10^{-4}/20 \times 10^{-3} = 0.68$ moles CTP/min per mole active
site (20×10^{-3} M and 27 min^{-1} are the values for K_m and
k_{cat} respectively for the ammonia reaction at pH 7.2). It
can therefore be safely concluded that the ammonia produced
from glutamine does not equilibrate with the solution in
the glutamine reaction but is used as nascent ammonia.
This conclusion is in agreement with the finding that no
ammonia is found in the reaction mixture when glutamine is
used as a substrate. The last point to be clarified is

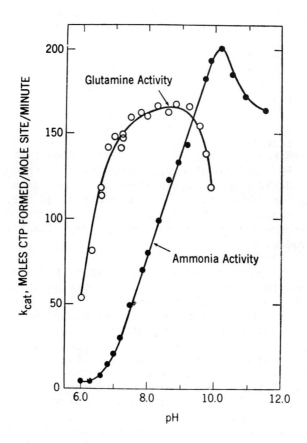

Figure 7:   pH-Activity Profile of CTP Synthetase

The glutamine activity and the ammonia activity were measured in the presence of saturating concentrations of ligands.

whether each subunit contains a single NH_3 site common to the external ammonia when the latter is used as a substrate and the nascent ammonia produced upon glutamylation of the enzyme when glutamine is used as a substrate. One can demonstrate that at pH 7.2 where the rate of the ammonia reaction is 18% of the glutamine reaction the two reactions are competitive. This finding is easily explained if one

remembers that at pH 7.2 the external ammonia is in the
form of NH_4^+. The latter species probably competes with
the nascent unprotonated ammonia released from glutamine
for a single ammonia site. It seems therefore that the reaction with NH_4^+ is slower than with unprotonated ammonia.
It is now possible to confirm the complete scheme for the
CTP-synthetase reactions suggested in Figures 5 and 6. Figure 6 describes the glutaminase step activated by GTP,
the latter facilitating the formation of the glutamyl-enzyme intermediate. Fig. 5 summarizes the steps following
the glutaminase reaction, which are the same also in the
ammonia reaction.

Effect of other ligands on the activation by GTP.- The
use of β,γ-NH-ATP as a substitute for ATP converts artifi-
cially the CTP synthetase to a glutaminase (Table I). It
is now possible to investigate further the role of other
ligands on the glutaminase step which is the one activated
by GTP. Table II summarizes the glutaminase activity of
the CTP synthetase as a function of added ligands. The
glutaminase activity of both the dimer and tetramer forms
of the enzymes is activated 11-fold by adding GTP. It is
also readily apparent that the dimer species has a lower
glutaminase activity than the tetramer. In order to produce the maximal glutaminase rate, ATP (or ADPNP) and UTP
must bind to the enzyme. In this respect the substrates
ATP and UTP are allosteric effectors with respect to the
glutaminase reaction; after this step these ligands function as substrates.

TABLE II: The Glutaminase Activity of CTP Synthetase.

Ligands added	Species of enzyme	k_{cat} [a]
Mg^{2+}	dimer	2.5
Mg^{2+} + GTP	dimer	31
Mg^{2+} + ADPNP + UTP	tetramer	14
Mg^{2+} + ADPNP + GTP + UTP	tetramer	145

The reaction mixture contained: 10 mM $MgCl_2$, 1 mM GTP,
1 mM ADPNP, 1.5 mM UTP, 10 mM glutamine in 0.02 M imidazole
-acetate pH 7.2 in a final volume of 1.0 ml. The glutaminase activity was measured using the sub-micro glutamate
dehydrogenase method described earlier (Levitzki, 1970).

[a] moles glutamine decomposed/min/mole active sites.

The duality of the GTP effect.- When CTP synthetase is labeled with the glutamine affinity label DON the glutamine activity is irreversibly inhibited. The DON-enzyme does however retain its full ammonia activity; furthermore, the response of the DON-enzyme towards other ligands is identical with that of the native enzyme (Table III).

TABLE III: Comparison of native CTP synthetase to DON-CTP synthetase

Ligand	$S_{0.5}$ (mM)		Hill Coefficient		k_{cat}	
	CTPS	DON-CTPS	CTPS	DON-CTPS	CTPS	DON-CTPS
ATP[a]	0.43	0.45	2.87	2.85		
UTP[b]	0.39	0.37	1.75	1.85		
UTP[c]	0.40	0.40	1.1	1.0		
NH_3	5.3	7.0	1.0	1.0	150[d]	150[d]

Assay solutions contained the standard assay mixture: 25 mM $(NH_4)_2SO_4$, 20 mM Tris-acetate pH 8.2, 10 mM $MgCl_2$ and UTP and ATP as specified, in a final volume of 1.0 ml. [a] [UTP]=0.375 mM (subsaturating); [b] [ATP]=0.375 mM (subsaturating); [c] [ATP]=0.85 mM (saturating). [d] Assay conducted at pH 9.25 (20 mM glycine-NaOH buffer).

When the effect of GTP on the ammonia activity of the DON-enzyme is compared with the GTP effect on the ammonia reaction of the native enzyme, a striking difference is observed. Whereas the ammonia activity of the native enzyme is not affected at all by GTP, the ammonia activity of the DON enzyme is inhibited by GTP (Fig. 8). Since the covalent compound DON-enzyme is an analog to the glutamyl-enzyme this finding means that GTP inhibits the consumption of external ammonia once the glutamyl enzyme intermediate is formed. This also means that at pH 7.2, the optimal pH for the glutamine activity, GTP prevents the competition of external NH_4^+ with the NH_3 released from glutamine, thus preventing the inhibition of the enzyme. It seems therefore that in the presence of both GTP and glutamine only the latter will be consumed by the enzyme.

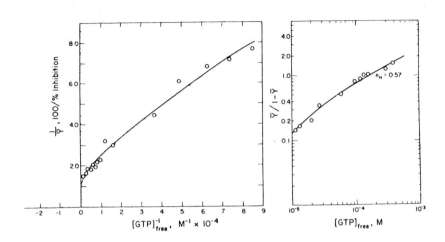

Figure 8:   Inhibition of the Ammonia Activity of
             DON-labeled CTP Synthetase by GTP

The ammonia activity was measured at pH 8.2 at saturating
ATP, UTP, $MgCl_2$ and ammonia under standard assay conditions
(38°) as a function of GTP concentration.   The data is
plotted as the double reciprocal plot (left) and a Hill
plot (right).   The response to GTP is typically negatively
cooperative (Levitzki and Koshland, 1969), with a Hill coefficient (n_H) of 0.57.

CONCLUSION

     The allosteric effector GTP facilitates the synthesis
of CTP by the enzyme CTP synthetase when glutamine is the
nitrogen source:

$$\text{Glu-NH}_2 + \text{UTP} + \text{ATP} \xrightarrow[\text{Mg}^{2+}]{\text{GTP}} \text{Glu+CTP+ADP+Pi}$$

GTP has no effect on the ammonia reaction when glutamine is
absent but inhibits the consumption of external ammonia
when glutamine is present.   The GTP was shown to facilitate
the formation of the glutamyl-enzyme and bound ammonia intermediate from the enzyme-glutamine Michaelis complex.

519

The enzyme bound ammonia is in the unprotonated form and is readily available for the subsequent steps which involve UTP amination, phosphorylation and dephosphorylation in sequence. The substrates ATP and UTP also act as allosteric effectors in the glutaminase step and their effect is additive to that of GTP.

The GTP effect is specifically felt in the glutamine subsite and the effector ligand has no effect on the response of the enzyme towards the other ligands involved in the reaction although their binding site is contiguous to that of the glutamine subsite (Figure 9).

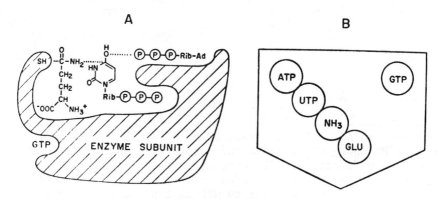

Figure 9: A Schematic Representation of the CTP Synthetase Subunit

The scheme (A) summarizes the known chemistry of CTP synthetase. There is one ammonia sub-site which binds either the ammonia released from glutamine or external ammonia. The competitiveness of the glutamine and the ammonia reactions indicate that the ammonia site is a part of the glutamine site (B). GTP is shown to affect only the glutamine binding site via an allosteric site functionally (and probably topographically) separated from the active site.

REFERENCES

Chakraborty, K.P., and Hurlbert, R.B. (1961). *Biochim. Biophys. Acta* 47, 607.

Fahien, L.A. and Shemisa, O. (1970). *Mol. Pharmacol.* 6,156.

Levitzki, A. and Koshland, D.E., Jr. (1969). *Proc. Natl. Acad. Sci.* 62, 1121.

Levitzki, A., and Koshland, D.E., Jr. (1971). *Biochemistry* 10, 3365.

Levitzki, A. and Koshland, D.E., Jr. (1972). *Biochemistry* 11, 247.

Levitzki, A., Stallcup, W.B. and Koshland, D.E., Jr. (1971). *Biochemistry* 10, 3371.

Lieberman, I. (1955). *J. Amer. Chem. Soc.* 77, 2661.

Lieberman, I. (1956). *J. Biol. Chem.* 222, 765.

Long, C.W., Levitzki, A. and Koshland, D.E.,Jr. (1970). *J. Biol. Chem.* 245, 80.

Long, C.W. and Pardee, A.B. (1967). *J. Biol. Chem.* 242, 4715.

Monod, J. Changeux, J.P. and Jacob, F. (1963). *J. Mol. Biol.* 6, 306.

Moos, C., Alpert, N. R. and Myers, T.C. (1960). *Arch. Biochem. Biophys.* 88, 183.

Yates, R.A. and Pardee, A.B. (1956). *J. Biol. Chem.* 221, 757.

Yount, R.B., Babcock, D., Ballantyne, W. and Ojata, D. (1971). *Biochemistry* 10, 2484.

521

ANTHRANILATE SYNTHETASE

Howard Zalkin

Department of Biochemistry, Purdue University,
Lafayette, Ind.

Abstract

Anthranilate synthetase catalyzes the first specific reaction of tryptophan biosynthesis and is subject to end product inhibition by tryptophan. Three patterns for aggregation of anthranilate synthetase with other enzymes of tryptophan synthesis have been recognized: type I, unaggregated; type II, aggregated with the second enzyme of the pathway, anthranilate-5-phosphoribosylpyrophosphate phosphoribosyltransferase (PR transferase); type III, associated with N-(5'-phosphoribosyl)anthranilate isomerase and indole glycerol 3-P synthetase or to just the latter. In all cases glutamine-dependent anthranilate synthetase activity requires participation of non identical polypeptide chains designated anthranilate synthetase Components I and II. The enzyme from Salmonella typhimurium contains two chains of each component. Component I alone catalyzes anthranilate synthesis with NH_3 but not with glutamine. There appears to be a common mechanism for glutamine utilization for the three types of enzyme. Glutamine binds to anthranilate synthetase Component II and the amide is transferred to the NH_3 site on component I. For type II enzyme the PR transferase chain is bifunctional and contains an NH_2-terminal segment of molecular weight approximately 20,000 (component II) covalently joined to the putative PR transferase segment of molecular weight about 40,000. Cooperative kinetics and cooperativity for tryptophan binding are observed with oligomeric anthranilate synthetase but not with free component I. End product inhibition of enzyme activity occurs when tryptophan binds to a regulatory site on component I and provokes or maintains a conformation having poor affinity for substrates.

523

Introduction

The pathway for tryptophan synthesis in microorganisms and plants is shown in Fig. 1. Anthranilate synthetase catalyzes the first reaction in the tryptophan biosynthetic pathway. In most organisms this enzyme is subject to end product inhibition by tryptophan. It is the purpose of this paper to review selected recent developments that may contribute to the eventual detailed understanding of the structure-function relationships and the mechanisms for catalysis and end product control by tryptophan.

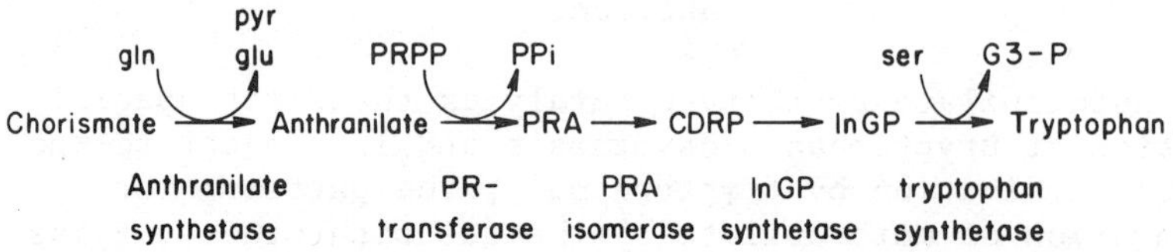

Fig. 1 Pathway for tryptophan biosynthesis. The abbreviations are: gln, glutamine; pyr, pyruvate; glu, glutamate; PRPP, 5-phosphoribosyl 1-pyrophosphate; PR transferase, anthranilate 5-phosphoribosylpyrophosphate phosphoribosyltransferase; PRA, N-(5'-phosphoribosyl) anthranilate; CDRP, 1-(o-carboxyphenylamino)-1-deoxyribulose phosphate; InGP, indole-3-glycerol 5-phosphate; ser, serine; G3-P, glyceraldehyde 3-phosphate.

Aggregation with Other Enzymes of Tryptophan Biosynthesis

Several patterns for association of anthranilate synthetase with other enzymes of tryptophan biosynthesis have been recognized (Zalkin, 1973). These are summarized in Table 1. In bacteria two types of anthranilate synthetase have been detected. In Bacillus subtilis, Chromobacterium violaceum, species of Pseudomonas and Serratia marcescens the enzyme appears not to be associated with other enzymes of tryptophan synthesis. This pattern we have designated type I. In Aerobacter aerogenes, Escherichia coli and Salmonella typhimurium anthranilate synthetase is aggregated to the second enzyme of the tryptophan pathway, PR transferase. This pattern is designated type II. A third type of anthranilate synthetase has been detected in yeast

Table 1

Patterns of Association of Anthranilate Synthetase
with Enzymes of Tryptophan Biosynthesis

Type	Activities aggregated to anthranilate synthetase	Organism
I	None	Bacillus subtilis Chromobacterium violaceum Pseudomonas putida Pseudomonas aeruginosa Pseudomonas acidovorans Pseudomonas testosteroni Serratia marcescens
II	PR transferase	Aerobacter aerogenes Escherichia coli Salmonella typhimurium
III	InGP synthetase or PRA isomerase and InGP synthetase	Saccharomyces cerevisiae Aspergillus nidulans Neurospora crassa

and fungi. In these organisms anthranilate synthetase is
associated with InGP synthetase. In cases where PRA isomerase is on the same protein chain with InGP synthetase it
is included in the aggregate with anthranilate synthetase.
Thus far tryptophan synthetase is the only enzyme of the
pathway not aggregated to anthranilate synthetase in some
organism. Patterns of enzyme association have been
deduced from gel filtration or zone centrifugation analyses
made on crude extracts. In some cases the results have
been verified following partial or complete purification.
It is of course possible that physiologically important
associations have been altered or destroyed upon cell dis-
ruption.

Our work has been with anthranilate synthetase from
Serratia marcescens (type I) and from Salmonella typhimurium (type II). J. A. DeMoss and co-workers (Hutter and
DeMoss, 1967; Gaertner and DeMoss, 1969; Arroyo-Begovich
and DeMoss, 1972) have studied the type III anthranilate

synthetase in fungi.

Requirement for Two Protein Components

Anthranilate synthetase catalyzes the reaction shown
below. Similar to other glutamine amidotransferases
(Meister, 1962; Zalkin, 1973), NH$_3$ can replace glutamine in
which case the products of the reaction are anthranilate,
pyruvate and H$_2$O.

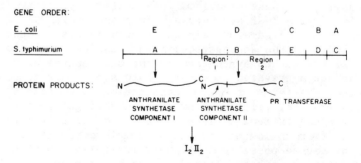

Ito and Yanofsky (1966) made the important observation
that glutamine-dependent anthranilate synthetase activity
in extracts of E. coli required participation of non identical subunits, anthranilate synthetase Component I, the
product of the E. coli trpE gene and PR transferase, the
product of trpD (Fig. 2). Because of its requirement for

Fig. 2 Gene-protein relationships for anthranilate
synthetase in E. coli and S. typhimurium. The product of
E. coli trpD or S. typhimurium trpB is bifunctional and
contains a component II segment corresponding to region 1
and a PR transferase segment from region 2. N and C refer
to amino and carboxyl terminal ends, respectively. The com-
pleted enzyme contains two chains of each protein (I$_2$II$_2$).

glutamine-dependent anthranilate synthetase activity, PR
transferase was designated anthranilate synthetase Component II. For type II anthranilate synthetase, the terms
anthranilate synthetase Component II and PR transferase
have been used interchangeably. The important point is
that the aggregated anthranilate synthetase can utilize
either glutamine or NH$_3$ whereas anthranilate synthetase
Component I by itself is active only with NH$_3$ and not with
glutamine. Bauerle and Margolin (1966) found the same
relationships for anthranilate synthetase in Salmonella
typhimurium. It appears that in extracts of wild type
cells from both organisms all of the anthranilate synthetase Component I and II is aggregated. Free component I or
II has not been detected in E. coli or S. typhimurium containing both proteins.

Anthranilate synthetase Component I from S. typhimurium was purified to homogeneity and partially characterized
(Nagano and Zalkin, 1970; Zalkin and Kling, 1968). Some of
its main properties are summarized in Table 2. Of interest
is the observation that a regulatory protein can be a

Table 2

Summarized Properties of Anthranilate Synthetase
Component I from S. typhimurium

Single polypeptide chain of molecular weight 64,000

Requires NH$_3$ and is inactive with glutamine

End product inhibition by tryptophan is competitive with
chorismate and noncompetitive with NH$_3$

Tryptophan causes decreased reactivity of enzyme sulfhydryl
groups with 5,5'-dithiobis(2-nitrobenzoate) corresponding
to extent of inhibition

Binds 1 mole of tryptophan per 64,000 g protein

No cooperativity for substrate saturation or inhibition

---

single polypeptide chain. Recently another case for a monomeric regulatory enzyme has been found (Panagou et al.,
1972). Tryptophan inhibition is competitive with chorismate
but there is no conclusive evidence that tryptophan and
chorismate compete for a single site. Based on decreased
sulfhydryl group reactivity we concluded that tryptophan
provokes a conformational change which could account for
end product inhibition (Nagano and Zalkin, 1970). In contrast to the results in Table 2, cooperative kinetics and
tryptophan binding were obtained with oligomeric anthranilate synthetase-PR transferase (Henderson et al., 1970;
Henderson and Zalkin, 1971). Coopertivity is therefore
associated with subunit interactions.

Mechanism of Glutamine Utilization

The glutamine analog 6-diazo-5-oxo-L-norleucine (DON)
was used to explore the mechanism of glutamine utilization
(Nagano et al., 1970). DON and certain other glutamine
analogs (O-diazoacetyl-L-serine [azaserine], 2-amino-4-oxo-
5-chloropentanoic acid and L-2-amino-3-ureidopropionic acid
[albizziin]) inactivate the glutamine-dependent activity of
several glutamine amidotransferases while having little or
no effect on the NH_3-dependent activity of many of these
enzymes (Meister, 1962; Zalkin, 1973). Similarly the differential inhibitory or inactivating effect of DON has been
noted earlier with anthranilate synthetase (Gibson et al.,
1967; Tamir and Srinivasan, 1969). We found that DON
specifically inactivated the glutamine-dependent activity
of the anthranilate synthetase aggregate from S. typhimurium but not anthranilate synthetase Component I (Nagano
et al., 1970). The effect of DON is shown in Fig. 3.
NH_3-dependent anthranilate synthetase activity was relatively stable and PR transferase activity was slightly
increased. Inactivation was irreversible. Activity was
not regained following 100-fold dilution, dialysis or gel
filtration. Glutamine afforded substantial protection
against inactivation by DON. These results indicate that
covalent attachment of DON to the glutamine site prevents
binding or reaction of glutamine but not NH_3 and thus provides evidence for distinct sites for these two ligands.
We concluded that anthranilate synthetase Component I is
unaffected by DON because it lacks a functional glutamine
binding site.

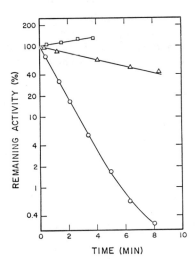

Fig. 3   Inactivation of glutamine-dependent (o—o)
anthranilate synthetase by DON. NH$_3$-dependent activity
(△—△) and PR transferase (□—□) were less affected.
From Nagano et al., 1970.

The rate of inactivation by low concentrations of DON
was stimulated 25-fold or more by chorismate (Table 3) suggesting ordered binding of first chorismate and then DON,
or by analogy glutamine. Mg^{2+} which is required for enzyme
activity had a relatively small effect on the reaction with
DON. Lack of effect of EDTA indicates that binding of
chorismate, DON and by analogy glutamine is not dependent
upon Mg^{2+}.

The conclusion that glutamine and NH$_3$ bind to distinct
sites was verified when anthranilate synthetase Component I
and PR transferase subunits of enzyme labeled with ^{14}C-DON
were separated by gel electrophoresis in 8 M urea. R$_f$
values relative to the bromophenol blue tracking dye were
0.16 and 0.13 for anthranilate synthetase Component I and
PR transferase, respectively. The data in Fig. 4 show that
radioactivity from ^{14}C-labeled DON is predominantly in the
region of the gel containing PR transferase. In 6 experiments, 72 to 94% of the radioactivity over the background
level was in the protein band corresponding to PR transferase.

Table 3

Effect of Chorismate and Mg^{2+} on Inactivation of Glutamine-Dependent Anthranilate Synthetase by DON

Pseudo first order rate constants for inactivation were determined from plots similar to those in Fig. 3. From Nagano et al., 1970.

Additions	k (min^{-1})
None	0.04
Chorismate	1.06
$MgCl_2$	0.06
Chorismate + $MgCl_2$	1.43
Chorismate + EDTA	1.06

Using the following two techniques it was shown that each PR transferase subunit binds one equivalent of DON. (a) Approximately two moles of ^{14}C-DON were incorporated per mole of enzyme aggregate. (b) The data in Fig. 5 show the extrapolated amount of DON required to completely inactivate the enzyme when enzyme is in excess over DON. Again the data show interaction of two DON per enzyme.

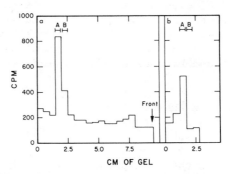

Fig. 4 Profiles from electrophoretic separations of ^{14}C-DON-labeled anthranilate synthetase-PR transferase. The regions designated A and B correspond to the PR transferase and anthranilate synthetase Component I protein bands, respectively. Two gel profiles are shown. The direction of electrophoresis was from left to right. From Nagano et al., 1970.

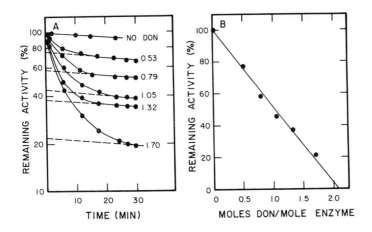

Fig. 5 Determination of the number of DON binding
sites. 5A shows the inactivation of glutamine-dependent
anthranilate synthetase with different molar ratios of DON
to enzyme (0 to 1.70). 5B shows the final remaining
activity plotted against the DON/enzyme ratio as obtained
from 5A. From Nagano et al., 1970.

Since the enzyme contains two chains of PR transferase
(Henderson and Zalkin, 1971) these data indicate binding of
one DON per PR transferase chain inactivates the glutamine-
dependent activity. We conclude that the glutamine site
for anthranilate synthetase is on the PR transferase chain
(anthranilate synthetase Component II).

Native anthranilate synthetase aggregate was found to
contain 8.6 exposed and 22.2 buried sulfhydryl groups as
determined by titration with 5,5'-dithiobis(2-nitrobenzo-
ate) (DTNB). After alkylation with DON the number of
exposed sulfhydryl groups was unchanged but two fewer
buried -SH groups were detected. It was concluded that DON
alkylates a specific cysteine residue of each PR transferase chain. Alkylation of cysteine residues by glutamine
analogs has been found for other glutamine amidotransferases (French et al., 1963; Mizobuchi and Buchanan, 1968;
Long et al., 1970; Trotta et al., 1971).

Binding of glutamine to anthranilate synthetase

Component II and chorismate to component I implies transfer of NH_3 between subunits. A glutaminase activity which could reflect the transfer mechanism was detected with anthranilate synthetase-PR transferase aggregate but not with either of the isolated components (Table 4). Glutaminase was assayed by measurement of the reaction glutamine to glutamate or by γ-glutamylhydroxamate formed in the presence of hydroxylamine. Of main importance is that

Table 4

Glutaminase Activity of Anthranilate Synthetase-
PR Transferase[a]

The complete reaction mixture contained 0.1 M Trischloride, 50 mM glutamine, 5 mM EDTA, 0.1 mM chorismate, 0.2 M hydroxylamine and 30 μg enzyme.

Conditions	Specific Activity
Complete	1200
+ Tryptophan	83
- Chorismate	57

[a] From Nagano et al., 1970.

glutaminase activity is dependent upon chorismate and is inhibited by tryptophan. The glutaminase activity therefore exhibits properties expected for a partial reaction of anthranilate synthetase. Somerville and Elford (1967) had earlier detected formation of a chorismate-dependent and tryptophan-inhibited hydroxamate in extracts of E. coli which is now explained as a glutaminase activity of E. coli anthranilate synthetase. We have proposed that the glutaminase activity may reflect the mechanism for transfer of the amide of glutamine from component II to the NH_3 site on anthranilate synthetase Component I. Fig. 6 summarizes our present view of the anthranilate synthetase mechanism. The scheme shows binding of chorismate to anthranilate synthetase Component I followed by binding of glutamine to anthranilate synthetase Component II. Glutamine is bound to a sulfhydryl group. Amide transfer leaves a γ-glutamyl thioester associated with component II. This thioester could react with hydroxylamine to yield γ-glutamylhydroxa-

532

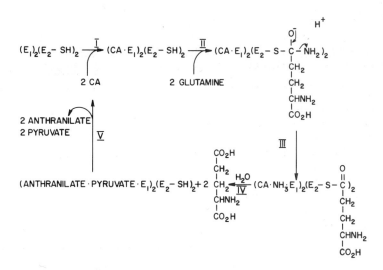

Fig. 6 Hypothetical reaction mechanism for glutamine-dependent anthranilate synthetase. E_1 and E_2, anthranilate synthetase Components I and II, respectively; CA, chorismate. The enzyme is shown as a tetramer containing two chains of each component. From Nagano et al., 1970.

mate. The final steps show formation and release of products from the enzyme. There is no evidence for the order of product release. This mechanism for glutamine utilization was first suggested for type II anthranilate synthetase from S. typhimurium (Nagano et al., 1970) but more recent experiments indicate an identical sequence for the type I anthranilate synthetase from Serratia marcescens (Zalkin and Hwang, 1971).

Subunit Composition of Anthranilate Synthetase from S. typhimurium and S. marcescens

Evidence pointing to a subunit composition of two chains each of anthranilate synthetase Components I and II for the enzyme from S. typhimurium is based on a consideration of the molecular weights of the isolated components and the intact aggregate, stoichiometry of ligand binding and densitometric analysis following electrophoretic separation and staining of subunits. Table 5 shows a summary of molecular weights and binding data which is consistent

Table 5

Summary of Molecular Weight and Ligand Binding for
Anthranilate Synthetase-PR Transferase of S. typhimurium

Molecular weight
    Anthranilate synthetase Component I 64,000
    PR transferase 63,000
    Anthranilate synthetase-PR transferase 285,000

Ligand binding per 285,000 g
    1.8 moles tryptophan
    1.9 moles chorismate
    2.0 moles DON

---

with the suggested stoichiometry. A densitometer tracing
from polyacrylamide electrophoresis in 8 M urea is shown in
Fig. 7. The average weight ratio of anthranilate synthetase Component I to PR transferase from 7 experiments was
1.16 ± 0.11. This value suggests that the two types of
subunits make equal weight contributions to the aggregate.
Our data therefore suggest that anthranilate synthetase-PR
transferase from S. typhimurium is a tetramer containing 2
protein chains of each type.

The composition of the type I anthranilate synthetase
from S. marcescens appears similar in several respects.
Although the enzyme from Serratia is not aggregated to
other enzymes of the tryptophan pathway it is an oligomer
containing non identical subunits. The present view is
that the Serratia enzyme contains 2 anthranilate synthetase
Component I chains of molecular weight approximately 60,000
and 2 anthranilate synthetase Component II chains of molecular weight approximately 21,000 (Zalkin and Hwang, 1971).
Treatment with DON specifically inactivates the glutamine-
dependent activity and results in covalent attachment of
one mole of DON per mole of anthranilate synthetase
Component II. As noted above, presently available data
indicate similar mechanisms for the type I and type II
anthranilate synthetases.

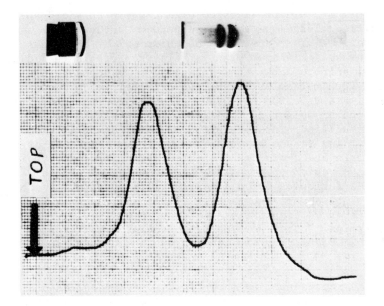

Fig. 7 Densitometer tracing of the subunits of
anthranilate synthetase-PR transferase separated by
electrophoresis in 8 M urea. Electrophoresis is from left
to right. The faster moving band is component I. From
Henderson and Zalkin, 1971.

Bifunctional Anthranilate Synthetase Component II
(PR Transferase) from Salmonella typhimurium

The PR transferase subunit from Salmonella typhimurium
is bifunctional because it is required for glutamine-
dependent anthranilate synthetase activity in association
with component I and it also provides PR transferase.
Recent results from two laboratories show that in vivo in
E. coli and S. typhimurium the anthranilate synthetase
Component II function is associated with an amino-terminal
segment of molecular weight approximately 20,000 whereas
the remaining two-thirds of the protein provides the PR
transferase activity (Yanofsky et al., 1971; Grieshaber and
Bauerle, 1972).

Our experiments (Hwang and Zalkin, 1971) have shown
that anthranilate synthetase-PR transferase from Salmonella

typhimurium could be digested with trypsin to yield an
unmodified glutamine- and NH_3-dependent anthranilate synthetase which is reduced in molecular weight and lacks PR
transferase activity. Disc gel electrophoresis in sodium
dodecyl sulfate shows that digestion of PR transferase
chains from molecular weight approximately 63,000 to
approximately 15,000 to 19,000 accounts for the reduction
in molecular weight of the aggregate. A highly speculative,
over-simplified, interpretation of this result is shown in
Fig. 8. The scheme proposes that the two functions of the
PR transferase chain are connected by an exposed trypsin-
sensitive region such that the PR transferase can be
released following brief digestion with trypsin. Recent
work by Grieshaber and Bauerle (1972) indicates that pro-
teolysis of PR transferase may occur stepwise to yield an
NH_2-terminal "core" peptide of molecular weight approximately 24,000 and not be limited to an exposed "hinge"
region.

The representation shown in Fig. 8 emphasizes one possibility for the evolution of multiple types of anthranilate synthetase. Monomeric, strictly NH_3-dependent,
anthranilate synthetase Component I may have been a primitive enzyme species. Upon evolution of glutamine and a

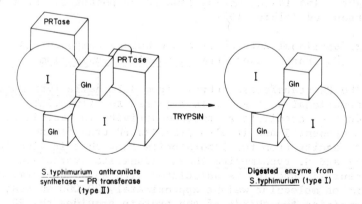

S. typhimurium anthranilate
synthetase – PR transferase
(type II)

TRYPSIN

Digested enzyme from
S. typhimurium (type I)

Fig. 8 Schematic representation for subunit arrangement of anthranilate synthetase-PR transferase from S.
typhimurium and the effect of digestion with trypsin. I,
anthranilate synthetase Component I; Gln, glutamine binding
segment of component II.

gene for a glutamine utilizing protein (anthranilate synthetase Component II) an oligomeric type I glutamine-
dependent anthranilate synthetase may have developed.
Fusion of the gene for anthranilate synthetase Component II
to the gene for PR transferase could have yielded type II
anthranilate synthetase-PR transferase. Summarized data in
Table 6 show the similarities between trypsin-digested
anthranilate synthetase aggregate from Salmonella
typhimurium and the Serratia enzyme. Proteolytic removal
of PR transferase from the Salmonella enzyme generates an
anthranilate synthetase remarkably similar to that found in
Serratia. The results suggest that covalent attachment of
PR transferase to anthranilate synthetase Component II of a
Serratia-like enzyme could have generated the Salmonella-
type anthranilate synthetase-PR transferase.

Table 6

Comparison of Anthranilate Synthetase Enzymes

Parameter	ASase-PRtase Salmonella	Trypsin-treated ASase-PRtase	ASase Serratia
Glutamine- and NH_3-ASase	Yes	Yes	Yes
PRtase	Yes	No	No
Oligomeric mol. wt.	285,000	≈141,000	≈141,000
Component I mol. wt.	64,000	≈64,000	≈60,000
Component II mol. wt.	≈63,000	≈15,000-19,000	≈21,000
Mole DON/mole enzyme	2	--	2
Composition[a]	I_2II_2	$I_2II_2^x$	I_2II_2
Glutaminase	Yes	Yes	Yes
K_m chorismate	3.7 μM	3.3 μM[b]	2.3 μM
K_m glutamine	0.67 μM	0.5 mM[b]	0.5 mM
n' tryptophan[c]	1.6	1.7	1.3-1.8

[a] I and II designate components I and II. Superscript x
   indicates digested with trypsin.
[b] From Tamir and Srinivasan (1969).
[c] Hill interaction coefficient.

On the Mechanism of Tryptophan Inhibition

The finding that tryptophan prevents inactivation of

glutamine-dependent anthranilate synthetase by DON (Nagano et al., 1970) provided an important clue to the mechanism of end product inhibition. This result localizes the action of tryptophan to one of the two binding steps: (a) binding of the first substrate chorismate or (b) the second substrate glutamine. It was later shown (Henderson and Zalkin, 1971) that 7.5 μM tryptophan inhibited by 90 to 100% the binding of chorismate to the S. typhimurium anthranilate synthetase aggregate.

More recent experiments (Zalkin and Chen, 1972) have shown that changes in conformation accompany binding of tryptophan and chorismate and that binding of one ligand excludes or nearly excludes binding of the other. Changes in conformation were estimated by reactivity of enzyme sulfhydryl groups under special conditions and by changes in fluorescence of enzyme-bound 1-anilinonapthalene-8-sulfonate (ANS). In the presence of 4 M urea chorismate and tryptophan were found to alter the rate of denaturation of anthranilate synthetase as measured by reactivity of sulfhydryl groups with DTNB. The data in Fig. 9 show that

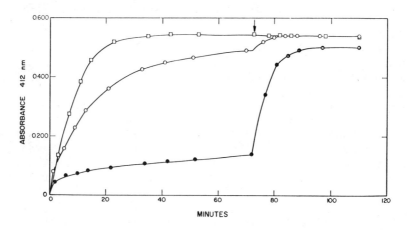

Fig. 9 Reaction of anthranilate synthetase sulfhydryl groups with DTNB in 4 M urea. Enzyme from S. marcescens with no addition (o—o); 0.5 mM chorismate (●—●) or 0.25 mM tryptophan (□—□). At the arrow, ↓, 0.1% sodium dodecyl sulfate was added to give complete denaturation.

tryptophan increased and chorismate decreased reactivity of sulfhydryl groups (absorbance 412 nm) under these conditions. All 15 to 17 -SH groups were titrated after addition of sodium dodecyl sulfate. We suggest that a possibly subtle conformational change resulting from binding of chorismate or tryptophan alters the rate of denaturation in 4 M urea and is thus greatly amplified. The tryptophan concentration dependence for change in the rate of DTNB reaction corresponds directly with that for inhibition of enzyme activity indicating the close relationship between the two parameters. With chorismate the half maximal change occurs at a concentration which is equal to the K_m.

Binding of ANS to anthranilate synthetase leads to a large enhancement of fluorescence as for many other proteins. As shown in Table 7 tryptophan and chorismate reduce the maximal fluorescence yield of enzyme-bound ANS

Table 7

ANS Binding to Serratia Anthranilate Synthetase

Addition	K_{app} (µM)	I_{max}
None	21	10.6
Tryptophan, 10 µM	25	1.9
Chorismate, 100 µM	23	4.1

without a change in the apparent dissociation constant for the fluorophore. Such results are consistent with chorismate- and tryptophan-dependent changes in conformation. Again, the concentration dependence for changes in conformation parallel the saturation of catalytic and inhibitory sites as shown in Fig. 10. Saturation of sites was determined by both activity and direct binding measurements.

Competitive binding experiments using equilibrium dialysis were conducted to examine the effect that binding of chorismate or tryptophan has on binding of the other. For competition of 2 ligands for identical non interacting sites or more generally, for mutually exclusive binding of two ligands the following equation applies (Curthoys and Rabinowitz, 1971):

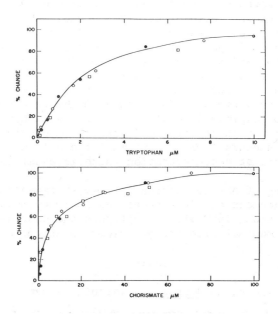

Fig. 10 (top) Effect of tryptophan concentration on inhibition of enzyme activity (●—●), ANS fluorescence (o—o) and tryptophan binding (□—□). (bottom) Effect of chorismate concentration on enzyme activity (●—●), ANS fluorescence (o—o) and chorismate binding (□—□).

$$A\frac{n-r}{r} = K_{D(A)} + B\frac{K_{D(A)}}{K_{D(B)}}$$

where A is the free concentration of fixed ligand, B the free concentration of varied ligand, n the number of enzyme sites for A, r the amount of A bound per enzyme and $K_{D(A)}$, $K_{D(B)}$ dissociation constants for A and B, respectively. A plot of $A(n-r/r)$ versus B yields a straight line with slope $K_{D(A)}/K_{D(B)}$ and ordinate intercept $K_{D(A)}$. Results of competitive binding experiments using equilibrium dialysis have shown that varied concentrations of one ligand interfere with the binding of a fixed concentration of the other. Summarized results are shown in Table 8. It is noted that dissociation constants determined from competitive binding experiments according to the binding

540

Table 8

Binding of Chorismate and Tryptophan to Anthranilate
Synthetase from Serratia marcescens

Fixed ligand (A)	Varied ligand (B)	$K_{D(A)}{}^a$	$K_{D(A)}{}^b$	$\dfrac{K_{D(A)}{}^a}{K_{D(B)}}$	$\dfrac{K_{D(A)}{}^b}{K_{D(B)}}$
		μM			
Tryptophan	Chorismate	4.7	2.1	0.24	0.31
Chorismate	Tryptophan	7.9	6.7	5.0	3.2

[a] Determined by competitive binding procedure.
[b] Determined from conventional one ligand binding experiments.

equation are in reasonable agreement with values obtained
from Scatchard plots in conventional one ligand binding
experiments. We conclude from the data in Figs. 9 and 10
and Tables 7 and 8:

(a) Changes in conformation accompany the binding of
chorismate and tryptophan to catalytic and regulatory
sites, respectively.

(b) Chorismate and tryptophan provoke or maintain
different conformational changes.

(c) The regulatory site for tryptophan must be distinct or at least partially distinct (partially overlapping) from the catalytic site since chorismate and tryptophan provoke different conformational changes.

(d) The conformational change that accompanies tryptophan binding to the regulatory site excludes the binding of
chorismate to the catalytic site.

(e) The conformational change that accompanies binding
of chorismate to the catalytic site facilitates binding of
the second substrate, glutamine.

If faced with a choice between concerted (Monod et
al., 1965) or sequential (Koshland et al., 1966) models for
the conformational transition the correspondence between
changes in saturation and conformation would favor a
sequential-type mechanism. Fig. 11 shows two hypothetical
schemes for the action of tryptophan and chorismate.
Three distinguishable enzyme conformations (E, E', E") are
written in Scheme 1 whereas two distinguishable conformations (E, E') are shown in Scheme 2. Based on changes in

HOWARD ZALKIN

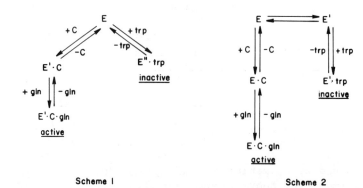

Scheme 1   Scheme 2

Fig. 11 Hypothetical schemes showing chorismate (C)-
and tryptophan (trp)-dependent conformation changes. From
a suggestion by K. G. Brandt.

fluorescence of enzyme bound ANS, chorismate and tryptophan
each promote a conformational state different from the free
enzyme fluorophore complex. On this basis we presently
favor Scheme 1 over Scheme 2. The data in Fig. 9 can be
accomodated by either scheme.

Acknowledgment

The author's research has been supported by grants
from the National Science Foundation and United States
Public Health Service (CA 11442).

References

Arroyo-Begovich, A., and DeMoss, J. A. (1972). In prepara-
tion.
Bauerle, R. H., and Margolin, P. (1966). Cold Spring
Harbor Symp. Quant. Biol. 31, 203.
Curthoys, N. P., and Rabinowitz, J. C. (1971). J. Biol.
Chem. 246, 6942.
French, T. C., Dawid, I. B., and Buchanan, J. M. (1963).
J. Biol. Chem. 238, 2186.
Gaertner, F. H., and DeMoss, J. A. (1969). J. Biol. Chem.
244, 2716.
Gibson, F., Pittard, J., and Reich, E. (1967). Biochim.
Biophys. Acta 136, 573.

Grieshaber, M., and Bauerle, R. H. (1972). Nature New Biology 236, 232.

Henderson, E. J., Nagano, H., Zalkin, H., and Hwang, L. H. (1970). J. Biol. Chem. 245, 1416.

Henderson, E. J., and Zalkin, H. (1971). J. Biol. Chem. 246, 6891.

Hutter, R., and DeMoss, J. A. (1967). J. Bacteriol. 94, 1896.

Ito, J., and Yanofsky, C. (1966). J. Biol. Chem. 241, 4112.

Koshland, D. E., Nemethy, G., and Filmer, D. (1966). Biochemistry 5, 365.

Long, C. W., Levitzki, A., and Koshland, D. E. (1970). J. Biol. Chem. 245, 80.

Meister, A. (1962). In "The Enzymes" (P. D. Boyer, H. Lardy and K. Myrback, eds.), Vol. VI, pp. 247-266. Academic Press, New York.

Mizobuchi, K., and Buchanan, J. M. (1968). J. Biol. Chem. 243, 4842.

Monod, J., Wyman, J., and Changeux, J.-P. (1965). J. Mol. Biol. 12, 88.

Nagano, H., and Zalkin, H. (1970). J. Biol. Chem. 245, 3097.

Nagano, H., Zalkin, H., and Henderson, E. J. (1970). J. Biol. Chem. 245, 3810.

Panagou, D., Orr, M. D., Dunstone, J. R., and Blakley, R. L. (1972). Biochemistry 11, 2378.

Somerville, R. L., and Elford, R. (1967). Biochem. Biophys. Res. Commun. 28, 437.

Tamir, H., and Srinivasan, P. R. (1969). J. Biol. Chem. 244, 6507.

Trotta, P. P., Burt, M. E., Haschemeyer, R. H., and Meister, A. (1971). Proc. Natl. Acad. Sci. U.S.A. 68, 2599.

Yanofsky, C., Horn, V., Bonner, M., and Stasiowski, S. (1971). Genetics 69, 409.

Zalkin, H. (1973). Advan. Enzymol., in press.

Zalkin, H., and Chen, S. H. (1972). J. Biol. Chem. 247, in press.

Zalkin, H., and Hwang, L. H. (1971). J. Biol. Chem. 246, 6899.

Zalkin, H., and Kling, D. (1968). Biochemistry 7, 3566.

BIOSYNTHESIS OF ANTHRANILIC AND p-AMINOBENZOIC ACIDS

P. R. Srinivasan

Department of Biochemistry
Columbia University
College of Physicians and Surgeons
New York, N.Y., 10032

INTRODUCTION

The aromatic pathway leading to the biosynthesis of the amino acids, phenylalanine, tyrosine and tryptophan and p-aminobenzoate, p-hydroxybenzoate, 2,3 dihydroxybenzoate, vitamin K and unbiquinone is one of the most complex metabolic pathways. Biochemical and genetic analysis of mutants of E. coli, Aerobacter aerogenes and Salmonella led to its elucidation and is shown in Fig. 1.

PATHWAY FOR AROMATIC BIOSYNTHESIS IN BACTERIA

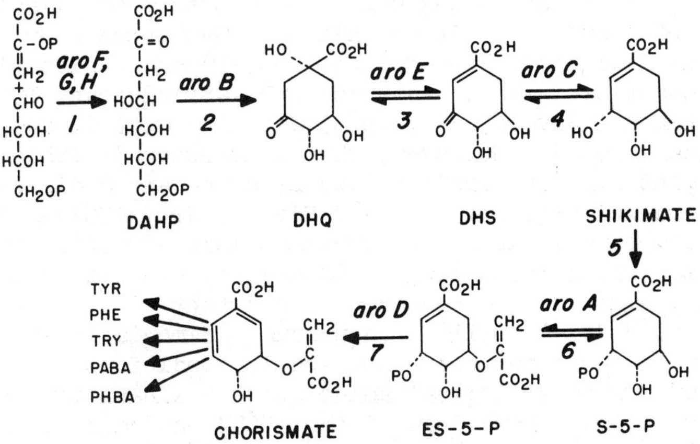

Fig. 1. The aromatic pathway.

For convenience, it could be divided into a common pathway comprising of intermediates required for the biosynthesis of all aromatic compounds and the branch pathways which

lead to the biosynthesis of the individual aromatic compounds. The common pathway, consisting of seven reactions,
is initiated by the condensation of the two intermediates
of carbohydrate metabolism, enolpyruvate phosphate and
erythrose-4-phosphate, to give the seven carbon compound,
3-deoxy-D-arabino-heptulosonic acid 7-phosphate (DAHP),
which is cyclized and converted in a series of six
reactions to chorismate, the branch point compound and the
last common intermediate in the pathway.

The conversion of chorismate to anthranilate and p-
aminobenzoate leads to the biosynthesis of tryptophan and
folic acid, while conversion to prephenate provides the
precursors for phenylalanine and tyrosine formation. The
other aromatic compounds are similarly derived from
chorismate (Gibson and Pittard, 1968)

BIOSYNTHESIS OF ANTHRANILIC ACID

Even before the identification of the branch point
compound as chorismate, it was possible to demonstrate
that the amino group of anthranilate is derived from the
amide N of glutamine using cell free extracts of E. coli
capable of converting either shikimate 5-phosphate or
enolpyruvylshikimate 5-phosphate (ES-5-P) and L-glutamine
almost quantitatively to anthranilate (Srinivasan, 1959;
Srinivasan and Rivera, Jr., 1963; and Rivera, Jr., and
Srinivasan, 1963). Moreover, this conversion is inhibited
by the glutamine antagonists, aza-L-serine and 6-diazo-5-
oxo-L-norleucine (Srinivasan and Rivera, Jr., 1963). The
latter compound is at least 200 times more effective than
the former glutamine analogue, aza-L-serine. With crude
extracts, ammonia is effective to an extent of 25 to 40%
at a pH of 8.2. As will be seen later, ammonia is an
effective amino donor under more basic conditions. The
nature of the physiological amino donor has been investigated by Gibson, Pittard and Reich(1967) who came to the
conclusion that glutamine is the "natural" substrate for
anthranilate synthesis. However, when E. coli is grown
in a medium containing NH_4^+, the level of glutamine synthetase is lowered and under these conditions of repression, if the ammonia concentration is high it could serve
as an efficient amino donor. This situation may be
applicable to most, if not all, transfer reactions involving glutamine as a substrate.

Early experiments on the incorporation of ^{14}C-glucose into aromatic amino acids in A. aerogenes (Rafelson,1955; Ehrensvärd, 1958) were at variance with the biosynthetic scheme outlined here and suggested that the ring of shikimate is probably not utilized as such i.e., intact, as a precursor of the aromatic amino acids. We studied this problem by comparing the labelling pattern of anthranilate isolated from culture filtrates of E. coli B-37 (a mutant blocked immediately after anthranilic acid) grown in the presence of (3,4-^{14}C) glucose with the pattern previously obtained in shikimate (Srinivasan, 1965). The results revealed that the carboxyl of shikimate becomes the carboxyl of anthranilic acid and the aromatization of the ring takes place without rearrangement (Fig. 2).

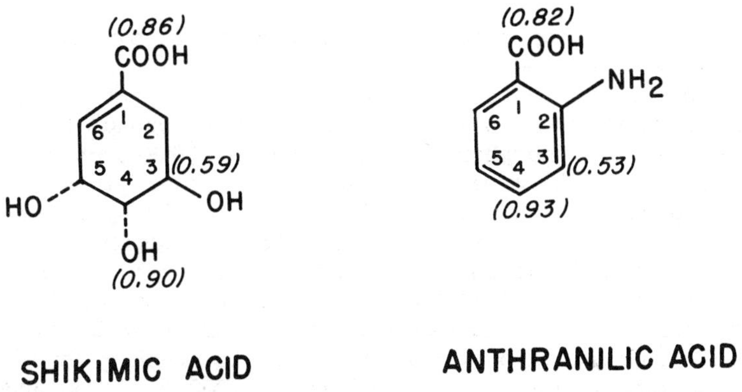

SHIKIMIC ACID   ANTHRANILIC ACID

Fig. 2. Incorporation of carbon atoms 3 and 4 of glucose into shikimic acid and anthranilic acid.

Furthermore, the methods employed for the degradation of anthranilic acid indicated that the amination occurs on carbon atom 2 rather than on carbon 6.

Properties of Anthranilate Synthase-Anthranilate synthase catalyzes the formation of anthranilate from chorismate and glutamine or chorismate and ammonia (Fig. 3).

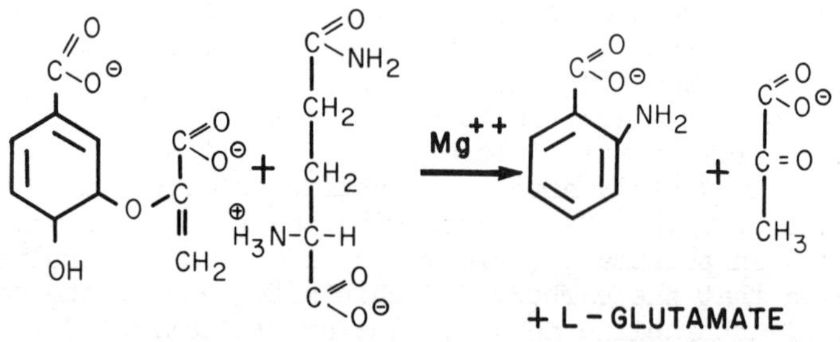

Fig. 3. Reaction catalyzed by anthranilate synthase.

The enzyme exists as an aggregate with the next enzyme in
the tryptophan pathway, anthranilate-5'-phosphoribosyl-
pyrophosphate phosphoribosyltransferase (P R transferase).
The aggregate is capable of utilizing either glutamine or
ammonia as an amino donor. The anthranilate synthase
protein devoid of the transferase termed Component I, can
utilize ammonia as an amino donor. The enzyme we have used
in our studies is derived from a mutant of Salmonella
typhimurium B-26, blocked immediately after anthranilic
acid. Our method of preparation yields a homogenous
enzyme with a molecular weight of 137,000 + 15,000 and a
sedimentation value of 6.75S (Tamir and Srinivasan, 1969).
This value is considerably lower than the values for the
molecular weight of the enzyme reported in the literature
(261,000). Since the anthranilate synthase was present
in the particulate fraction of the extract (105, 000 x g
pellet), we solubilized the enzyme with a crude lipase
preparation and to explain the discrepancy in the molecular
weight data we had suggested that our method of solubilization may have removed a part of the complex not essential for the activity of the enzyme. In the light of more
recent investigations by Hwang and Zalkin (1971) our interpretation for the differences in molecular weight seems
to be correct. It appears that our preparation has an
intact Component I (M.W. = 62,000) and partially digested
PR transferase (15,000 to 19,000) which still contains a
functional glutamine site and interacts with Component I to

give a functional glutamine dependent anthranilate synthase activity indistinguishable from that of the native enzyme (Fig. 4).

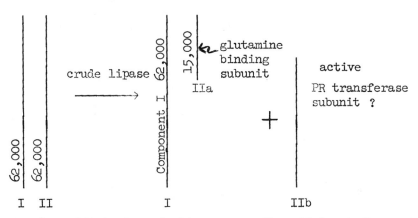

Fig. 4. Effect of crude lipase on anthranilate synthase.

In passing it is interesting to point out that the glutamine binding subunit of <u>Pseudomanas putida</u> has a molecular weight of 18,000 and is devoid of PR transferase activity (Queener and Gunsalus, 1970). Thus the PR transferase subunit of <u>E. coli</u> and <u>S. typhimurium</u> is the translational product of a gene which has evolved by fusion of two genes, a gene coding for the glutamine binding subunit and a gene coding for the PR transferase activity.

<u>Effect of Inhibitors</u>:The various compounds that inhibit anthranilate synthase can be divided into three groups based on the nature of their inhibition (Tamir and Srinivasan, 1969). The first group affects the activity of the enzyme with respect to both glutamine and ammonia as amino donors. The second group affects the activity of the enzyme only with glutamine as amino donor, and the third group only with ammonia as amino donor (Table I). p-Chloromercuribenzene sulfonic acid completely inhibited the enzymatic activity with both NH_3 and glutamine as amino donors and this inhibition can be reversed by thiol reagents. The glutamine analogs at low concentrations exerted their effect only with glutamine as an amino donor. Under these conditions, the inhibitory effect with ammonia

Table I

Effect of inhibitors on activity of anthranilate synthase

Group	Compound	Concentration	Inhibition	
			Gluta-mine	NH₄Cl
		M	%	
I	p-Chloromercuribenzene	2.5×10^{-6}	62	64
	sulfonic acid	1×10^{-5}	95	95
	Hg^{++}	1×10^{-5}	100	100
II	6-Diazo-5-oxo-L-norleucine	5.0×10^{-4}	100	21
	Azaserine	5.0×10^{-3}	100	15
	5-Chloro-4-oxo-L-norvaline	5.0×10^{-4}	100	14
	Hydroxylamine	2.0×10^{-1}	60	27
III*	Urea	2.0	0	50
	Dimethylurea	7.5×10^{-1}	15	67
	Tetramethylurea	5.0×10^{-1}	15	85
	Guanidine	5.0×10^{-1}	29	82

as the amino donor is minimal, the inhibition ranging from 15 to 20%. However, when the concentration of NH_4Cl is below its Km the enzymatic reaction with chorismate and NH_4Cl is abolished. These results can be interpreted on the basis of two different reactive sites being involved: one for glutamine, amino donor group, and the other for NH_3. This is unlikely from the results of the experiment presented in Table II.

Table II

Effect of glutamine addition on rate of synthesis of anthranilate from NH₄Cl and chorismate

Amino donor	Concentration	Anthranilic acid
	μmoles/ml	*μmole*
Glutamine	5	0.11
	20	0.14
	50	0.16
Glutamine plus 100 μmoles of NH₄Cl	5	0.19
	20	0.18
	50	0.17
NH₄Cl	100	0.20

In the presence of excess chorismate, the addition of varying amounts of glutamine to a saturating level of NH_4Cl
did not materially alter the synthesis of anthranilate.
The counterpart of this experiment, in which a large excess
of NH_4Cl was added to excess glutamine and chorismate,
also did not change the rate of the reaction. Thus, it is
reasonable to conclude that only a single site is involved
in the actual amination reaction. This is supported by an
examination of the kinetics of inhibition of 5-chloro-4-
oxo-L-norvaline on the enzymatic reactions of chorismate
with glutamine (Fig. 5, a and b)

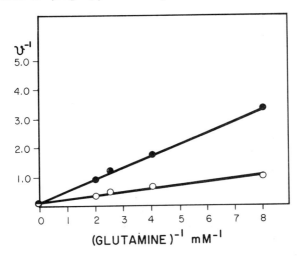

$$v^{-1}$$

(GLUTAMINE)$^{-1}$ mM^{-1}

Fig. 5a: Double reciprocal plots of initial velocity
against glutamine concentration in the presence and absence
of 5-chloro-4-oxo-L-norvaline. o———o, no inhibitor;
●———●, with 2.5 x 10^{-4} M inhibitor.

and chorismate with ammonia. In both cases the inhibition
is competitive. Similar results were obtained with the
other glutamine analogue, 6-diazo-5-oxo-L-norleucine.
Moreover, the Michaelis constants for glutamine and ammonia
at their respective pH optima are close. The V_{max} for both
amino donors at pH 7.9 is the same.

The third group of inhibitors exerted their major
effect on the enzymatic reaction of chorismate and ammonia
at concentrations ranging from 0.5 to 0.75 are the derivatives of urea. It is interesting that these derivatives,

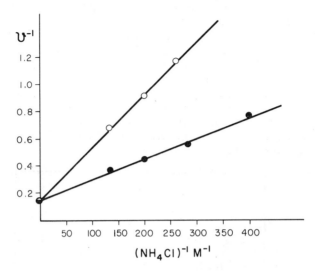

Fig. 5b: Double reciprocal plots of initial velocity against NH₄Cl concentration in the presence and absence of 5-chloro-4-oxo-L-norvaline. ●——●, no inhibitor; ○——○ , with 2.5 x 10⁻⁴ M inhibitor.

though really amides, did not affect the reaction with glutamine as amino donor at low concentrations. Moreover, the inhibition with group III compounds is not immediate but required incubation of the enzyme with inhibitors.

Effect of Hydroxylamine and N-Methylhydroxylamine-Somerville and Elford, (1967) have shown that NH_2OH and even more so, its derivative, CH_3-NHOH, are inhibitors of anthranilate synthase, the inhibitions resulting in the formation of unidentified hydroxamates. The fact that hydroxamate formation is completely dependent on chorismate, glutamine, and enzyme and is inhibited by the end product tryptophan, suggests that hydroxylamine may function by interfering with the overall reaction. Zalkin and Kling working with Component I of anthranilate synthase, observed the formation of the hydroxamate only in the presence of CH_3-NHOH (Zalkin and Kling, 1968). Moreover, it was dependent only on chorismate and enzyme whereas the amino donors, ammonia or glutamine had no effect. In view of these findings, it was felt that the identification of the hydroxamate may shed light on the mechanism of anthran-

ilate synthase reaction (Tamir and Srinivasan, 1971). The
effect of variation of NH_2OH and CH_3-NHOH concentration on
hydroxamate formation is illustrated in Fig. 6 a and b.

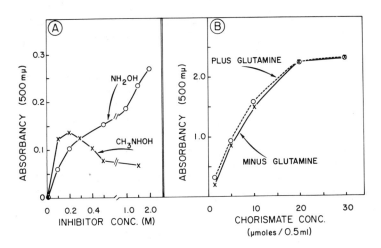

Fig. 6a: Effect of inhibitor concentration (conc.) on
hydroxamate formation.
b: Effect of chorismate concentration on hydroxamate formation with CH_3-NHOH.

At a concentration of 0.1 M, CH_3-NHOH is twice as effective
as NH_2OH, whereas very high concentrations of CH_3-NHOH
markedly reduces the formation of hydroxamate. NH_2OH,
on the other hand, behaves differently, increasing the
concentration yields a biphasic curve. At high concentrations of NH_2OH the formation of hydroxamate is probably
nonenzymatic (Meister et al. 1955) and this may explain
the biphasic curve. In agreement with the observations of
Somerville and Elford, Mg^{++} was not necessary for the
formation of hydroxamate with either NH_2OH or CH_3-NHOH.
While studying the effect of chorismate concentration on
the formation of hydroxamate with CH_3-NHOH it was found that
the addition of glutamine at high chorismate concentration
did not have any influence; even at low concentrations of
chorismate there was only a slight stimulation of hydroxamate formation as measured by the $FeCl_3$ color reaction
(Fig. 6). However, with NH_2OH as an inhibitor, the
presence of glutamine was essential for hydroxamate
formation.

553

These results suggested that the mechanism of inhibition, as well as the hydroxamate formed with CH_3-NHOH may be different from that with NH_2OH.

The hydroxamate formed in the presence of chorismate, glutamine and NH_2OH by anthranilate synthase was identified as γ-glutamylhydroxamate based on the following evidence:

a) Uniformly labelled [^{14}C]chorismate failed to label the hydroxamate isolated from the enzymatic reaction.

b) [^{14}C]Glutamine gave labelled hydroxamate.

c) The isolated hydroxamate had the same mobility as authentic γ-glutamylhydroxamate in two solvent systems (n-butanol, acetic acid, H_2O; ethanol, NH_4OH, H_2O) with two spray reagents: $FeCl_3$ and ninhydrin.

d) The infrared spectra of the enzymatic hydroxylamine product and of the authentic γ-glutamylhydroxamate were identical in all respects.

The hydroxamate formed with CH_3-NHOH and chorismate in the anthranilate synthase reaction was also studied and found to be an adduct of the enolpyruvyl moiety of chorismate. The evidence obtained suggests that the structure of this adduct is α-carboxy-α, N-dimethylnitrone (Tamir and Srinivasan, 1971).

$$CH_3 - \underset{\underset{\displaystyle CH_3 - N \longrightarrow O}{||}}{C} - COOH$$

Recently Pabst and Somerville (1971), while substantiating our findings on the identity of hydroxamates have suggested that the formation of the nitrone with CH_3-NHOH may be nonenzymatic in nature. In our studies with CH_3-NHOH we have omitted Mg^{++} and have used a large concentration of EDTA to prevent the formation of pyruvate which could result from the enzymatic conversion chorismate to anthranilate.

The identification of the reaction product with NH_2OH as γ-glutamylhydroxamate suggests that anthranilate synthase in the presence of chorismate can form an acyl-enzyme through the γ-carboxyl of glutamic acid and can exhibit glutaminase activity. Evidence for such an activity has recently been provided by Nagano et al.(1970). Glutaminase activity can be demonstrated only in the presence of chorismate. Although chorismate is essential for demonstrating glutaminase activity of Mg^{++} free anthranilate synthase, increasing the chorismate concentration had a non-linear effect on the glutaminase activity (Fig. 7).

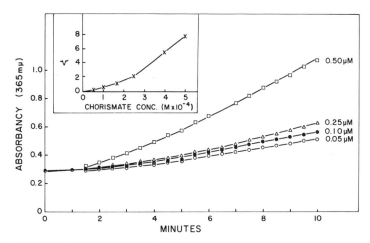

Fig. 7. Effect of chorismate concentration (conc.) on glutaminase activity.

Addition of hydroxylamine to the reaction mixture containing enzyme, chorismate and glutamine results in a decrease of glutamic   acid formed with the concomitant production of γ-glutamylhydroxamate (Tamir and Srinivasan, 1971).

Evidence for an Acyl Enzyme:Hydroxylamine inhibits the formation of anthranilate from chorismate and glutamine as well as the glutaminase activity exhibited by anthranilate synthase in the absence of Mg^{++}. To explore the relationship of the partial reaction to the complete reaction (i.e., anthranilate formation), the effect of hydroxylamine on these two reactions was examined (Tamir and Srinivasan, 1972) (Table III). Increasing the concentration of hydroxylamine decreases the rate of formation of anthranilate as well as the rate of hydrolysis of glutamine and the degree of inhibition at the different concentrations of NH_2OH is similar for both reactions. These results could imply that a common intermediate may be involved in these reactions, and that hydroxylamine effectively competes

555

Table III

Effect of NH₂OH on the rate of formation of anthranilate by anthranilate synthase and on the glutaminase activity of anthranilate synthase

NH₂OH concentration	Anthranilate[a] formed	Glutamate[a] release
M		
0.0	1.0	1.0
0.05	0.66	0.72
0.10	0.47	0.41
0.20	0.24	0.23

with chorismate or H_2O for the "enzyme complex." That the partial reaction of glutaminase activity and the overall reaction of anthranilate formation may go through a common intermediate is also supported by the finding that the end product tryptophan inhibits both reactions.

The production of an acyl intermediate between glutamine and enzyme during the complete reaction could explain the formation of γ-glutamylhydroxamate with NH_2OH. Such a hypothesis could be tested by changing the rate of formation of γ-glutamylhydroxamate by increasing the concentration of NH_2OH in the overall reaction. This should cause a proportional decrease in the amount of anthranilate formed. An experiment of this type is presented in Table IV. Changes in the NH_2OH concentration results in a concomitant alteration in the rates of synthesis of γ-glutamylhydroxamate and anthranilate. However, the sum of the two products, γ-glutamylhydroxamate and anthranilate, at the various NH_2OH concentrations appears to be constant and this indicates that the rate limiting step of the reaction is probably the formation of the "acyl-intermediate" rather than its subsequent decomposition. Such a conclusion may be valid because by varying the concentration of one of the acceptors, i.e., NH_2OH, the partitioning of the "intermediate" to products changed. The constant sum corresponds to the rate of formation of the intermediate.

If a glutamyl enzyme is an intermediate then γ-glutamylhydroxamate may be expected to be hydrolyzed by

Table IV

Effect of NH₂OH on the rate of formation of γ-glutamylhydroxamate and anthranilate

NH₂OH concentration	γ-Glutamyl-hydroxamate	Anthranilate[a]	Anthranilate plus γ-glutamyl hydroxamate
M		μmole	
None		0.478	0.478
0.4	0.187	0.266	0.453
0.5	0.227	0.245	0.472
0.6	0.240	0.229	0.469
0.7	0.280	0.213	0.493
0.8	0.290	0.186	0.470

anthranilate synthase and this was found to be the case. In Fig. 8 are given the Lineweaver-Burk plots for the hydrolysis of γ-glutamylhydroxamate and glutamine. The slopes of the curves (Vmax/Km) for both compounds appear to be very close and suggests that the two reactions may be subjected to the same rate limiting step. It should be

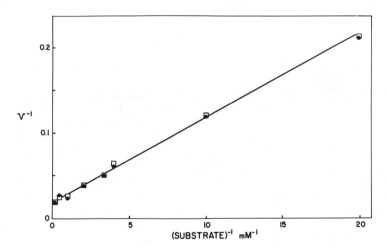

Fig. 8. Effect of variation of glutamine (●) and γ-glutamylhydroxamate (◻) concentration on glutaminase activity of anthranilate synthase.

557

stressed that the hydrolysis of either glutamine or γ-glut-amylhydroxamate is absolutely dependent on the presence of chorismate.

If the postulated "acyl-intermediate" such as a glutamyl enzyme is formed by the enzyme and glutamine then the anthranilate synthase which exhibits a glutaminase activity might also show an esterase activity as well. This was explored with γ-ethylglutamate as substrate and the results are presented in Fig. 9.

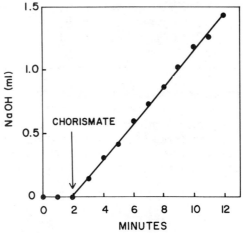

Fig. 9. Esterase activity of anthranilate synthase.

γ-Ethylglutamate is hydrolyzed by anthranilate synthase only in the presence of chorismate. Addition of hydroxyl-amine results in the formation of a hydroxamate with this substrate further suggesting the involvement of a glutamyl enzyme in the hydrolytic reaction. The yield of hydrox-amate is, however, poor with γ-ethylglutamate as compared to glutamine as substrate. The Km and V_{max} with this substrate was studied using the alcohol dehydrogenase assay for the ethanol formed in this reaction (Fig. 10). Although there is no change in the V_{max} (0.08 μmole per min) with this substrate as compared to the other substrates glutamine or γ-glutamylhydroxamate, the Michaelis-Menten constant (Km = 2.1×10^{-3} M) is considerably higher than with the other two substrates. The Km for glutamine as well as γ-glutamylhydroxamate is 5×10^{-4} M. These

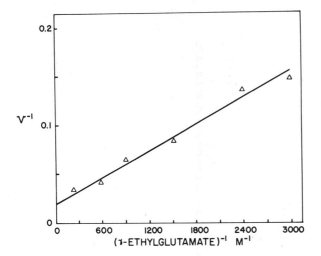

Fig. 10. Effect of variation of γ-ethylglutamate concentration on esterase activity of anthranilate synthase.

studies strongly suggest the participation of an acyl enzyme, i.e., a γ-glutamylenzyme in the reaction catalyzed by anthranilate synthase. Definitive proof for the existence of such an intermediate awaits the actual isolation of such an intermediate.

On the basis of some of the evidence presented here, the following scheme for the enzymatic reactions catalyzed by anthranilate synthase can be advanced (Fig. 11). Anthranilate synthase reacts with chorismate to form enzyme-chorismate complex. This complex can combine with ammonia in the presence of Mg^{++} to give anthranilate, pyruvate and free enzyme or react with CH_3-NHOH to yield the adduct shown in the figure. The enzyme chorismate complex also interacts with glutamine. This ternary complex can undergo hydrolysis with H_2O to give glutamate, NH_3, chorismate and free enzyme or react with NH_2OH to give γ-glutamylhydroxamate, chorismate, ammonia and free enzyme or form anthranilate, pyruvate, glutamate and free enzyme in the presence of Mg^{++}.

In his discussion of anthranilate synthase, Dr.Zalkin will present experimental results which indicate that the

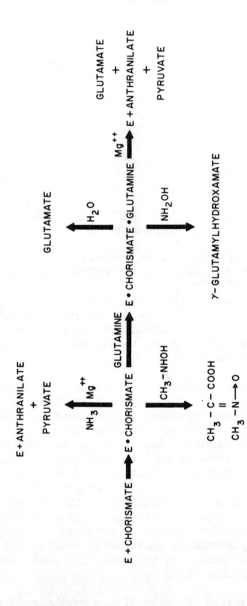

Fig. 11. Scheme for the various reactions catalyzed by anthranilate synthase.

"enzyme-chorismate-glutamine complex" is a thioester complex. The formulation of the "enzyme-chorismate-glutamine tetrahedral complex," while clarifying the glutaminase activity and the formation of γ-glutamylhydroxamate avoids the release of the amino group as ammonia in the presence of Mg^{++} (Fig. 12).

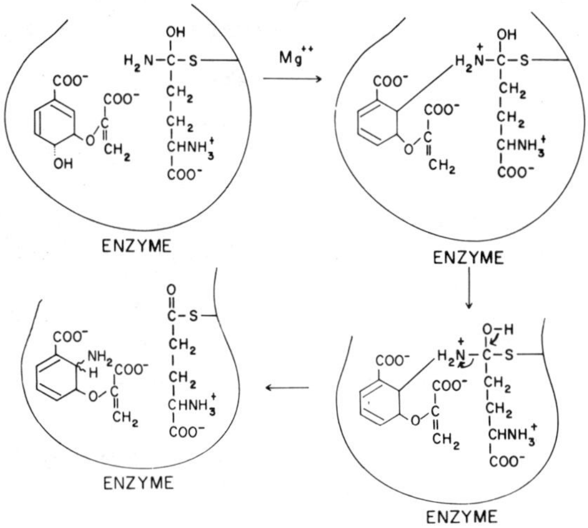

Fig. 12. Mechanism of anthranilate synthase reaction.

Subsequently the amination of chorismate and the rearrangement of the tetrahedral complex to yield a glutamyl complex eventually leads to the formation of glutamate, anthranilate, pyruvate and free enzyme. However, these ideas do not really explain the basic mechanism of anthranilate synthesis. In the aromatization of chorismate, a hydroxyl group and an enolpyruvyl group must be eliminated. Elimination of the enolpyruvyl group is accompanied by protonation to form pyruvate. The source of this proton has been investigated in our laboratories by performing the enzymatic reaction in 99.7% D_2O. The isolated pyruvate contained close to an atom of deuterium in the methyl group. High resolution mass spectra also revealed that about 6% of the deuterio-pyruvate contains a CHD_2 species (Tamir and Srinivasan, 1970).

Enolpyruvyl esters are high energy compounds and the

free energy of hydrolysis of the enolpyruvyl group can be deduced from other reactions to be greater than 10 kilo-calories.

$$\text{Enolpyruvate-P}^{3^-} + H_2O \longrightarrow \text{pyruvate}^- + H_3PO_4^{3^-}$$
$$\Delta F' = -13,300 \text{ cal.}$$

$$\text{ES-5-P}^{4^-} + H_2O \longrightarrow \text{shikimate 5-P}^{3^-} + \text{pyruvate}^-$$
$$\Delta F' = -11,300 \text{ cal.}$$

(Levin and Sprinson, 1964)

Although the formation of the tetrahedral complex might facilitate amination of chorismate at carbon 2 resulting in a rearrangement of the double bonds and elimination of the hydroxyl group, it appears wasteful that the final step of aromatization which is itself favored due to a large gain in resonance energy should require the elimination of an enolpyruvyl group. These considerations lead us to suggest the following mechanism (Fig. 13).

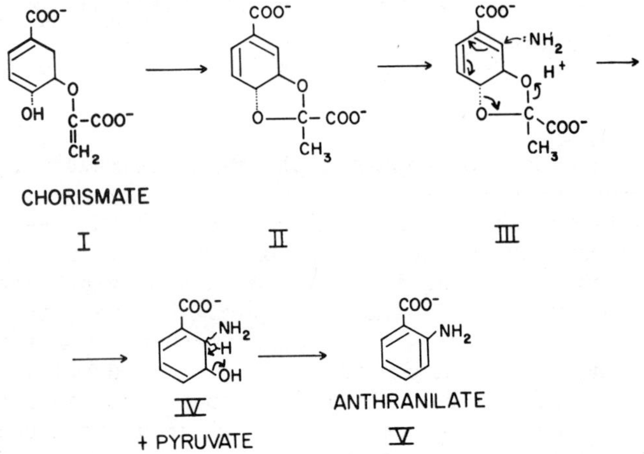

Fig. 13. Postulated scheme for anthranilate synthase reaction.

Reaction of the hydroxyl group on carbon 4 of chorismate with the double bond of the enolpyruvyl group results in the formation of the cyclic intermediate (II). This compound can then undergo amination by NH_3 or by the amino group from the "tetrahedral complex" discussed earlier

562

resulting in the release of the pyruvate group to yield compound IV. Aromatization then yields anthranilate. Since there are no free intermediates in the anthranilate synthase reaction, we must assume that the postulated compounds II and IV are enzyme bound. It is feasible to test this hypothesis with the aid of 4-^{18}O-chorismate. Such a compound should yield ^{18}O-pyruvate which can be trapped as lactate to prevent exchange. These studies are in progress.

BIOSYNTHESIS OF p-AMINOBENZOIC ACID

Earlier investigations revealed that the formation of p-aminobenzoic acid also proceeds by the common pathway discussed in the preceding section. Cell free extracts of Baker's yeast were capable of converting shikimate 5-phosphate and L-glutamine to p-aminobenzoic acid (Weiss and Srinivasan, 1959). NH_4Cl was a poor amino donor in these extracts at a pH of 8.2 in contrast to anthranilate biosynthesis in crude extracts where NH_4Cl was effective to an extent of 25 to 40% as an amino donor at this pH. Using L-(amide-^{15}N) glutamine we were able to demonstrate that the amino group of p-aminobenzoic acid was derived from the amide-N of glutamine (Srinivasan and Weiss,1961). The glutamine analogues, 6-diazo-5-oxo-L-norleucine and aza-L-serine, abolished the synthesis of p-aminobenzoic acid. The former inhibitor was 200 times less inhibitory in this system compared to its effect in the anthranilate forming system. The earlier observation of Davis (1955) that mutants blocked immediately after 3-enolpyruvyl shikimate 5-phosphate required p-aminobenzoic acid in addition to the aromatic amino acids implicated chorismate as an immediate precursor. Cell free extracts of yeast (Hendler and Srinivasan, 1967) and A. aerogenes 62-1 (Gibson et al. 1964) were found to convert chorismate to p-aminobenzoic acid in the presence of L.glutamine.

Cell free extracts of yeast can be further fractionated to yield two fractions, both of which were required for the conversion of chorismate to p-aminobenzoate (Table V). The admixture of Fraction I with heated Fraction II or vice versa was not effective. These results suggest that two protein fractions are needed for the conversion of chorismate to p-aminobenzoate (Hendler and Srinivasan,1967).

563

Table V

CONVERSION OF CHORISMATE TO p-AMINOBENZOATE BY VARIOUS
ENZYMATIC FRACTIONS OF BAKER'S YEAST

Enzyme fraction	Protein in reaction mixture (mg)	p-Aminobenzoate formed (mμmoles)
Crude extract	4.0	2.4
40 to 60% $(NH_4)_2SO_4$	4.0	13.7
Fraction I	2.0	0
Fraction II	0.8	0
Fractions I plus II	2.8	12.2

A similar situation also exists in E. coli. Mutant
strains requiring p-aminobenzoate have been isolated in
E. coli by Gibson and his colleagues and these mutants map
at two distinct loci. The genes concerned with p-aminobenzoate synthesis in E. coli are designated the pab A and
pab B genes (Huang and Pittard, 1967). These mutants have
no other growth requirements which suggests that the metabolic lesion must be between chorismate and p-aminobenzoate.
No cross feeding could be demonstrated between these p-
aminobenzoate auxotrophs (huang and Gibson, 1970). However,
the cell extracts complement each other. Chromatography of
the cell extracts on Sephadex G-100 gave two proteins,
fractions A and B, from the pab A mutant and pab B mutant
respectively. Components A and B while incapable of converting chorismate and L-glutamine to p-aminobenzoate individually, can form p-aminobenzoate from the substrates if
they were mixed together. The molecular weights of the
two components A and B have been analyzed by gel filtration
techniques and found to be 9,000 and 48,000 respectively
(Huang and Gibson, 1970). Probably, the small component
of 9,000 molecular weight from pab A strain is the glutamine binding subunit and should show glutaminase activity
in the presence of component B and chorismate. We have
also studied a mutant of E. coli 107-14 which requires p-
aminobenzoic acid and histidine for growth and this mutant
does not seem to accumulate intermediates which could be
converted to p-aminobenzoate. These studies indicate that
there may not be any free intermediates in the biosynthesis
of p-aminobenzoate from chorismate in E. coli and yeast.

Recently Altendorf, Gilch and Lingens (1971) have reported on the accumulation of an intermediate from the

culture medium of A. aerogenes 62-1 AC, a p-aminobenzoate auxotroph. This intermediate (Compound A) has been isolated and characterized by these authors to have the structure shown in Fig. 14. Compound A is utilized by

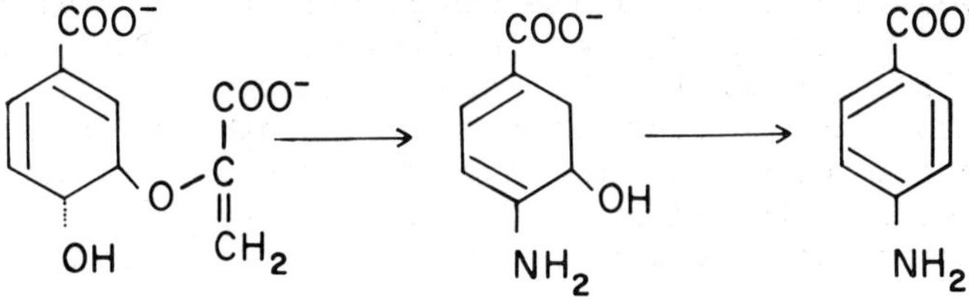

CHORISMATE   'COMPOUND A'   p- AMINO-
                                 BENZOATE

Fig. 14. Biosynthesis of p-amiobenzoate (Altendorf et al. 1971).

---

p-aminobenzoate auxotrophs of E. coli K12 and is converted to p-aminobenzoic acid by prolonged incubation at pH 3.5.

Genetic analysis of p-aminobenzoate requiring strains of Neurospora crassa led Drake (1956) to conclude that these mutants were associated with two loci. Further evidence that there may be free intermediates in the biosynthesis of p-aminobenzoate has come from analysis of these mutants (Table VI). When a sterile portion of the culture

Table VI

GROWTH RESPONSE OF NEUROSPORA CRASSA MUTANTS

The organisms were grown with aeration on Medium N supplemented with 2% sucrose and the indicated additions at 26°.

Strain	Additions	Comments
830	$2 \cdot 10^{-2}$ ug p-aminobenzoate	100 mg of mycelial mat in 4 days
H 193	$2 \cdot 10^{-2}$ ug p-aminobenzoate	100 mg of mycelial mat in 4 days
830	5 ml growth filtrate* of H 193	100 mg of mycelial mat in 4 days
H 193	5 ml growth filtrate* of 830	No growth even after 10 days

* The filtrate was found to be free of p-aminobenzoate by bioassay with E. coli 107-14.

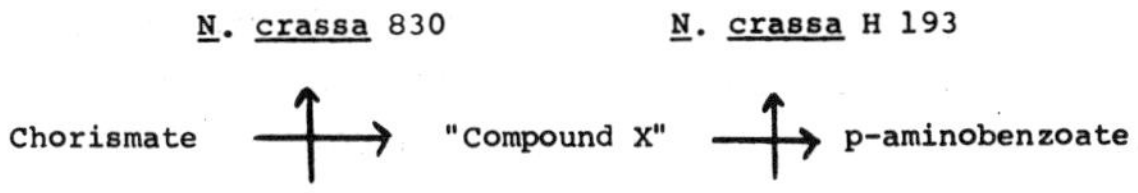

N. crassa 830   N. crassa H 193

Chorismate ─┤→ "Compound X" ─┤→ p-aminobenzoate

565

filtrate of a 15 day growth of N. crassa H193, grown in Vogel's medium (Vogel, 1956) containing a limiting amount of p-aminobenzoate (2×10^{-2} µg per ml) was subsequently inoculated with N. crassa mutant 830, maximum growth was observed in 4 days. On the other hand, the culture filtrate of N. crassa 830 was unable to replace the p-aminobenzoate requirement. This cross feeding experiment suggests that the conversion of chorismate to p-aminobenzoate involves at least two steps and that N. crassa 830 is blocked in the first step while N. crassa mutant H193 is blocked in the second step (Hendler and Srinivasan, 1967).

The identity of the proposed intermediate Compound X is under investigation. It is interesting to point out that the culture filtrates of the N. crassa mutants do not support the growth of the p-aminobenzoate requiring E. coli mutant 107-14.

In conclusion, the two enzymes, anthranilate synthase and p-aminobenzoate synthase, from E. coli and S. typhimurium, which catalyze the biosynthesis of anthranilate and p-aminobenzoate respectively, posses many common features. They utilize the same substrates:chorismate and glutamine. The enzymes are composed of two subunits, a larger subunit which binds chorismate and a smaller subunit (or a part of the second subunit) which binds glutamine. Except for anthranilate and p-aminobenzoate, the same products are produced. The interrelationship between these two enzymes goes a level higher in Bacillus subtilis where the same glutamine binding subunit, trp x is shared by anthranilate synthase and p-aminobenzoate synthase (Kane et al.1972). However, only tryptophan regulates the synthesis of the glutamine binding subunit trp x. Therefore, addition of tryptophan to the growth medium results in a progressive inhibition of growth and either p-aminobenzoate or folate completely reverses the inhibition of growth produced by tryptophan.

ACKNOWLEDGMENT S

The investigations reported from my laboratory were supported by a grant from the National Institutes of Health. I want to express my deep appreciation and thanks to my associates Drs. Bernard Weiss, Americo Rivera Jr.,

Sheldon Hendler, and Hadassah Tamir for their active role in
the development of the various aspects of our work reported in this article.

REFERENCES

Altendorf, K.H., Gilch, B. and Lingens, F. (1971). FEBS
    Letters, 16, 95.
Davis, B.D. (1955). Advances in Enzymology, 16, 287.
Drake, B. (1956). Genetics, 41, 640.
Ehrensvärd, G. (1958). Chemical Society Symposia (Bristol,
    1958), Special Publication No. 12, London, The
    Chemical Society, P. 17
Gibson, F., Gibson, M., and Cox, G.B. (1964). Biochim
    Biophys. Acta 82, 637.
Gibson, F., Pittard, J. and Reich, E. (1967). Biochim.
    Biophys. Acta 136, 573.
Gibson, F., and Pittard, J. (1968). J. Bacteriol. Rev.
    32, 465.
Hendler, S., and Srinivasan, P.R. (1967). Biochim. Biophys.
    Acta 141, 656.
Huang, M., and Pittard, J. (1967). J. Bacteriol. 93, 1938.
Huang, M., and Gibson, F. (1970). J. Bacteriol. 102, 767.
Hwang, L.H., and Zalkin, H. (1971). J. Biol. Chem. 246,
    2338.
Kane, J.F., Holmes, W.M., and Jensen, R. (1972). J. Biol.
    Chem. 247, 1587.
Levin, J.G., and Sprinson, D.B. (1964). J. Biol. Chem.
    239, 1142.
Meister, A., Levintow, L., Greenfield, R.E., and
    Abendschein, P. (1955). J. Biol. Chem. 215, 441.
Nagano, H., Zalkin, H., and Henderson, E.J. (1970). J.
    Biol. Chem. 245, 3810.
Pabst, M.J., and Somerville, R. (1971). J. Biol. Chem.
    246, 7214.
Queener, S.F., and Gunsalus, I.C. (1970). Proc. Nat. Acad.
    Sci. U.S.A. 67, 1225.
Rafelson, M.E., Jr. (1955). J. Biol. Chem. 212, 953: 213,
    479.
Rivera, A., Jr., and Srinivasan, P.R. (1963). Biochemistry
    2, 1063.
Somerville, R.L., and Elford, R. (1967). Biochim. Biophys.
    Res. Commun. 28, 437.
Srinivasan, P.R. (1959). J. Amer. Chem. Soc. 81, 1772.

Srinivasan, P.R., and Weiss, B. (1961). Biochim. Biophys. Acta. 51, 597.

Srinivasan, P.R., and Rivera, A. Jr. (1963). Biochemistry 2, 1059.

Srinivasan, P.R. (1965) Biochemistry 4, 2860.

Tamir, H., and Srinivasan, P.R. (1969). J. Biol. Chem. 244, 6507.

Tamir, H., and Srinivasan, P.R. (1970). Proc. Nat. Acad. Sci. U.S.A. 66, 547.

Tamir, H., and Srinivasan, P.R. (1971). J. Biol. Chem. 246, 3024.

Tamir, H., and Srinivasan, P.R. (1972). J. Biol. Chem. 247, 1153.

Vogel, H. (1956). Microb. Genet. Bull., 13, 42.

Weiss, B., and Srinivasan, P.R. (1959). Proc. Nat. Acad. Sci. U.S.A. 45, 1491.

Zalkin, H., and Kling, D. (1968). Biochemistry 7, 3566.

THE GLUTAMINE (OR AMMONIUM) REQUIREMENT
FOR HISTIDINE BIOSYNTHESIS

Bruce N. Ames
Biochemistry Department
University of California
Berkeley, California 94720

Salmonella typhimurium uses ten enzymic steps to convert phosphoribosylpyrophosphate and ATP to histidine. The biosynthetic steps are illustrated in Fig. 1. Any mutant lacking one of the enzymic activities will grow normally when supplied with exogenous histidine; hence, the pathway has no branch points leading to other metabolites required for growth. Several thousand mutations leading to a requirement for histidine have been described in the cluster of genes specifying the enzymes of histidine biosynthesis (Hartman et al., 1971): these genes, then, are both necessary and sufficient to produce the enzymes to synthesize histidine. The enzymology of histidine biosynthesis has been recently reviewed in detail (Martin et al., 1971).

Glutamine was shown by isotopic work to be involved in the histidine pathway as a donor of amide nitrogen for the synthesis of the imidazole ring (Neidle and Waelsch, 1959). This was first shown in a cell free system by Moyed and Magasanik (1960). The enzymology and genetics were worked out by Smith and Ames (1964) who showed that the conversion of BBMIII to IGP required either glutamine or ammonia. They showed that the enzyme coded for by the F gene was necessary and sufficient for conversion if high ammonia was used, and for the utilization of glutamine the product of the H gene was required as well. As glutamine was required in lower amounts and could be used at a more physiological pH, it appeared to be the natural nitrogen donor.

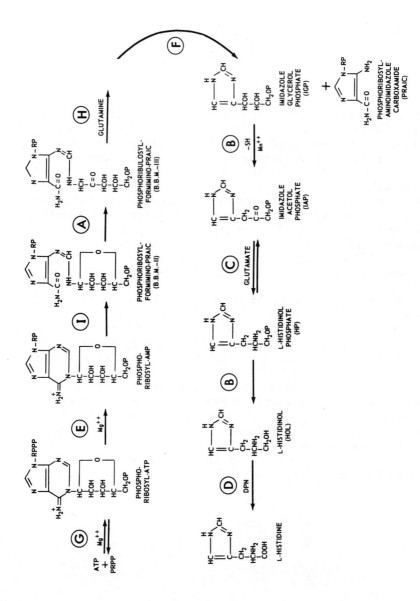

Figure 1. Pathway of histidine biosynthesis

REFERENCES

Hartman, P. E., Hartman, Z., Stahl, R. C. and Ames, B. N. (1971). Advances in Genetics 16, 1.

Martin, R. G., Berberich, M. A., Ames, B. N., Davis, W. W., Goldberger, R. F., and Yourno, J. D. (1971). In "Methods in Enzymology 17B" (H. and C. W. Tabor, eds.), Academic Press, New York.

Moyed, H. S., and Magasanik, B. (1960). J, Biol. Chem. 235, 149.

Neidle, A., and Waelsch, H. (1959). J. Biol. Chem. 234, 586.

Smith, D. W. E., and Ames, B. N. (1964).J. Biol. Chem. 239, 1848.

UTILIZATION OF GLUTAMINE FOR THE BIOSYNTHESIS OF ASPARAGINE

Bernard Horowitz and Alton Meister

Department of Biochemistry, Cornell University Medical College
New York, N.Y. 10021

ABSTRACT

This paper describes studies on the properties of glutamine-dependent asparagine synthetase isolated from mouse leukemia cells (RADA1). This work developed from earlier observations in this laboratory (Horowitz et al., 1968), which showed (in part) that certain tumors that are resistant to therapy with asparaginase exhibit very high asparagine synthetase activity. Purified asparagine synthetase was found to catalyze the following reaction:

$$\text{L-Glutamine} + \text{L-Aspartate} + \text{ATP} + H_2O \xrightarrow{\text{Mg}^{++} \text{ or Mn}^{++}}$$

$$\text{L-Asparagine} + \text{L-Glutamate} + \text{AMP} + \text{PPi}$$

Glutamine may be replaced by NH_4^+ as the amide nitrogen donor. β-L-Aspartyl adenylate was synthesized and direct evidence was obtained for its participation as an intermediate in asparagine synthesis. This enzyme seems to differ in at least two significant respects from other glutamine amidotransferases: (1) The enzyme exhibits glutaminase activity both in the presence and absence of the other substrates required for asparagine synthesis. (2) Both the glutaminase and glutamine dependent (but not the ammonium-dependent) synthetase require chloride ion. Chloride can be replaced by certain other anions, but with much less activity. The enzyme is inhibited by L-2-amino-4-oxo-5-chloropentanoic acid, aminomalonic acid, N-methyl-DL-aspartate, and several other compounds.

INTRODUCTION

The central significance of the amide nitrogen atom of glutamine in a variety of biosynthetic processes (e.g., the synthesis of GMP, DPN, CTP, histidine, tryptophan, carbamyl phosphate, purines) has been emphasized by many of the papers presented at this symposium. Of the several glutamine amidotransferases that are involved in these reactions, glutaminase (which catalyzes a reaction in which water is the acceptor) was the first to be recognized, while the synthesis of asparagine has been a relatively recent addition to this category of enzymes (see (Meister, 1962) for a review of this area). The synthesis of asparagine catalyzed by certain bacterial enzymes takes place by conversion of aspartate and ammonia to asparagine in a reaction in which ATP is cleaved to AMP and inorganic pyrophosphate (Ravel et al., 1962; Cedar and Schwartz, 1969; Burchall et al., 1964). Earlier reports that the synthesis of asparagine in animal tissues and in certain plants takes place by a reaction analogous to that catalyzed by glutamine synthetase in which ATP is cleaved to ADP and inorganic phosphate do not seem to have been correct (Meister, 1965); there have apparently been no additional publications on such a pathway. The first indication that the amide nitrogen atom of glutamine is directly utilized for asparagine biosynthesis came from the studies of Levintow (1957), who found that when HeLa cells were grown on media containing [^{15}N-amide] glutamine, substantial amounts of isotopic nitrogen were incorporated into the protein-asparagine; similar studies with ^{15}NH$_3$ did not reveal significant incorporation of ^{15}N into asparagine. Later work in which cell free preparations of asparagine synthetase were studied have demonstrated that mammalian asparagine synthetase catalyzes the following reaction:

$$\text{L-Aspartate + L-Glutamine + ATP + H}_2\text{O} \xrightarrow{\text{Mg}^{++}\text{ or Mn}^{++}}$$

$$\text{L-Asparagine + L-Glutamate + AMP + PPi}$$

The enzyme also catalyzes the synthesis of asparagine when

574

glutamine is replaced by ammonium ion; in this connection, it should be noted that preparations of bacterial asparagine synthetase have been reported to be active only with ammonia and to be in-active when ammonia is replaced by glutamine (Ravel et al., 1962; Burchall et al., 1964; Cedar and Schwartz, 1969).

The present work on glutamine-dependent asparagine synthe-tase developed from earlier studies in this laboratory on the aspara-gine synthetase activities of a series of mouse leukemia cells (Horowitz et al., 1968). In this work it was found that normal mouse tissues exhibit asparagine synthetase activity, that leukemias which are sensitive to treatment with asparaginase exhibit either no asparagine synthetase or very low levels of this activity, and that leukemias resistant to asparaginase therapy exhibit moderate to high asparagine synthetase activities. Thus, a striking correlation was found between resistance to asparaginase treatment and possession of asparagine synthetase activity on the one hand, and sensitivity to asparaginase and absence of asparagine synthetase on the other. These and related studies therefore appear to provide an explanation at the enzymatic level for the therapeutic effectiveness (or lack of effectiveness) of asparaginase. Thus, tumors that do not have asparagine synthetase must obtain asparagine from the extracellular fluid and when this process is prevented by asparagin-ase therapy, the tumor cells cannot survive. These observations indicate that asparagine synthetase is of central significance in relation to the effectiveness of asparaginase therapy of certain tumors.

In the course of this work it was found that certain aspara-ginase resistant tumors exhibit very high levels of asparagine synthetase; such tumors serve as good sources for the isolation of this enzyme. A partially purified preparation of asparagine synthe-tase was obtained from Novikoff hepatoma (Patterson and Orr, 1968); this preparation was found to catalyze the stoichiometric formation of L-asparagine, AMP, and PPi, but no data were report-ed on the utilization of glutamine or the formation of glutamate in this reaction. In this paper, we will review studies carried out in our laboratory (Horowitz and Meister, 1972) on the characterization of asparagine synthetase from RADA1 cells, an asparaginase-

575

resistant mouse leukemia. Experiments were performed on the mechanism of action of the enzyme and direct evidence has been obtained for the intermediate participation of β-aspartyl adenylate. A detailed study of the stoichiometry of the reaction unexpectedly revealed that the isolated asparagine synthetase preparation exhibits considerable glutaminase activity. It is notable that the glutaminase activity of asparagine synthetase is much higher than the glutaminase activities exhibited by other glutamine amidotransferases, for example, carbamyl phosphate synthetase. Another finding of interest is that both the glutamine dependent asparagine synthetase activity and the glutaminase activity exhibited by the enzyme require chloride. A number of compounds were studied as possible inhibitors of asparagine synthetase, and several were found which inhibit the enzyme by competing with glutamine for attachment to the glutamine binding site of the enzyme.

EXPERIMENTAL PROCEDURES

Asparagine synthetase was determined by following the conversion of labeled L-aspartate to L-asparagine. That the asparagine formed was of the L-configuration was shown by studies in which guinea pig serum L-asparaginase was employed. The details of the assay procedures used have been given elsewhere (Horowitz and Meister, 1972).

RADA1 mouse leukemia was carried by intraperitoneal transplantation in strain A mice (Old et al., 1963). The cells were collected from the ascitic fluid and homogenized with a Potter-Elvehjem homogenizer. After centrifugation of the homogenate, the supernatant solution was treated with ATP, magnesium chloride and L-asparagine. The solution was heated at 54° for 13 minutes and then cooled and centrifuged to remove denatured protein. The supernatant solution was subjected to ammonium sulfate fractionation and the active fraction was chromatographed on a column of Sephadex G-100. This material was subjected to an additional ammonium sulfate fractionation. The details of the purification procedure have been published (Horowitz and Meister, 1972). About 8 mg of purified enzyme were obtained from the tumor cells harvested from the ascitic fluid of 600 mice. Although a consider-

576

able purification of the enzyme was achieved (about 170-fold), the enzyme preparation was not homogeneous. It is estimated that the most purified preparation of the enzyme is about 30% pure. The molecular weight of the enzyme was estimated by gel filtration to be about 105,000.

General Catalytic Properties of the Enzyme

The purified enzyme catalyzed conversion of L-[4-^{14}C] aspartate to L-[4-^{14}C] asparagine in reaction mixtures containing ATP, magnesium ions, and either L-glutamine or ammonium ion.

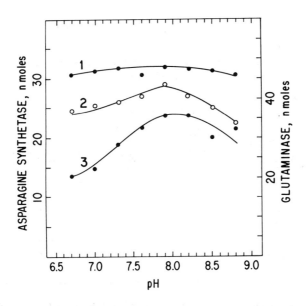

Fig. 1 pH activity curves. Curve 1: glutaminase; Curve 2: glutamine-dependent asparagine synthetase; Curve 3: ammonium-dependent asparagine synthetase (Horowitz and Meister, 1972).

No asparagine was formed when ATP, magnesium ions, or glutamine (or ammonia) was separately omitted. The apparent Km values for L-aspartate, ATP, L-glutamine, and ammonium chloride were, respectively, 0.9, 0.2 , and 9 mM. As described in Figure 1, glutamine-dependent asparagine synthetase activity varied little over the pH range 6.7 to 8.8. In contrast, ammonium-dependent asparagine synthetase activity exhibited a much greater increase as the pH was raised from 6.7 to 8.8. Asparagine synthesis was not observed when L-glutamine was replaced by D-glutamine or L-asparagine. Replacement of glutamine by 0.1 M hydroxylamine led to formation of β-aspartylhydroxamate. D-Aspartate did not inhibit the synthesis of asparagine from L-aspartate. When magnesium ions were replaced by manganese ions, much lower asparagine synthetase activity was observed. Similarly, replacement of ATP by GTP, CTP, TTP, or UTP led to marked reduction of asparagine synthesis.

Glutaminase Activity of the Enzyme

Studies on the stoichiometry of the reaction are summarized in Table I. The formation of asparagine was accompanied by equi-

TABLE I
Stoichiometry of Product Formation*

Exp.	Product Formed (nmoles)			
	L-Asparagine	PPi	AMP	L-Glutamate
1a	53	52	56	–
b	55	56	58	
2a	71	–	–	144
b	75			153
3a	103	–	–	268
b	105			262

*The reaction mixtures contained L-[4-^{14}C] aspartate (1.5 μmoles; 1.67 Ci/mole), ATP (3 μmoles), MgCl$_2$ (7.5 μmoles), L-glutamine (30 μmoles), Tris-HCl buffer (150 μmoles; pH 7.6), and enzyme (0.11, 0.15 and 0.20 unit, respectively, in exps. 1, 2, and 3), in a final volume of 1.5 ml; the mixtures were incubated for 30 minutes at 37°. The formation of L-asparagine, PPi, AMP, and L-glutamate was determined on separate aliquots as described in the text. The values given for AMP and PPi were corrected by sub-tracting blank values for the formation of AMP and PPi in the absence of L-aspartate; the blank values were about 5 nmoles for AMP and 16 nmoles for PPi.

---

molar formation of AMP and inorganic pyrophosphate, but the formation of L-glutamate was much higher than that of L-asparagine. This finding initially suggested the presence in the enzyme prepara-tion of contaminating glutaminase activity. Subsequent studies revealed however, that the formation of L-glutamate in the com-plete asparagine synthetase system was equivalent to the formation of L-glutamate in similar reaction mixtures lacking L-aspartate. Thus, in exp. 2 (Table I), L-glutamate formation was 155 nmoles when L-aspartate was omitted. In exp. 3, the formation of L-glutamate was 260 nmoles in the absence of L-aspartate. Additional studies showed that in the complete glutamine-dependent asparagine synthetase system, the formation of L-glutamate was equal to the sum of the formation of L-asparagine and ammonium. The free ammonium formed in this reaction cannot account for the observed synthesis of L-asparagine since substitution of glutamine by this concentration of ammonium does not promote appreciable synthesis of asparagine. It may thus be concluded that the glutaminase activity is associated with asparagine synthetase. This conclusion is further supported by the observation that glutaminase and asparagine synthetase moved together on gel filtration and also on acrylamide gel electrophoresis. Thus, as shown in Figure 2, the ratio of glutaminase to asparagine synthetase activity was constant during elution from the Sephadex column. Figure 3 describes the results obtained on acrylamide gel electrophoresis; the gel was sectioned after electrophoresis and each section was examined for glutaminase and glutamine-dependent asparagine synthetase activity. Both activities were found in the same sections of the gel, which contained a major protein component.

Chloride-Dependence of Glutamine-Dependent Asparagine Synthetase

The discovery that chloride is required for glutamine-

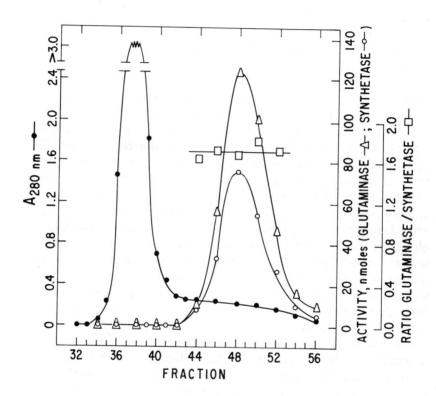

Fig. 2 Elution of the enzyme from Sephadex G-100. Step 5 of the purification procedure (see Horowitz and Meister, 1972).

dependent asparagine synthetase activity was made during the course of studies on glutaminase activity of the enzyme. Several experiments were carried out in which chloride was omitted from the reaction mixture and under these conditions no glutaminase activity was observed. As indicated in Figure 4, the relationship between glutaminase activity and chloride concentration follows a hyperbolic curve; the apparent Km value for chloride calculated from these data is 2 mM. While bromide and iodide can substitute

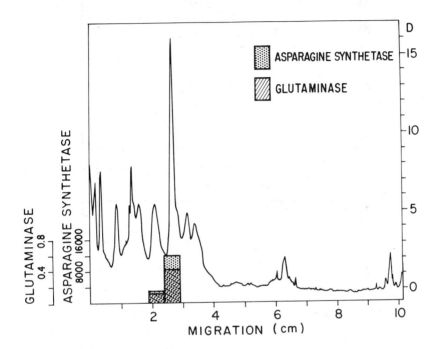

Fig. 3 Migration of glutamine-dependent asparagine synthetase and glutaminase activities on acrylamide gel electrophoresis;
from (Horowitz and Meister, 1972).

in part for chloride (Table II), no glutaminase activity was found
when chloride was replaced by phosphate, sulfate, citrate, and
acetate. The accidental finding that chloride is required for
glutaminase activity suggested that asparagine synthesis might
exhibit a similar requirement. As indicated by the data given in
Table II, this proved to be the case; although some asparagine
synthesis took place in the absence of added chloride, the possibility

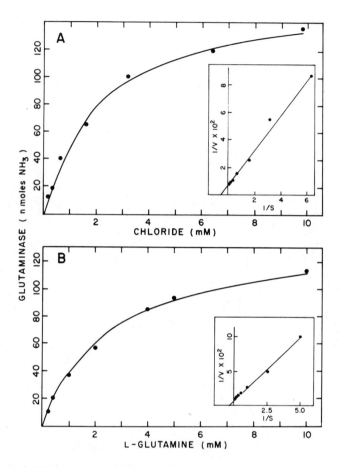

Fig. 4 Effect of chloride (A) and L-glutamine (B) on glutaminase activity (From Horowitz and Meister, 1972).

that small amounts of chloride were present as contaminants in the
several components of the synthesis system cannot be excluded. It
is notable that the ammonium-dependent synthesis of asparagine does
not require chloride. The finding that both the glutaminase and
glutamine-dependent synthetase activities require chloride gives
additional support to the conclusion that the observed glutaminase
activity is catalyzed by the synthetase itself rather than by a contaminating enzyme. It is notable also that the apparent Km value

TABLE II

Effect of Anions on Glutaminase and Asparagine Synthetase*

Anion	Relative Activities		
Added	Glutaminase	Asparagine Synthetase	
		With L-Glutamine	With NH_4^+
None	0	16	91
Chloride	[100]	[100]	[100]
Sulfate	0	18	
Citrate	0	17	
Acetate	0	16	
Bicarbon-ate	0	15	
Fluoride	0	0	
Bromide	80	91	
Iodide	46	66	

*Enzyme at step 6 of the purification procedure was passed through a Sephadex G-25 column (22 x 0.9 cm) equilibrated and eluted with 0.05 M sodium phosphate buffer (pH 7.2) containing 2 mM DTT and 0.5 mM EDTA. Glutaminase activity was measured by adding the enzyme to a mixture (final volume, 0.5 ml) containing sodium phosphate buffer (20 μmoles; pH 7.2), glutamine (5 μmoles), and the potassium salt of the appropriate anion (2.5 μmoles); incubated for 30 minutes at 37°. Glutamate formation was determined. Glutamine-dependent asparagine synthetase activity was measured by adding the enzyme to a mixture (final volume 0.5 ml) consisting of potassium phosphate buffer (20 μmoles, pH 7.2), sodium ATP (0.5 μmole), magnesium phosphate (0.833 μmole), L-[4-^{14}C]aspartate (0.25 μmole; 1.67 Ci/mole), and the potassium salt of the appropriate anion (2.5 μmoles); incubated for 20 minutes at 37°. Asparagine formation was determined. Ammonium-dependent asparagine synthetase activity was determined in the same way except that ammonium phosphate (16.7 μmoles) was added in place of glutamate.

for L-glutamine (calculated from the data given in Figure 4, B) is about 1 mM, which is about the same as the apparent Km value determined for L-glutamine in the asparagine synthetase reaction.

We have begun to consider the mechanism by which chloride stimulates asparagine synthetase activity. There is evidence that anions can stimulate enzymatic activity by increasing their stability under the conditions of assay (Muus et al., 1956; Muus, 1953; Kakiuchi and Tomizawa, 1964; Skeggs et al., 1954; Johnson, 1941; McDonald et al., 1966; Unemoto and Hayashi, 1969; Webb and Morrow, 1959; Massey, 1953; Askari, 1966; Hughes and Williamson, 1952; Carter and Greenstein, 1947; Errera and Greenstein, 1949; Klingman and Handler, 1958), by shifting the pH dependence to a more basic region (Muus et al., 1956; Unemoto and Hayashi, 1969; Myrback, 1926; Caldwell and Kung, 1953; Schuberth, 1966; Reyes and Huennekens, 1967), and by increasing the affinity for substrate (Unemoto and Hayashi, 1969; Schuberth, 1966; Sayre and Roberts, 1958). Since the ammonium-dependent synthesis of asparagine is not dependent on chloride, it would appear that chloride functions only in amide nitrogen transfer. Chloride does not apparently influence the stability of the enzyme under the conditions of assay since asparagine synthesis is linear with time both in the presence and in the absence of added chloride (Figure 5). We have found that asparagine synthetase activity is not very greatly affected by pH changes in the range 6.5-8.8. Thus, as shown in Figure 6, addition of 5 mM chloride produces a small shift in the pH activity profile toward the basic region, but it seems un-likely that this shift is directly associated with the chloride-stimulated activity. On the other hand, chloride was found to exhibit a pronounced effect on the affinity of the enzyme for L-glutamine (Figure 7). Thus, in the presence of 1 mM chloride ion, the apparent Km value for L-glutamine was found to be 2 mM, while in the presence of 5 mM chloride ion the apparent Km value for L-glutamine was 0.8 mM. While these data are consistent with the conclusion that chloride may be required for the binding of gluta-mine, they do not exclude the possibility that chloride may function in the utilization of glutamine. As discussed below, L-asparagine effectively inhibits glutamine-dependent and ammonium-dependent asparagine synthetase activities; the data indicate that asparagine inhibits competitively with respect to both glutamine and ammonium.

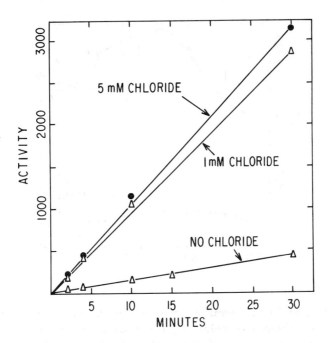

Fig. 5 Time course of asparagine synthesis in the presence and absence of chloride. The enzyme (0.06 unit) was added to assay mixtures (final volume, 1 ml) containing potassium phosphate buffer (40 μmoles; pH 7.2), magnesium phosphate (1.65 μmole), ATP (2 μmoles), L-glutamine (10 μmoles), L-[4-^{14}C] aspartate (1 μmole; 1.67 Ci/mole), and potassium chloride at the indicated concentrations. Aliquots (0.15 ml) were removed at intervals and added to 20 μl of 15% trichloroacetic acid. The amount of [^{14}C] asparagine formed was determined.

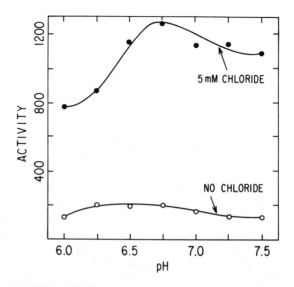

Fig. 6 Effect of pH on asparagine synthetase activity in the presence and absence of chloride. The assay mixtures (final volume 0.5 ml) contained enzyme (0.02 unit), potassium phosphate buffer (40 μmoles; pH as indicated), magnesium phosphate (0.835 μmole), ATP (1 μmole), L-glutamine (5 μmoles), and L-[4-^{14}C] aspartate (0.5 μmole; 1.67 Ci/mole) in the presence or absence of potassium chloride (2.5 μmoles). The amount of [^{14}C] asparagine formed in 30 minutes was determined.

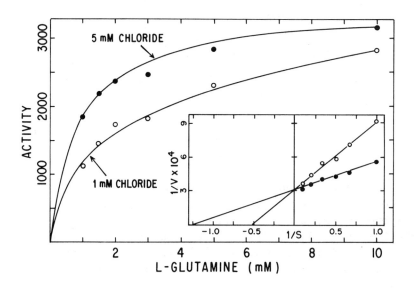

Fig. 7 Effect of chloride on the apparent Km value for L-glutamine. The enzyme was added to assay mixtures (final volume, 0.5 ml) containing potassium phosphate buffer (20 μmoles; pH 7.2), magnesium phosphate (0.835 μmole), ATP (1 μmole), L-[4-^{14}C] aspartate (0.5 μmole; 1.67 Ci/mole), potassium chloride (1 mM or 5 mM), and L-glutamine (as indicated). The asparagine formed in 30 minutes was determined.

TABLE III
Effect of Chloride on the Inhibition by L-Asparagine of the
Ammonium-Dependent Synthesis of Asparagine*

Chloride	L-Asparagine	NH_4^+-Dependent Asparagine Synthesis (c.p.m.)	Inhibition (%)
+	–	1025	[0]
+	+	325	68%
–	–	930	[0]
–	+	425	54%

*Enzyme was added to assay mixtures (final volume, 0.5 ml) containing potassium phosphate buffer (20 µmoles; pH 7.2), magnesium phosphate (0.835 µmole), ATP (1 µmole), $(NH_4)_3PO_4$ (16.5 µmoles, pH 7.2), L-[4-^{14}C] aspartate (0.5 µmole; 1.67 Ci/ mole), and where indicated KCl (2.5 µmoles) and L-asparagine (7.5 µmoles). The amount of [^{14}C] asparagine formed after 30 minutes was determined.

Therefore, if chloride affects the binding of glutamine to the enzyme, it might also be expected to affect the binding of L-asparagine. Since the ammonium-dependent synthesis of asparagine does not require chloride, it is possible to determine whether asparagine inhibits ammonium-dependent synthesis of asparagine in the absence of chloride. An experiment designed to answer this question is described in Table III. The data show that L-asparagine produces substantial inhibition of ammonium-dependent asparagine synthesis in the absence of added chloride, although the inhibition was not as great as in the presence of chloride. This result suggests that the binding of asparagine to the enzyme may be only slightly affected by chloride. This experiment would thus seem to add weight to the possibility that the decrease in the apparent Km value for L-glutamine observed in the presence of chloride reflects an effect on the utilization of glutamine rather than solely on the binding of glutamine to the enzyme.

Evidence for a β-L-Aspartyl-AMP Intermediate

Preparations of the enzyme were found to catalyze an L-aspartate-dependent ATP-PPi exchange (Table IV). The extent of

TABLE IV
Aspartate-Dependent Incorporation of PPi into ATP*

Enzyme (units)	[^{32}P] ATP formed (nmoles)
0	0
0.06	0.22
0.12[a]	1.2
0.12[a]	1.2
0.12[a,b]	0
0.18	1.5
0.24[b]	2.4
0.24[b]	0

*Enzyme was added to a mixture consisting of Tris-HCl buffer (50 μmoles; pH 7.6), MgCl$_2$ (2.5 μmoles), ATP (1 μmole), L-aspartate (0.5 μmole) and [^{32}P] PPi (0.24 μmole; 1.3 Ci/mole); incubated at 37° for 30 minutes. The incorporation of radioactivity into ATP was measured. [a] tRNA (0.25 mg) added; [b] L-aspartate omitted.

incorporation of [^{32}P] PPi into ATP was dependent on the amount of enzyme, and no incorporation occurred when either enzyme or L-aspartate was omitted. The addition of mouse liver aspartyl-tRNA had no effect on the incorporation of inorganic pyrophosphate into ATP; this renders unlikely the possibility that the observed incorporation is due to the presence of an α-aspartate activating enzyme.

We have obtained direct evidence indicating intermediate participation of β-L-aspartyl adenylate in the asparagine synthetase reaction. In this work, β-L-aspartyl adenylate was prepared by organic synthesis as described elsewhere (Horowitz and Meister,

1972). As indicated in the experiment described in Figure 8, when

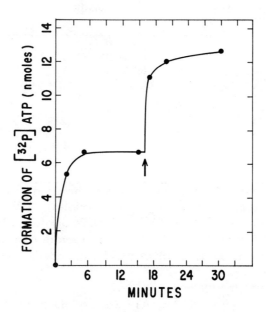

Fig. 8 Synthesis of ATP from β-L-aspartyl-AMP and PPi. β-L-Aspartyl-AMP (1.3 mg) was added initially and after 16 minutes to a mixture (final volume, 1.2 ml) containing enzyme (1 unit), Tris-HCl buffer (150 μmoles; pH 7.0), sodium phosphate buffer (20 μmoles; pH 7.2), magnesium chloride (10 μmoles), and [^{32}P] PPi (0.6 μmole; 1.3 Ci/mole). Aliquots were withdrawn at intervals and the amount of labeled ATP formed was determined. From (Horowitz and Meister, 1972).

β–L–aspartyl adenylate was added to a reaction mixture containing enzyme, Tris buffer, sodium phosphate, magnesium chloride, and [^{32}P] inorganic pyrophosphate, [^{32}P] ATP was formed. The formation of ATP stopped within 5 minutes under these conditions; this reflects the considerable instability of β–L–aspartyl adenylate in neutral solution. Addition of a further amount of β–L–aspartyl adenylate to this reaction mixture led to additional ATP synthesis. When β–L–aspartyl adenylate was added to the reaction mixture 10 minutes prior to addition of enzyme, no labeled ATP was formed; furthermore, no labeled ATP was formed in controls in which β–L–aspartyl adenylate was omitted. Thus, the data indicate that the formation of labeled ATP is dependent on the presence of both the enzyme and β–L–aspartyl adenylate, and the findings exclude the possibility that the formation of labeled ATP is due to ATP–pyrophosphate exchange or to direct synthesis of ATP from AMP and inorganic pyrophosphate. The findings, which are analogous to earlier results on the α–aminoacyl RNA synthetases (DeMoss et al., 1956; Berg, 1958; Krishnaswamy and Meister, 1960), offer strong support for the formation of enzyme–bound β–L–aspartyl adenylate in the reaction.

Inhibition of Asparagine Synthetase

Asparagine synthetase activity is inhibited by three of the products of the synthetase reaction, i.e., L–asparagine, inorganic pyrophosphate, and AMP. L–Glutamate, however, is not an effective inhibitor. It is of interest that L–asparagine inhibits noncompetitively with respect to L–aspartate but competitively with respect to both L–glutamine and ammonium (Figure 9). It was also found that inorganic pyrophosphate inhibited competitively with respect to ATP; on the other hand, AMP inhibited noncompetitively with respect to ATP (Horowitz and Meister, 1972).

A number of studies have been carried out on the inhibition of asparagine synthetase by L–2–amino–4–oxo–5–chloropentanoic acid. This compound was designed and synthesized in this laboratory as a potential glutamine antagonist in the glutamine–dependent carbamyl phosphate synthetase system (Khedouri et al., 1966). When carbamyl phosphate synthetase is incubated with this chloro–

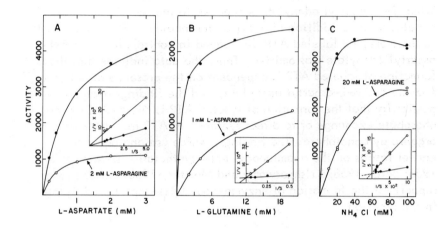

Fig. 9 Inhibition of asparagine synthetase by asparagine. The enzyme was added to the standard synthetase assay mixtures containing varying amounts of L-aspartate (A), L-glutamine (B), or ammonium chloride (C). From (Horowitz and Meister, 1972).

ketone analog of glutamine, there is substantial loss of the glutamine-dependent synthetase activity, but the treated enzyme is fully active in catalyzing carbamyl phosphate synthesis from ammonia (Khedouri et al., 1966; Pinkus and Meister, 1972). In the present studies, when 0.12 mM chloroketone was added to the

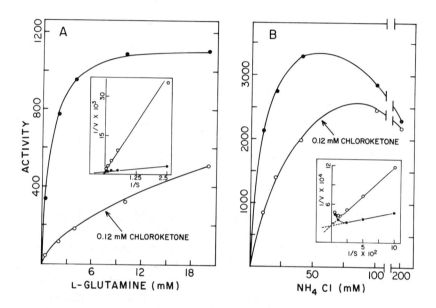

Fig. 10 Inhibition of glutamine-dependent and ammonium-dependent asparagine synthetase activities by L-2-amino-4-oxo-5-chloropentanoic acid. From (Horowitz and Meister, 1972).

standard asparagine synthetase reaction mixture, both glutamine-dependent and ammonium-dependent asparagine synthetase activities were inhibited (Figure 10); the findings are consistent with competitive inhibition. However, when asparagine synthetase was preincubated with 0.167 mM chloroketone and then assayed (after dilution) for asparagine synthetase activity, findings analogous to those made on carbamyl phosphate synthetase were obtained. In the experiment

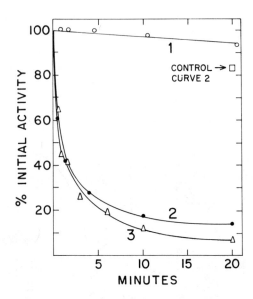

Fig. 11 Effect of preincubating the enzyme with L-2-amino-4-oxo-5-chloropentanoic acid (0.167 mM). Aliquots of the reaction mixtures were removed at intervals and added to asparagine synthetase assay mixtures containing either L-glutamine (Curve 2) or ammonium chloride (Curve 1). Aliquots were also added to standard glutaminase assay mixtures (Curve 3). The L-asparagine (Curves 1 and 2) and L-glutamate formed in 30 minutes was determined (Curve 3). The effect of including L-glutamine (33.3 mM) during the preincubation period on the inactivation of the glutamine-dependent synthetase was determined (single point, control).

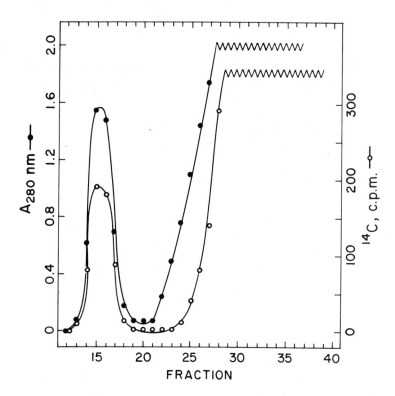

Fig. 12 Binding of L-2-amino-4-oxo-5-chloro-[5-^{14}C]-pentanoic acid to the enzyme. The enzyme (2.4 units; 4.16 mg) was incubated at 20° with [5-^{14}C] chloroketone (0.35 μmole; 1.13 Ci/mole) in a final volume of 2.2 ml. After 8 minutes this solution was chilled to 4° and passed through a Sephadex G-25 column (37 x 1.1 cm), which was equilibrated and eluted with 0.06 M Tris-HCl buffer (pH 7.6) containing 1 mM DTT and 0.5 mM EDTA. The absorbance at 280 nm (●) and the radioactivity (O) were determined. From (Horowitz and Meister, 1972).

described in Figure 11, there was rapid disappearance of glutamine-dependent asparagine synthetase activity after brief incubation of the enzyme with the chloroketone (Curve 2). The glutaminase activity also disappeared, following a very similar course (Curve 3). However, the ammonium-dependent asparagine synthetase activity (Curve 1) was only slightly inhibited under these conditions. When the enzyme was preincubated with chloroketone in the presence of L-glutamine, there was much less inhibition of the glutamine-dependent asparagine synthetase activity (control, Curve 2; Fig. 11). These studies show that when the enzyme is incubated with chloroketone there is irreversible inactivation of the enzyme; the findings are similar to those obtained with carbamyl phosphate and suggest that the chloroketone binds to the enzyme. Binding of the chloroketone to the enzyme was demonstrated in an experiment in which the enzyme was incubated with L-2-amino-4-oxo-5-chloro-[5-^{14}C]-pentanoic acid. The reaction mixture was then placed on a column of Sephadex G-25 as indicated in Figure 12. In this study, a considerable amount of radioactivity emerged from the column with the protein. Experiments of this type indicate the binding of about 8 nanamoles of chloroketone per milligram of protein; similar experiments, but carried out in the presence of 0.1 M L-glutamine showed about a 50% decrease in the binding of chloroketone. The findings are therefore in accord with the conclusion that the chloroketone binds to the glutamine binding site of the enzyme. However, under the conditions of these experiments, there is extraneous binding to other protein sites. Such extraneous binding may be attributed to the ability of the chloroketone to serve as a general acylating agent (Pinkus and Meister, 1972).

Glutamine-dependent asparagine synthetase activity is also inhibited by aminomalonate; studies described in detail elsewhere (Horowitz and Meister, 1972) indicate that aminomalonate acts as a competitive inhibitor with respect to L-aspartate. It is of interest, that aminomalonate was found to be about 12% as effective as L-aspartate in stimulating the incorporation of pyrophosphate into ATP. This finding suggested the possibility that aminomalonate might serve as a substrate and therefore be converted to aminomalonamate.

However, studies with $[2-^{14}C]$ aminomalonate failed to reveal the formation of $[^{14}C]$ aminomalonamate.

A number of other compounds were tested as potential inhibitors of asparagine synthetase activity; a summary of these data is given elsewhere (Horowitz and Meister, 1972). L-Albizziin, which is known to be an irreversible inhibitor of formylglycinamide ribonucleotide amidotransferase (Li and Buchanan, 1971)inhibits asparagine synthetase to some extent. However, preincubation of asparagine synthetase with albizziin did not lead to irreversible inactivation of the enzyme.

DISCUSSION

The present data show that the asparagine synthetase isolated from RADA1 cells is a glutamine amidotransferase that catalyzes the coupled synthesis of asparagine and the cleavage of ATP to AMP and inorganic pyrophosphate in accordance with the following scheme:

$$ENZ + ATP + L\text{-Aspartate} \rightleftharpoons ENZ \ [\beta\text{-}L\text{-Aspartyl-AMP}] + PPi$$

$$ENZ \ [\beta\text{-}L\text{-Aspartyl-AMP}] + L\text{-Glutamine} + H_2O \longrightarrow ENZ + $$

$$AMP + L\text{-Asparagine} + L\text{-Glutamate}$$

The data indicate that asparagine synthetase can catalyze the transfer of the amide group of L-glutamine to water as well as to aspartyl adenylate, and that the glutaminase activity exhibited by the enzyme is a function of the synthetase itself. It would be of interest to learn whether the very high glutaminase activity of this enzyme is characteristic of other asparagine synthetases; the glutamine-dependent asparagine synthetase isolated from Novikoff hepatoma (Patterson and Orr, 1968) was apparently not investigated in this respect. Several other glutamine amidotransferases e.g., anthranilate synthetase (Nagano et al., 1970; Tamir and Srinivasan, 1971), carbamyl phosphate synthetase (Tatibana and Ito, 1969; Wellner et al., in preparation), cytidine triphosphate synthetase

Levitzki and Koshland, 1971), and formylglycinamidine ribonucleotide synthetase (Li and Buchanan, 1971) have been reported to
exhibit glutaminase activity. However, in contrast to the present
results on asparagine synthetase, in each of the cases cited,
stoichiometry was observed between the formation of L-glutamate
and the other products. It therefore appears that the present preparation of asparagine synthetase is unusual in that there is uncoupling
between the step involving transfer of the amide nitrogen atom from
the rest of the reaction. Such uncoupling may possibly be due to
modification of the enzyme during isolation. A somewhat analogous
phenomenon has been observed in our laboratory in studies on
carbamyl phosphate synthetase; when this enzyme is placed at pH 9
at 0°, the glutaminase activity, which is initially quite low,
increases dramatically and the synthetase activity decreases (Pinkus
et al., 1972). Another interpretation of the uncoupling phenomenon
observed with glutamine-dependent asparagine synthetase is that it
reflects some type of regulatory mechanism. However, such a
mechanism for the control of asparagine synthesis would be rather
costly in terms of energy utilization. In this respect it is interesting to note that the synthesis of asparagine is more costly in terms
of energy than the synthesis of the homologous amide, glutamine.
While the synthesis of glutamine from glutamate and ammonia
requires the cleavage of only one high energy phosphate bond of
ATP, glutamine-dependent synthesis of asparagine requires the
hydrolysis of three high energy phosphate bonds, assuming that the
pyrophosphate liberated in the reaction is subsequently cleaved to
orthophosphate.

The present studies reveal an additional interesting property
of asparagine synthetase, namely that the transfer of the amide
nitrogen atom of L-glutamine to either β-aspartyl adenylate or to
water requires chloride. Since the ammonium-dependent asparagine
synthetase activity does not require chloride, while the glutaminase
activity does, it may be concluded that only the amidotransferase
function is chloride-mediated. Although we have found no reports
in the literature indicating that other glutamine amidotransferases
may be significantly stimulated by anions, it should be mentioned
that our finding that asparagine synthetase requires chloride was
serendipitious. The possibility exists that one or more of the other

amidotransferases may also require chloride. The effect of anions
has of course been examined on a number of other enzymes,
including amylases (Muus et al., 1956; Muus, 1953; Myrback,1926;
Trotta et al., 1971), certain peptidases (Kakiuchi and Tomizawa,
1964; Skeggs et al., 1954; Johnson, 1941; McDonald et al., 1966),
cyclophosphodiesterase (Unemoto and Hayashi, 1969), sulphatase
(Webb and Morrow, 1959), fumarase (Massey, 1953), AMP deaminase (Askari, 1966), and interestingly a bacterial glutaminase
(Hughes and Williamson, 1952). Our studies on the mechanism by
which chloride stimulates asparagine synthetase activity indicate
that chloride does not function by affecting the stability of the
enzyme under the conditions of assay, nor does it appreciably alter
the pH-activity profile. It is of considerable interest that chloride
markedly affects the apparent Km value for L-glutamine. The
present studies on the inhibition by L-asparagine of the ammonium-
dependent synthesis of asparagine indicate that L-asparagine (which
is a competitive inhibitor with respect to ammonium and L-glutamine)
is almost as good an inhibitor in the absence as it is in the presence
of chloride. This finding is consistent with the belief that chloride
influences the utilization of glutamine rather than only its binding
to the enzyme. It is possible that chloride produces a change in
the conformation of the enzyme such that the position of the nucleophilic residue responsible for displacement of the amide group of
glutamine becomes shifted.

    The experiments on the inhibition of glutamine-dependent
asparagine synthetase by L-2-amino-4-oxo-5-chloropentanoic acid
are in accord with the view that this chloroketone derivative binds
to the glutamine binding site of the enzyme by forming a covalent
linkage. In similar studies on carbamyl phosphate synthetase, the
chloroketone was found to bind to a cysteine residue at the glutamine binding site of the light subunit of this enzyme (Pinkus and
Meister, 1972). When either asparagine synthetase or carbamyl
phosphate synthetase are incubated with the chloroketone, enzymes
are obtained which can no longer function with glutamine but which
can still use ammonia. In the studies on carbamyl phosphate synthetase, it was established that the glutamine binding site of the enzyme
is located on a separate subunit (Trotta et al., 1971). It will be of
interest to determine whether asparagine synthetase also possesses

599

a separate glutamine binding subunit. The bacterial asparagine synthetases which have been studied have been reported to be inactive with glutamine; however, in other respects the reaction catalyzed is analogous to that catalyzed by the glutamine–dependent asparagine synthetase. It seems likely that the bacterial asparagine synthetases do not have active glutamine binding subunits, possibly because a metabolic need for such additional catalytic equipment did not develop during the course of their evolution; alternatively, the glutamine binding subunit of these enzymes may have been lost or inactivated during isolation. It is of possible interest in this connection that early studies on the synthesis of CTP by a purified fraction obtained from E. coli led to isolation of an enzyme that was active only with ammonia and which did not function with glutamine, see (Meister, 1962). More recent studies indicate that glutamine is required for CTP synthesis. Similar phenomena have been observed in studies on other glutamine amidotransferases. The possibility that the glutamine binding subunit of a glutamine amidotransferase may be removed or destroyed during enzyme purification seems quite plausible, and indeed the uncoupling phenomenon observed in the present studies on glutamine-dependent asparagine synthetase may reflect a similar event.

REFERENCES

Askari, A. (1966), Molec. Pharmacology, 2, 518.

Berg, P. (1958), J. Biol. Chem., 233, 601.

Burchall, J.J., Reichelt, E.C., and Wolin, M.J. (1964), J. Biol. Chem., 239, 1794.

Caldwell, M.L., and Kung, J.T. (1953), J. Amer. Chem. Soc., 75, 3132.

Carter, C.E., and Greenstein, J.P. (1947), J. Natl. Cancer Inst., 7, 433.

Cedar, H., and Schwartz, J.H. (1969), J. Biol. Chem., 244, 4112.

Cole, S.W. (1904), J. Physiology, 30, 202.

DeMoss, J.A., Genuth, S.M., and Novelli, G.D. (1956), Proc. Natl. Acad. Sci. U.S., 42, 325.

Errera, M., and Greenstein, J.P. (1949), J. Biol. Chem., 178, 495.

Horowitz, B., Madras, B.K., Meister, A., Old, L.J., Boyse, E. A., and Stockert, E. (1968), Science, 160, 533.

Horowitz, B., and Meister, A. (1972), J. Biol. Chem., 247,

Hughes, D.E., and Williamson, D.H. (1952), Biochem. J., 51, 45.

Johnson, M.J., (1941), J. Biol. Chem., 137, 575.

Kakiuchi, S., and Tomizawa, H.H. (1964), J. Biol. Chem., 239, 2160.

Khedouri, E., Anderson, P.M., and Meister, A. (1966), Biochemistry, 5, 3552.

Klingman, J.D., and Handler, P. (1958), J. Biol. Chem., 232, 369.

Krishnaswamy, P.R., and Meister, A. (1960), J. Biol. Chem., 235, 408.

Levintow, L. (1957), Science, 126, 611.

Levitzki, A., and Koshland, D.E. Jr. (1971), Biochemistry, 10, 3365.

Li, H.-C., and Buchanan, J.M. (1971), J. Biol. Chem., 246, 4713.

McDonald, J.K., Ellis, S., and Reilly, T.J. (1966), J. Biol. Chem., 241, 1494.

Massey, V. (1953), Biochem. J., 53, 67.

Meister, A. (1965), "Biochemistry of the Amino Acids", Second Edition, Vol. I, pp. 457–460. Academic Press, New York.

Meister, A. (1962), In "The Enzymes", Second Edition, 6, 247. (P.D. Boyer, H. Lardy and K. Myrback, eds.), Academic Press, New York.

Muus, J. (1953), Comptes Rendus des Travaux du Laboratoire Carlsberg Serie Chemique, 28, 317.

Muus, J., Brockett, F.P., and Connelly, C.C. (1956), Arch. Biochem. Biophys., 65, 268.

Myrback, K. (1926), Hoppe-Seyler's Z. Physiol. Chem., 159, 1.

Nagano, H., Zalkin, H., and Henderson, E.J. (1970), J. Biol. Chem., 245, 3810.

Old, L.J., Boyse, E.A., and Stockert, E. (1963), J. Natl. Cancer Inst., 31, 977.

Patterson, M.K., and Orr, G.R. (1968), J. Biol. Chem., 243, 376.

Pinkus, L.M., and Meister, A. (1972), J. Biol. Chem., 247,

Pinkus, L.M., Wellner, V.P., and Meister, A. (1972), Federation Proc., 31, 474.

Ravel, J.M., Norton, S.J., Humphreys, J.S., and Shive, W. (1962), J. Biol. Chem., 237, 2845.

Reyes, P., and Huennekens, F.M. (1967), Biochemistry, 6, 3519.

Sayre, F.W., and Roberts, E. (1958), J. Biol. Chem., 233, 1128.

Schuberth, J. (1966), Biochim. Biophys. Acta, 470, 122.

Skeggs, L.T. Jr., Marsh, W.H., Kahn, J.R., and Shumway, N. P. (1954), J. Exptl. Med., 99, 275.

Tamir, H., and Srinivasan, P.R. (1971), J. Biol. Chem., 246, 3024.

Tatibana, M., and Ito, K. (1969), J. Biol. Chem., 244, 5403.

Trotta, P.P., Burt, M.E., Haschemeyer, R.H., and Meister, A. (1971), Proc. Natl. Acad. Sci. U.S., 68, 2599.

Unemoto, T., and Hayashi, M. (1969), Biochim. Biophys. Acta, 171, 89.

Webb, E.C., and Morrow, P.F.W. (1959), Biochem. J., 73, 7.

Wellner, V.P., Anderson, P.M., and Meister, A., in preparation.

Appendix: Nomenclature for the Enzymes of Glutamine Metabolism*

E. C. Number	Systematic Name	Trivial Names	Reaction
1. 6.3.1.2	L-Glutamate: ammonia ligase (ADP)	Glutamine synthetase	ATP + L-glutamate + NH_3 = ADP + orthophosphate + L-glutamine
2. 1.4.1.X	L-Glutamate: $NADP^+$ oxidoreductase (deaminating, glutamine forming)	Glutamate synthase / Glutamate synthetase / Glutamine amide-2-oxoglutarate amino-transferase (oxidoreductase, NADP) (GOGAT)	2-Oxoglutarate + L-glutamine + reduced NADP = 2 glutamate + NADP
3. 1.4.1.4	L-Glutamate: NADP oxidoreductase	Glutamate dehydrogenase	L-glutamate + H_2O + NADP = 2-oxoglutarate + NH_3 + reduced NADP
4. 2.6.1.15	L-Glutamine: 2 oxoacid aminotransferase	Glutamine-ketoacid aminotransferase / Glutamine-ketoacid transaminase / Glutamine transaminase	L-glutamine + a 2-oxoacid = 2-oxoglutarate + an amino acid
5. 3.5.1.2	L-Glutamine aminohy-drolase	Glutaminase	L-glutamine + H_2O = L-glutamate + NH_3
6.		Glu-t-RNA^{Gln}amidotransferase / Glutamyl-t-RNA (glutamine specific) amidotransferase	Glutamyl-t-RNA^{Gln} + ATP + L-glutamine = glutaminyl-t-RNA^{Gln} + ADP + orthophosphate + L-glutamate
7. 2.6.1.16	L-Glutamine: D-fructose-6-phosphate amino-transferase	Glutamine-fructose-6-phosphate amidotransferase / Hexosephosphate aminotransferase	L-Glutamine + D-fructose-6-phosphate = 2-amino-2-deoxy-D-glucose-6-phosphate + L-glutamate
8. 2.4.2.14	Ribosylamine-5-phosphate: pyrophosphate phos-phoribosyltransferase (glutamate amidating)	Glutamine phosphoribosylpyrophosphate amidotransferase / Amido phosphoribosylpyrophosphate transferase / PRPP amidotransferase	β-D-Ribosylamine 5-phosphate + pyro-phosphate + L-glutamate = L-glutamine + 5'-phospho-α-D-ribosyl-pyrophosphate + H_2O

*The enzyme nomenclature used above is taken, wherever possible, from the recommendations of the International Union of Biochemistry on the Nomenclature and Classification of Enzymes as stated in "Enzyme Nomenclature" (1965), Elsevier Publ., Amsterdam.

Appendix (*continued*)

E. C. Number	Systematic Name	Trivial Names	Reaction
9. 6.3.5.3	5'-Phosphoriboxyl-formyl-glycine-amide: L-glutamine amido-ligase (ADP)	Phosphoriboxyl-formylglycineamidine synthetase; Formylglycinamide ribonucleotide amido-transferase	ATP + 5'-phosphoribosyl-formyl glycine-amide + L-glutamine + H_2O = ADP + orthophosphate + 5'-phosphoribosyl-formyl-glycineamidine + L-glutamate
10. 6.3.5.2	Xanthosine-5'-phosphate: L-glutamine amido-ligase (AMP)	GMP synthetase; XMP aminase	ATP + xanthosine-5'-phosphate + L-glutamine + H_2O = AMP + pyrophosphate + GMP + L-glutamate
11. 6.3.5.1	Deamido-NAD: L-gluta-mine amido-ligase (AMP)	NAD synthetase	ATP + deamido-NAD + L-glutamine + H_2O = AMP + pyrophosphate + NAD + L-glutamate
12. 2.7.2.X	ATP: carbamate phospho-transferase (carbon dioxide: L-glutamine amido-ligase)	Carbamylphosphate synthetase; Carbamate kinase	2ATP + L-glutamine + CO_2 + H_2O = 2ADP + orthophosphate + carbamylphosphate + L-glutamate
13. 6.3.5.X	UTP: L-glutamine amido-ligase (ADP)	CTP synthetase	ATP + UTP + L-glutamine = ADP + orthophosphate + CTP + L-glutamate
14. 2.6.1.X		Anthranilate synthetase; Anthranilate synthase	Chorismate + L-glutamine = anthranilate + L-glutamate + pyruvate
15.		p-Aminobenzoate synthase	Chorismate + L-glutamine = p-aminobenzoate + L-glutamate + pyruvate
16.		H amidotransferase	N-(5'-phospho-D-ribulosylformimino)-5-amino-1-(5''-phosphoribosyl)-4-imidazolecarboxamide + L-glutamine = imidazole glycerol phosphate + amino-imidazole carboxamide ribonucleotide + L-glutamate
17. 6.3.5.X	Aspartate: L-glutamine amido-ligase (AMP)	Asparagine synthetase	ATP + L-aspartate + L-glutamine = AMP + pyrophosphate + L-asparagine + L-glutamate

SUBJECT INDEX